Lecture Notes in Physics

Volume 819

For further volumes:
http://www.springer.com/series/5304

The Lecture Notes in Physics

The series Lecture Notes in Physics (LNP), founded in 1969, reports new developments in physics research and teaching—quickly and informally, but with a high quality and the explicit aim to summarize and communicate current knowledge in an accessible way. Books published in this series are conceived as bridging material between advanced graduate textbooks and the forefront of research and to serve three purposes:

- to be a compact and modern up-to-date source of reference on a well-defined topic
- to serve as an accessible introduction to the field to postgraduate students and nonspecialist researchers from related areas
- to be a source of advanced teaching material for specialized seminars, courses and schools

Both monographs and multi-author volumes will be considered for publication. Edited volumes should, however, consist of a very limited number of contributions only. Proceedings will not be considered for LNP.

Volumes published in LNP are disseminated both in print and in electronic formats, the electronic archive being available at springerlink.com. The series content is indexed, abstracted and referenced by many abstracting and information services, bibliographic net-works, subscription agencies, library networks, and consortia.

Proposals should be sent to a member of the Editorial Board, or directly to the managing editor at Springer:

Christian Caron
Springer Heidelberg
Physics Editorial Department I
Tiergartenstrasse 17
69121 Heidelberg/Germany
christian.caron@springer.com

Doru S. Delion

Theory of Particle and Cluster Emission

Springer

Doru S. Delion
Theoretical Physics Department
Institute of Physics and Nuclear Engineering
Atomistilor 407
077125 Bucharest-Magurele
Romania
e-mail: delion@theory.nipne.ro

ISSN 0075-8450
e-ISSN 1616-6361

ISBN 978-3-642-14405-9
e-ISBN 978-3-642-14406-6

DOI: 10.1007/978-3-642-14406-6

Springer Heidelberg Dordrecht London New York

Library of Congress Control Number: 2010933484

Cover design: eStudio Calamar, Berlin/Figueres

Printed on acid-free paper

Springer is part of Springer Science+Business Media (www.springer.com)

To my wife Daniela and my son Daniel

Preface

This book has two main purposes. Firstly, we review the main theoretical methods describing decay processes induced by the strong interaction. Thus, most of the book is addressed to a broad audience within the nuclear physics community.

Secondly, this book is an attempt to clarify some fundamental aspects connected with the microscopic α-like emission theory and given parts, like Chap. 10, are addressed to nuclear theorists interested in the α-decay theory.

Nowadays experimental nuclear physics pushes its limits towards highly unstable nuclei. The theoretical description of proton-rich and neutron-rich nuclei or superheavy elements has become an important part of the modern nuclear physics. The main tool to investigate such unstable nuclei concerns radioactive decays. The family of radioactive processes triggered by the strong interaction contains various decays, namely particle (proton or neutron) emission, two-proton emission, α-decay, heavy cluster emission and binary or ternary fission. Other decay processes are induced by electromagnetic (γ-decay) or weak forces (β-decays). We will investigate only the first type of fragmentations, where the emitted fragments are left in ground or low-lying excited states. We call these decays cold emission processes. They are presently among important tools to study nuclei far from the stability line. Nuclei close to the proton drip line (proton-rich nuclei) are investigated through proton emission, while the neutron drip line region (neutron-rich nuclei) is probed by cold fission processes. Superheavy nuclei are exclusively detected by α-decay chains.

In the first part we review the phenomenological theoretical framework to investigate nuclear structure, from proton emission up to the cold fission. The main assumption is that the dynamics of emitted fragments is fully described by the potential picture. We analyze the most general method to estimate the interaction potential between the emitted fragments, given by the double folding approach. We also introduce the significant observables, characterizing emission processes from deformed nuclei, namely partial decay widths and angular distribution, in terms of the stationary coupled channels formalism and Gamow resonant states. We then extensively describe the coupled channels approach for axially symmetric and triaxial nuclei with even–even or odd-mass structure. We discuss coupled channels techniques like numerical integration, diagonalization method, analytic continuation method, distorted waves approach and two potential method. The semiclassical approach within the Cluster Model, Super Asymmetric Fission Model, Effective Liquid Drop Model and Fragmentation Theory is extensively analyzed. We also investigate the coupling between collective excitations (rotations and vibrations) of emitted fragments and the relative motion, in terms of the so called core-angular harmonics. It turns out that partial decay widths to excited states of emitted fragments are very sensitive to the relative wave function in the region of the nuclear surface, which is important especially for very unstable exotic nuclear systems.

In the second part we review mainly the α-like microscopic approaches of emission processes. Initially we discuss various methods to describe the emission of clusters from nuclei, like time dependent approach, resonating group method, Feshbach reaction theory, R-matrix approach and multi-step shell model. The description of the preformation amplitude for α-particles, as well as for heavier clusters like ^{8}Be, ^{12}C and ^{14}C is given. We apply the multi-step technique to describe α-like states above ^{208}Pb and ^{40}Ca. Pairing approach to estimate the α-particle preformation amplitude is extensively analyzed. Then we analyze the two harmonic oscillator method, describing α-clustering properties, and perform a systematic analysis of selfconsistency in α-decay theory. We give a description of the α-decay fine structure in vibrational nuclei within the Quasiparticle Random Phase Approximation. Finally, we present a short introduction into the Two Center Shell Model.

Bucharest, April 2010

Doru S. Delion

Acknowledgments

Part of this book is based on original contributions written during the last two decades in a close cooperation with professors R. J. Liotta and R. Wyss (Stockholm), A. Insolia (Catania), J. Suhonen (Jyväskylä), G. G. Dussel (Buenos Aires) A. Săndulescu (Bucharest) and W. Greiner (Frankfurt/Main). I also mention the fruitful discussions with professors P. Schuck (Orsay) and K. A. Gridnev (St. Petersbourg) on various aspects of the α-clustering. Important contributions were given by Dr. S. Peltonen (Jyväskylä) in the field of the α-decay fine structure and Dr. M. Mirea (Bucharest) in the Chapter devoted to the Two Center Shell Model. I am grateful to Mrs. Marlena Pintilie for the careful reading of the manuscript.

Contents

Part I Phenomenological Description of Emission Processes

Part I
Phenomenological Description of Emission Processes

The importance of the interplay between the Coulomb barrier and the energy of the emitted particle (Q-value) on the α-decay was one of the most important discoveries made by Gamow [1] in the early days of the nuclear physics. It explained the exponential dependence of half lives upon the Q-value, evidenced experimentally by the Geiger–Nuttall [2, 3] law. The proposed physical picture was very simple, but contradicting the classical intuition, namely a preformed α-particle inside the nucleus penetrates quantum mechanically the repulsive Coulomb barrier.

In a phenomenological description one supposes that the dynamics of the decaying systems is fully described by a potential acting between the emitted fragments. It contains two main components, namely the nuclear attraction in the internal region, surrounded by a Coulomb repulsion. Indeed, systematic calculations of half lives for α-particle emitters have shown that experimental values are well described using an equivalent local potential [4]. The attractive depth and the radius determine the energy and wave function of the decaying state, understood as a narrow resonance [5, 6].

In this first part we will review different approaches, all of them based on the phenomenological description of emission processes, involving ground or low-lying excited states in both parent and daughter nuclei. We will introduce the main tool to describe emission processes, namely Gamow resonances. Then we will describe the experimental material in terms of the Geiger–Nuttall law for different emission processes. The double folding procedure, as the most general method to compute the inter-fragment potential, is extensively analyzed.

It turns out that half lives of α-decays or heavy cluster emission processes, predicted by these potentials, are too short and this feature is a signature that such clusters do not exist as free components on the nuclear surface. The problem of how the binary system is born from the initial nucleus concerns the microscopic description of the decay process. The so-called spectroscopic amplitude, computed by the overlap between the initial and final configurations, can explain experimentally measured half lives. We will discuss this concept in the second part, devoted to microscopic approaches.

We will introduce the general form of the wave functions describing various emission processes in terms of the so-called core-angular harmonics. Then we will analyze the methods to integrate the coupled channels system of differential equations describing emission processes, namely numerical integration, diagonalization, analytical continuation and distorted wave approach. A very important approach to compute the decay width is the semiclassical method. We will extensively analyze its application within various models in the literature.

The fine structure of the α-decay is described by considering rotational and vibrational degrees of freedom in emitted fragments, axial and triaxial symmetry of the nuclei. Then we will investigate a more general case, namely the double fine structure in the binary cold fission process.

Ternary emission processes are also reviewed. We give the general framework to describe the two proton emission in terms of hyperspherical harmonics. Then we describe the ternary fission process in terms of spheroidal harmonics.

References

1. Gamow, G.: Zur Quantentheorie des Atomkernes. Z. Phys. **51**, 204–212 (1928)
2. Geiger, H., Nuttall, J.M.: The ranges of the α particles from various substances and a relation between range and period of transformation. Philos. Mag. **22**, 613–621 (1911)
3. Geiger, H.: Reichweitemessungen an α-Strahlen. Z. Phys. **8**, 45–57 (1922)
4. Buck, B., Merchand, A.C., Perez, S.M.: Half-lives of favoured alpha decays from nuclear ground states. At. Data Nucl. Data Tables **54**, 53–73 (1993)
5. Breit, G.: Theory of resonant reactions and allied topics. Springer, Berlin (1959)
6. Fröman, P.P.: Alpha decay from deformed nuclei. Mat. Fys. Skr. Dan. Vid. Selsk. **1**, 3 (1957)

Chapter 1
Introduction

1.1 Binding Energy and Q-Value

According to our daily experience, by weighing two objects of one kilogram each, one obtains exactly two kilograms. But in this case the obvious relation 1 + 1 = 2 is actually an approximation, and this amazing fact was revealed by Albert Einstein at the beginning of twentieth century using his well known relation, connecting the energy and mass $E = mc^2$, where c is the speed of the light in vacuum. The order of magnitude for electromagnetic forces between atoms is 1 electron volt (eV), while the mass of the lightest atom (hydrogen) is about 1 GeV $= 10^9$ eV and practically it is concentrated in the nucleus. Its dimension, of about 1 fermi (fm) = 10^{-15} m, is by five orders of magnitude smaller than the whole atom, i.e. 1 Ångström (Å) = 10^{-10} m. This means that the matter we feel is actually almost "empty" and its basic constituents (atoms) are weekly bound in comparison with their masses. Thus, the usual addition rule of masses remains valid with a high accuracy at the atomic level.

The situation completely changes by "weighing" for instance a deuteron, made of one proton and one neutron. The deuteron mass $M_d c^2$ is smaller than the sum of its constituents $M_p c^2 + M_n c^2$, because the magnitude of strong forces, binding together the two nucleons, i.e. 1 MeV $= 10^6$ eV, is comparable with their masses.

A basic concept of the nuclear physics is the binding energy, defined as the difference between the mass of the nucleus and the masses of its constituents, i.e. Z protons and N neutrons

$$B(Z,N) = Mc^2 - ZM_p c^2 - NM_n c^2. \tag{1.1}$$

This quantity is a negative number and defines the amount of energy binding together the nucleons inside the nucleus. The range of nuclear forces between nucleons is comparable with their sizes, i.e. about 1 fm, and therefore only neighboring nucleons interact with each other. At distances larger than 0.2 fm the interaction is attractive, while for shorter distances it becomes repulsive. The binding energy per nucleon is approximately constant, $B(Z,N)/A \approx 8\,\text{MeV}$ and

D. S. Delion, *Theory of Particle and Cluster Emission*, Lecture Notes in Physics, 819,
DOI: 10.1007/978-3-642-14406-6_1, © Springer-Verlag Berlin Heidelberg 2010

this property reflects the saturation property of nuclear forces, i.e the nucleus behaves like a liquid drop.

Actually the binding energy per nucleon has minima along a valley called stability valley. These values are plotted in Fig. 1.1 versus the mass number. Due to the electrostatic repulsion between protons the binding energy per nucleon becomes less negative by increasing the mass number. In order to compensate the effect of the Coulomb repulsion the number of neutrons increases faster than that of protons and therefore the stability line $Z(N)$ lies below the first bisectrice of the (N, Z) plane. Actually the nuclei with $Z > 92$ (transuranic nuclei) are unstable and were synthesized artificially.

The surface $B(Z, N)/A$ has another remarkable property. It has local minima along the stability valley for some proton (neutron) numbers, called magic numbers with $Z(N) = 2, 8, 20, 28, 50, 82, 126$. Double magic nuclei like ${}^{4}_{2}\mathrm{He}_{2}$, ${}^{16}_{8}\mathrm{O}_{8}$, ${}^{40}_{20}\mathrm{Ca}_{20}$, ${}^{56}_{28}\mathrm{Fe}_{28}$, ${}^{132}_{50}\mathrm{Sn}_{82}$, ${}^{208}_{82}\mathrm{Pb}_{126}$ are the most stable nuclei in nature (we used the notation ${}^{A}_{Z}\mathrm{X}_{N}$). This property is explained by the so-called shell model of the nucleus, where a given nucleon moves in the field created by all other nucleons. The nucleons are fermions and their spin $s = \frac{1}{2}$ strongly interacts with the angular momentum l by a spin-orbit force. A single particle state in this selfconsistent mean field is labeled by the isospin index $\tau = p, n$ (for protons or neutrons), energy level ε, angular momentum l, total spin j and its projection m, i.e.

$$a^{\dagger}_{\tau\varepsilon ljm}|0\rangle \to \psi_{\tau\varepsilon ljm}(\mathbf{r}, \mathbf{s}), \tag{1.2}$$

where $\mathbf{r}$ denotes the spatial and $\mathbf{s}$ the spin coordinate. The above notation means that the operator $a^{\dagger}$ acts on the vacuum state $|0\rangle$ in order to create a single particle state. The conjugate operator $a_{\tau\varepsilon ljm}$ annihilates this state.

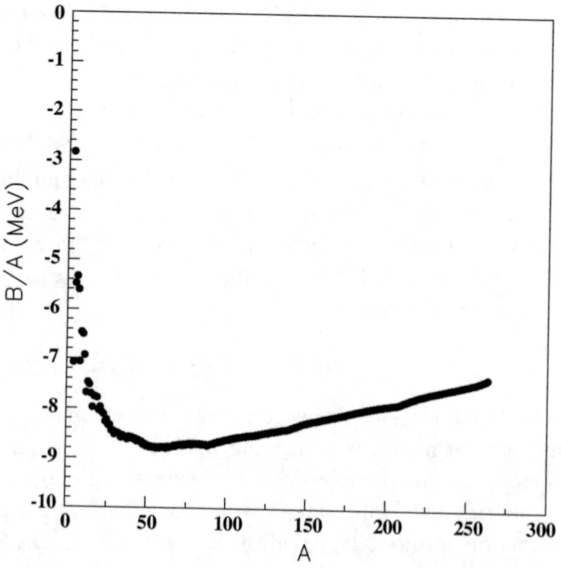

Fig. 1.1 Binding energy per mass number B/A versus the mass number along the stability valley

Due to these two properties the single particle levels, created in this selfconsistent potential, are grouped in several regions separated by larger gaps, corresponding to the above mentioned magic numbers. In the first approximation protons and neutrons fill these levels. Two nucleons cannot occupy the same single particle state and this property is mathematically described by the antisymmetric character of a two-particle state, i.e.

$$a^{\dagger}_{j_1m_1}a^{\dagger}_{j_2m_2}|0\rangle \rightarrow \frac{1}{\sqrt{2}}\left[\psi_{j_1m_1}(\mathbf{r}_1,\mathbf{s}_1)\psi_{j_2m_2}(\mathbf{r}_2,\mathbf{s}_2) - \psi_{j_2m_2}(\mathbf{r}_1,\mathbf{s}_1)\psi_{j_1m_1}(\mathbf{r}_2,\mathbf{s}_2)\right], \quad (1.3)$$

where the short-hand notation $(\tau\varepsilon lj) \rightarrow j$ is used. On each level with given quantum numbers (ε, l, j) one finds two nucleons of the same kind with opposite spin projections (m, $-m$). These nucleons interact with each other via residual two-body forces, the most important one being the pairing interaction. In this way a nuclear state becomes a coherent superposition of several pair states

$$a^{\dagger}_{jm}a^{\dagger}_{j-m}|0\rangle \rightarrow \frac{1}{\sqrt{2}}\left[\psi_{jm}(\mathbf{r}_1,\mathbf{s}_1)\psi_{j-m}(\mathbf{r}_2,\mathbf{s}_2) - \psi_{j-m}(\mathbf{r}_1,\mathbf{s}_1)\psi_{jm}(\mathbf{r}_2,\mathbf{s}_2)\right], \quad (1.4)$$

and the nucleons loose their individuality.

The nuclei out of the stability valley are in general unstable versus the splitting process into two or more fragments. To be more specific, the decay process is allowed as soon as the final configuration is energetically more favourable (i.e. more bound) than the initial one. For transitions between ground states this means that the binding energy of the initial nucleus is lower (more positive) than for the final configuration. Thus, the so-called Q-value of the splitting process into $1, 2, \ldots$ components is positive, i.e.

$$Q = B_i - B_1 - B_2 - \ldots > 0. \quad (1.5)$$

1.2 Strong Emission Processes

The instability of nuclei is called radioactivity. Nuclear physics was born when the natural spontaneous radioactivity of Uranium was discovered by Henry Becquerel in 1896. The new radiations were called by Ernst Rutherford using the first Greek letters, namely α, β and γ rays in 1899. Soon it was recognized that the α-radioactivity consists in emission of charged particles from atomic nuclei, β^- particles were identified with electrons, while γ rays with the electromagnetic radiation of much higher frequency than the usual light. Physicists realized that α-decay involves a new type of short range force called strong interaction, while β-decay is connected with the so-called weak interaction.

In 1928 George Gamow proposed a simple explanation for the exponential dependence of half-lives in α-decays upon Q-values [1], evidenced experimentally by the Geiger–Nuttall law [2, 3]. Gamow conceived the α-particle as a small ball, although composed of six particles, namely four protons and two electrons, moving in the mother nucleus which, through bouncing upon the nuclear surface, eventually penetrates quantum mechanically the surrounding Coulomb barrier. Only after James Chadwick has discovered the neutron in 1932 nuclear physicists realized that the α-particles are very bound clusters, made of two protons and two neutrons. By denoting the parent nucleus by ${}^{A}_{Z}P_{N}$, this process can be written as follows

$$ {}^{A}_{Z}P_{N} \rightarrow {}^{A-4}_{Z-2}D_{N-2} + \alpha. \tag{1.6} $$

Here by D with the corresponding labels we denoted the daughter nucleus. Of course α-decay is mainly allowed by the large binding energy of ${}^{4}_{2}$He, compared with neighboring nuclei. The nuclei connected by α-decays are situated on a line parallel with the first bisectrice in the (N, Z) plane. This line is called α-line.

Several important theoretical achievements in the field of the nuclear physics are directly connected with the α-decay theory. Let us mention here the Breit–Wigner theory of nuclear resonances [4, 5], the R-matrix approach of reactions and emission processes [6–9], Feshbach theory of nuclear reactions [10], the description of α-decays from deformed nuclei [11] and the shell model estimate of the α-particle spectroscopic factor [12–14].

Nowadays nuclear physics uses α-decays to investigate nuclear structure. The so-called α-spectroscopy gives an important experimental information concerning the nuclear structure of collective low-lying states. There are a lot of high precision data concerning α-decay intensities not only to ground, but also to excited states. Secondly, the α-decay chains are the only tool to investigate superheavy nuclei, which are very exotic systems, lying at the border of the nuclear stability.

In 1980, long time after the discovery of the α-decay, based on the large binding energy of ^{208}Pb, the spontaneous emission of heavier clusters like ^{14}C, ^{24}Ne, and ^{28}Mg was predicted [15, 16]. This new kind of natural radioactivity was soon experimentally found [17]. Now there are about 20 nuclei emitting heavy clusters. Their half-lives are much larger in comparison with typical α-decay half-lives. The corresponding decay process can be written as follows

$$ {}^{A}_{Z}P_{N} \rightarrow {}^{A-a}_{Z-z}D_{N-n} + {}^{a}_{z}C_{n}. \tag{1.7} $$

The main feature is that the heavy fragment is close or coincides with some Pb isotope and this is the reason why this kind of decay is also called "magic radioactivity". It lies between α-decay and fission.

In the last years, the study of the exotic very unstable nuclei has become the central subject of the nuclear physics. Proton rich nuclei are such systems, which can be investigated exclusively by proton emission using radioactive beams. The first discovery of the proton emission from ^{53}Co was made in 1970 [18] and today about 50 proton emitters are known [19]. A very exotic decay, namely two proton emission, is supposed to exist for some nuclei, where this process is energetically allowed. It was predicted in 1960 [20], but only relative recently it was detected in ^{45}Fe [21] and ^{48}Ni [22]. These two processes can be represented as follows

$$^{A}_{Z}P_N \rightarrow ^{A-k}_{Z-k}D_N + kp, \quad k = 1,2. \tag{1.8}$$

On the other extreme the nuclear fission is an important and active field in studying neutron rich nuclei. They are born during binary or ternary fission processes. Nuclear fission accompanied by the emission of neutrons was discovered by O. Hahn and F. Strassmann in 1939,

$$^{A}_{Z}P_N \rightarrow ^{A_1}_{Z_1}D_{N_1} + ^{A_2}_{Z_2}D_{N_2} + kn, \tag{1.9}$$

where

$$Z_1 + Z_2 = Z, \quad N_1 + N_2 + kn = N. \tag{1.10}$$

Only in 1962 some fission experiments reported binary fission fragments having very large kinetic energy [23]. In this case the low excitation energy of fragments shows that they are close to their ground states [24]. Therefore, this process was called cold (neutronless, $k = 0$) fission. Let us also mention the magic character of the cold fission, because one of the fragments is close or coincides with some Sn isotope. The first direct observation of the cold binary fission of ^{252}Cf was reported at Oak Ridge National Laboratory in 1994 [25, 26]. The so-called ternary fission is also possible

$$^{A}_{Z}P_N \rightarrow ^{A_1}_{Z_1}D_{N_1} + ^{A_2}_{Z_2}D_{N_2} + ^{a}_{z}C_n. \tag{1.11}$$

It is characterized by the emission of two heavy fragments, together with the equatorial emission of different light clusters, like α-particles, ^{10}Be, $^{12-14}$C, etc.

All these decay processes have a common feature: the emitted fragments are in their ground or low-lying states. This is the reason why they are called cold emission processes. As we already mentioned, another characteristic is that they are triggered by the strong interaction.

1.3 Electro-Weak Emission Processes

By considering nuclei with $Z + N =$ constant (this curve is also called β-line) one obtains for the binding energy a parabolic dependence. The nuclei along both sides of this parabola decay by β-decays until they rich the bottom. The β^--decay corresponds to neutron rich nuclei and consists of the emission of two leptons, namely an electron plus an antineutrino

$$^{A}_{Z}P_N \rightarrow ^{A}_{Z+1}D_{N-1} + e^- + \tilde{\nu}. \tag{1.12}$$

The energy spectrum of this process (Q-value) is continuous due to its three body character. The β^+-decay corresponds to the emission of a positron and a neutrino

$$^{A}_{Z}P \rightarrow ^{A}_{Z-1}D_{N+1} + e^{+} + \nu. \tag{1.13}$$

It is related to another process called electron-capture decay from the lowest atomic s-orbital

$$^{A}_{Z}P + e^{-} \rightarrow ^{A}_{Z-1}D_{N+1} + \nu. \tag{1.14}$$

These decay processes are triggered by the weak interaction. The two involved leptons are fermions with the spin $S = \frac{1}{2}$. Thus, in β-decays the final state can have total spin $S = 0$ or $S = 1$. In the electron capture the initial proton and electron can couple to $j \pm \frac{1}{2}$ and the final neutron and neutrino to $j \pm \frac{1}{2}$ or $j \mp \frac{1}{2}$. Thus, in all cases lepton spins can change the total nuclear angular momentum J by 0 or 1. Beta decays with no angular momentum change are called Fermi transitions, and those with an angular momentum change by one unit are called Gamow–Teller transitions.

Another process is also possible, namely double beta decay, involving the emission of two electrons and two antineutrinos

$$^{A}_{Z}P_{N} \rightarrow ^{A}_{Z+2}D_{N-2} + 2e^{-} + 2\tilde{\nu}. \tag{1.15}$$

This process occurs due to the fact that the pairing force renders the even-even nuclei more stable than odd–odd nuclei with broken pairs. Thus, the single β-decay process becomes energetically forbidden and $2\nu\beta\beta$-decay is the only possible channel.

Due to the small value of the coupling constant the decay rate can be estimated in a similar way to the γ-decay process, as a result of the transition from an excited to the ground state, i.e.

$$^{A}_{Z}X^{*} \rightarrow ^{A}_{Z}X_{N} + \gamma. \tag{1.16}$$

$\beta^{(\pm)}$-decays are collective processes involving many single particle orbitals. Thus, the corresponding transition operators can be written as a superposition of a neutron/proton annihilation and proton/neutron creation single particle operators (1.2), i.e.

$$\begin{aligned} \hat{\beta}^{(-)}_{\lambda} &= \Sigma_{if} \langle pf | T_{\lambda\mu} | ni \rangle a^{\dagger}_{pf} a_{ni} \\ \hat{\beta}^{(+)}_{\lambda} &= \Sigma_{if} \langle nf | T_{\lambda\mu} | pi \rangle a^{\dagger}_{nf} a_{pi}, \end{aligned} \tag{1.17}$$

where i/f denotes the quantum numbers εljm of the initial/final state. The meaning of the above expression is simple: in each term a neutron/proton initial state is annihilated and a proton/neutron final state is created. The coefficients are the β-decay matrix elements of the transition operator with the multipolarity $\lambda = 0, 1$.

Fermi transition operator is obviously the unity operator $T_{00} = 1$, while Gamow–Teller decay is described by the Pauli matrix $T_{1\mu} = \sigma_{1\mu}$.

Electromagnetic transitions can be described in a similar way, but with one kind of single particle operators

$$\hat{\gamma}_{\lambda} = \Sigma_{\tau=p,n}\Sigma_{if} e_{\tau}\langle\tau f|T_{\lambda\mu}|\tau i\rangle a^{\dagger}_{\tau f} a_{\tau i}, \tag{1.18}$$

where e_{τ} is the effective electric charge. The electric transition operator is proportional to the spherical harmonics defined below $T_{\lambda\mu} = r^{\lambda} Y_{\lambda\mu}$.

These decay processes, in which the lepton number is conserved, are described within the so-called Standard model of the electro-weak interaction formulated by Glashow [27], Weinberg [28] and Salam [29].

Let us mention here that another hypothetical double β-decay process is possible, namely neutrino-less $0\nu\beta\beta$-decay

$${}^{A}_{Z}P_{N} \rightarrow {}^{A}_{Z+2}D_{N-2} + 2e^{-}, \tag{1.19}$$

within the so-called Grand Unification Theory, if neutrino is a Majorana particle, i.e. it coincides with antineutrino, and the lepton-number conservation is violated.

References

1. Gamow, G.: Zur Quantentheorie des Atomkernes. Z. Phys. **51**, 204–212 (1928)
2. Geiger, H., Nuttall, J.M.: The ranges of the α particles from various substances and a relation between range and period of transformation. Philos. Mag. **22**, 613–621 (1911)
3. Geiger, H.: Reichweitemessungen an α-Strahlen. Z. Phys. **8**, 45–57 (1922)
4. Breit, G., Wigner, E.P.: Capture of slow neutrons. Phys. Rev. **49**, 519–531 (1936)
5. Breit, G.: Theory of Resonant Reactions and Allied Topics. Springer, Berlin (1959)
6. Kapur, P.L., Peirls, R.: The dispersion formula for nuclear reactions. Proc. Roy. Soc. A**166**, 277–295 (1938)
7. Teichmann, T., Wigner, E.P.: Sum rules in the dispersion theory of nuclear reactions. Phys. Rev. **87**, 123–135 (1952)
8. Thomas, R.G.: A formulation of the theory of alpha-particle decay from time-independent equations. Prog. Theor. Phys. **12**, 253–264 (1954)
9. Lane, A.M., Thomas, R.G.: R-Matrix theory of nuclear reactions. Rev. Mod. Phys. **30**, 257–353 (1958)
10. Feshbach, H.: Unified theory of nuclear reactions. Ann. Phys. (NY) **5**, 357–390 (1958)
11. Fröman, P.P.: Alpha decay from deformed nuclei. Mat. Fys. Skr. Dan. Vid. Selsk. **1**, no. 3 (1957)
12. Mang, H.J.: Calculation of α-transition probabilities. Phys. Rev. **119**, 1069–1075 (1960)
13. Săndulescu, A.: Reduced widths for favoured alpha transitions. Nucl. Phys. A **37**, 332–343 (1962)
14. Soloviev, V.G.: Effect of pairing correlations on the alpha decay rates. Phys. Lett. **1**, 202–205 (1962)
15. Săndulescu, A., Poenaru, D.N., Greiner, W.: Fiz. Elem. Chastits At Yadra **11**, 1334 (1980)
16. Săndulescu, A., Poenaru, D.N., Greiner, W.: New type of decay of heavy nuclei intermediate between fission and α decay. Sov. J. Part. Nucl. **11**, 528 (1980)

17. Rose, H.J., Jones, G.A.: A new kind of natural radioactivity. Nature (London) **307**, 245–246 (1984)
18. Jackson, K.P., et. al.: $^{52}Co^{m}$: a Proton-Unstable Isomer. Phys. Lett. B **33**, 281–283 (1970)
19. Woods, P.J., Davids, C.N.: Nuclei beyond the Proton Drip-Line. Ann. Rev. Nucl. Part. Sci. **47**, 541–590 (1997)
20. Goldansky, V.I.: On neutron deficient isotopes of light nuclei and the phenomena of proton and two-proton radioactivity. Nucl. Phys. **19**, 482–495 (1960)
21. Kryger, R.A., et. al.: Two-proton emission from the ground state of ^{12}O, Phys. Rev. Lett. **74**, 860–863 (1995)
22. Grigorenko, L.V., Johnson, R.C., Mukha, I.G., Thomson, I.J., Zhukov, M.V.: Two-proton radioactivity and three-body decay: General problems and theoretical approach. Phys. Rev. C **64**, 054002/1–12 (2001)
23. Milton, J.C.D., Fraser, J.S.: Time-of-Flight fission studies on ^{233}U, ^{235}U, and ^{239}Pu. Can. J. Phys. **40**, 1626–1663 (1962)
24. Guet, C., Ashgar, M., Perrin, P., Signarbieux, C.: A method to separate the mass and fission fragments with high kinetic energy. Nucl. Instrum. Methods **150**, 189–193 (1978)
25. Hamilton, J.H., et al.: Zero neutron emission in spontaneous fission of ^{252}Cf: a form of cluster radioactivity. J. Phys. G **20**, L85–L89 (1984)
26. Ter-Akopian, G.M., et al.: Neutron multiplicities and yields of correlated Zr–Ce and Mo–Ba fragment pairs in spontaneous fission of ^{252}Cf. Phys. Rev. Lett. **73**, 1477–1480 (1994)
27. Glashow, S.L.: Partial symmetries of weak interaction. Nucl. Phys. **22**, 579–588 (1961)
28. Weinberg, S.: A model of Leptons. Phys. Rev. Lett. **19**, 1264–1266 (1967)
29. Salam, A.: Elementary particle theory: relativistic groups and analyticity. In: Svartholm, N. (ed.) Proceedings of 8th Nobel Symposium, p. 367. Almqvist and Wiksell, Stockholm (1968)

Chapter 2
Binary Emission Processes

2.1 General Remarks

Most of nuclei decay by emitting a particle like proton, neutron, electron, positron or a composite cluster as deuteron, α-particle, Be, C, O, Mg, Ne, Si. Heavy nuclei can also fission into two fragments with comparable sizes. All these decays are called binary emission processes and they can be written as follows

$$P(J_iM_i) \rightarrow D_1(J_1M_1) + D_2(J_2M_2), \tag{2.1}$$

where J_iM_i is the initial spin and its projection. We suppose that other quantum numbers, like parity, are also embedded into this notation. The final spins J_k, $k = 1, 2$ satisfy the triangle rule

$$|J_1 - J_2| \leq J_i \leq J_1 + J_2, \tag{2.2}$$

and the initial parity equals the product of fragment parities.

As we already pointed out, our presentation concerns the description of decay processes induced only by the strong interaction. An important feature of these decays is connected with the large Coulomb barrier preventing the two fragments from moving apart. In the phenomenological description, we suppose that this barrier is extended in the internal region, so that the dynamics of the two fragments is fully described by a potential, defined for all distances. Such interaction is shown for the proton emission in Fig. 2.1, where it is given the spherical part of the Woods–Saxon potential, describing proton emission from ^{131}Eu. The energy of the system (dot-dashed line) is smaller that the height of the barrier, but the two fragments can penetrate it, due to the very small, but different from zero, wave function outside the barrier. We will show below, that this property can be described in terms of the so-called penetrability.

The main difficulty that one encounters when studying decay processes, not only from an experimental point of view, but also theoretically, is the instability of the initial nucleus. One may expect that these decay processes cannot be considered stationary and one has to use the time-dependent Schrödinger equation

D. S. Delion, *Theory of Particle and Cluster Emission*, Lecture Notes in Physics, 819,
DOI: 10.1007/978-3-642-14406-6_2, © Springer-Verlag Berlin Heidelberg 2010

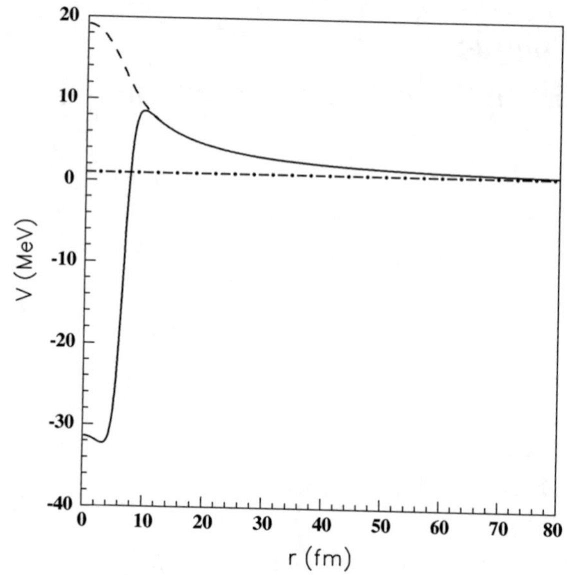

Fig. 2.1 Spherical part of the nuclear plus Coulomb interaction (*solid line*) as a function of the radius in ^{131}Eu. The Coulomb part (*dash line*) and the proton Q-value (*dot-dashed line*) are also shown

$$i\hbar\frac{\partial \Phi(t,\mathbf{r})}{\partial t} = \mathbf{H}\Phi(t,\mathbf{r}). \tag{2.3}$$

Anyway, due to large Coulomb barriers the probability of a cluster to escape from nucleus is very small and short lived states actually correspond to very narrow decay widths of such states, which are called resonant states. For instance, the shortest measurable half life is about $T_{1/2} = 10^{-12}$ s for proton emission, and the corresponding decay width is, according to the uncertainty relation

$$\Delta E \Delta t \sim \hbar, \tag{2.4}$$

$\Gamma \sim \Delta E = 6.6 \times 10^{-10}$ MeV. Since the characteristic nuclear time is $T_N \approx 10^{-22}$ s, the nucleus lives a long time before decaying in this energy-time scale and, therefore, the decay process may be considered stationary.

The wave function in the region of a resonance, with the width $\Delta E \sim \Gamma$, lying at an energy $E^{(0)}$ can be factorized in terms of the energy depending Lorentzian distribution as follows [1]

$$\begin{aligned}\Phi(E,\mathbf{r}) &= \frac{\Gamma/2}{\pi\left[(E-E^{(0)})^2+(\Gamma/2)^2\right]}\Psi(\mathbf{r}) \\ &= \frac{1}{2\pi i}\left[\frac{1}{E-(E^{(0)}-i\Gamma/2)} - \frac{1}{E-(E^{(0)}+i\Gamma/2)}\right]\Psi(\mathbf{r}).\end{aligned} \tag{2.5}$$

The corresponding evolution in time is given by the Fourier transform, which can easily be estimated by using the Cauchy theorem

$$\Phi(t,\mathbf{r}) = \int_{-\infty}^{\infty} \Phi(E,\mathbf{r}) e^{-iEt/\hbar} dE = \Psi(\mathbf{r}) e^{-iE_r t/\hbar}, \tag{2.6}$$

and therefore it has the form of a stationary state, but with a complex energy

$$E_r = E^{(0)} - \frac{i}{2}\Gamma, \tag{2.7}$$

as first proposed by Gamow in his paper explaining α-decay as a quantum penetration process, thus imposing the probabilistic interpretation of the quantum mechanics [2]. In this way, a narrow resonant state behaves almost like a bound state. In the same year, a similar explanation was given by Condon and Gurwey in Ref. [3].

Gamow put forward this idea, novel for his time and bold even today, that since the time dependence of the wave function, corresponding to the decaying resonance, should have the stationary form (2.6), the resonance energy could be considered complex with the form (2.7).

This idea proved to be of great significance in the study of all resonant processes. One of its important consequences is that, by going to the complex energy plane, the theory becomes dubious and difficult, but it has the great feature of transforming a time dependent process into a stationary one [4].

2.2 Angular Momentum Representation

The general framework to describe emission processes is the stationary scattering theory. The main tool is the angular momentum representation of solutions. Firstly we will give a textbook version of the formalism for spherical nuclei emitting structureless fragments. Then we will generalize the formalism to deformed nuclei emitting fragments with a given structure.

2.2.1 Spherical Boson Emitters

Let us consider that the fragments are structureless bosons emitted from a spherical nucleus. The typical case is the α-decay, connecting ground states of even–even spherical nuclei. The stationary Schrödinger equation describing such a process is written as follows

$$\left[-\frac{\hbar^2}{2\mu}\Delta + V_0(r)\right]\Psi(\mathbf{r}) = E\Psi(\mathbf{r}), \tag{2.8}$$

where μ denotes the reduced mass of the α-daughter system. The potential, $V_0(r)$ between the α-particle and daughter nucleus has two components: the short range nuclear part $V_N(r)$ and the Coulomb interaction $V_C(r)$. The wave function can be expanded in partial waves in terms of spherical harmonics

$$\Psi(\mathbf{r}) = \sum_l \frac{f_l(r)}{r} Y_{lm}(\hat{r}), \tag{2.9}$$

where $\mathbf{r} = (r, \hat{r}) \equiv (r, \theta, \phi)$. The spherical harmonics are factorized as follows

$$Y_{ln}(\theta, \phi) = \Theta_{lm}(\theta)\Phi_m(\phi), \quad \Phi_m(\phi) = \frac{e^{im\phi}}{\sqrt{2\pi}}, \tag{2.10}$$

where each component is orthonormal

$$\begin{aligned} \int_0^{\pi} \Theta^*_{lm}(\theta)\Theta_{l'm'}(\theta)\sin\theta d\theta &= \delta_{ll'} \\ \int_0^{2\pi} \Phi^*_m(\phi)\Phi_{m'}(\phi)d\phi &= \delta_{mm'}. \end{aligned} \tag{2.11}$$

The Laplacian in spherical coordinates has the following form

$$\Delta = \frac{1}{r}\frac{\partial^2}{\partial r^2} r - \frac{\hat{L}^2}{r^2}, \tag{2.12}$$

where the angular part of the operator acts on spherical harmonics as follows

$$\hat{L}^2 Y_{lm}(\theta, \phi) = l(l+1) Y_{lm}(\theta, \phi). \tag{2.13}$$

We insert the expansion in partial waves (2.9) in (2.8) and then we multiply it with $Y^*_{lm}(\theta, \phi)$. By using the orthogonality of spherical harmonics (2.11), one obtains the following equations for radial components

$$\frac{d^2 f_l(r)}{dr^2} = \left\{ \frac{l(l+1)}{r^2} + \frac{2\mu}{\hbar^2}[V_0(r) - E] \right\} f_l(r). \tag{2.14}$$

This system of decoupled equations can also be written in terms of the dimensionless reduced radius $\rho = \kappa r$, depending on the momentum $\kappa = \sqrt{2\mu E}/\hbar$, as follows

$$\left[-\frac{d^2}{d\rho^2} + \frac{V_l(r)}{E} - 1 \right] f_l(r) = 0, \tag{2.15}$$

where we introduced the angular momentum-dependent potential,

$$\frac{V_l(r)}{E} = \frac{l(l+1)}{\rho^2} + \frac{V_0(r)}{E}. \tag{2.16}$$

Since the nuclear interaction V_N is of a finite range, one has $V_N(r) = 0$ beyond some radius. Therefore at large distances only the spherical Coulomb interaction is active and the ratio between the interaction potential and emission energy can be expressed as follows

$$\frac{V_0(r)}{E} \to \frac{Z_1 Z_2 e^2}{rE} = \frac{\chi}{\rho}, \tag{2.17}$$

where Z_k are the charges of emitted fragments. Here we introduced the Coulomb parameter (twice the Sommerfield parameter)

$$\chi = 2\frac{Z_1 Z_2 e^2}{\hbar v}, \tag{2.18}$$

with the asymptotic velocity defined as follows

$$v = \sqrt{\frac{2E}{\mu}} = \frac{\hbar \kappa}{\mu}. \tag{2.19}$$

The independent solutions of the Coulomb equation

$$\left[-\frac{d^2}{d\rho^2} + \frac{l(l+1)}{\rho^2} + \frac{\chi}{\rho} - 1 \right] f_l(r) = 0, \tag{2.20}$$

are the standard regular and irregular Coulomb functions [5]. They are real functions of ρ. The regular solution $F_l(\chi, \rho)$ vanishes at the origin and increases as a function of the distance inside the Coulomb barrier, while the irregular solution $G_l(\chi, \rho)$ diverges at the origin but decreases with distance inside the Coulomb barrier.

At large distances both solutions oscillate, i.e. their asymptotic behaviour is given by

$$\begin{aligned} f_l(\chi, \rho) = F_l(\chi, \rho) &\to_{\rho\to\infty} \sin\left(\rho - \frac{1}{2} l\pi + \sigma_l\right), \\ G_l(\chi, \rho) &\to_{\rho\to\infty} \cos\left(\rho - \frac{1}{2} l\pi + \sigma_l\right). \end{aligned} \tag{2.21}$$

where σ_l is the Coulomb phase shift

$$\sigma_l = \arg\Gamma\left(l + 1 + i\frac{\chi}{2}\right) - \frac{1}{2}\chi \ln 2\rho, \tag{2.22}$$

with Γ being the Euler Gamma-function. In the above definition we also included the term depending upon the logarithm of the reduced radius.

If one of the emitted fragments is neutral, like for instance the neutron, then $\sigma_l = 0$ and the above spherical waves are proportional to spherical Bessel functions

$$\begin{aligned} f_l(\rho) = \rho j_l(\rho) &\to_{\rho\to\infty} \sin\left(\rho - \frac{1}{2}l\pi\right) \\ = \rho n_l(\rho) &\to_{\rho\to\infty} \cos\left(\rho - \frac{1}{2}l\pi\right). \end{aligned} \tag{2.23}$$

The outgoing/ingoing Coulomb–Hankel waves are defined in terms of the above Coulomb functions as follows

$$H_l^{(\pm)}(\chi,\rho) = G_l(\chi,\rho) \pm iF_l(\chi,\rho) \to_{\rho\to\infty} exp\left[\pm i\left(\rho - \frac{1}{2}l\pi + \sigma_l\right)\right]. \tag{2.24}$$

These waves become usual Hankel functions for neutral particles, i.e.

$$\rho h_l^{(\pm)}(\rho) = \rho[n_l(\rho) + ij_l(\rho)] \to_{\rho\to\infty} exp\left[\pm i\left(\rho - \frac{1}{2}l\pi\right)\right]. \tag{2.25}$$

2.2.2 Spherical Fermion Emitters

In the case of fermion (proton or neutron) emission from spherical nuclei the central potential entering Schrödinger equation (2.8) contains also the spin-orbit part, i.e.

$$V_0(r,\mathbf{s}) = V_0(r) + V_{so}(r)\mathbf{l}.\boldsymbol{\sigma}, \tag{2.26}$$

where $V_0(r)$ is the central nuclear plus Coulomb potential and $\boldsymbol{\sigma} = 2\mathbf{s}$. The wave function is expanded

$$\Psi(\mathbf{r},\mathbf{s}) = \sum_l \frac{f_{lj}(r)}{r}\mathcal{Y}_{jm}^{(l)}(\hat{r},\mathbf{s}), \tag{2.27}$$

in terms of spin-orbital harmonics

$$\mathcal{Y}_{jm}^{(l)}(\hat{r},\mathbf{s}) = \left[Y_l(\hat{r}) \otimes \chi_{\frac{1}{2}}(\mathbf{s})\right]_{jm} \equiv \sum_{m_1+m_2=m} \langle lm_1;\frac{1}{2}m_2|jm\rangle Y_{lm_1}(\hat{r})\chi_{\frac{1}{2}m_2}(\mathbf{s}), \tag{2.28}$$

where the bracket symbol denotes Clebsch–Gordan recoupling coefficients corresponding to the angular momentum addition $\mathbf{j} = \mathbf{l} + \mathbf{s}$. By using the same manipulations as in the previous case one obtains the system of equations for radial components

$$\left[-\frac{d^2}{d\rho^2}+\frac{l(l+1)}{\rho^2}+\frac{V_0(r)+V_{so}(r)\langle\mathbf{l}.\boldsymbol{\sigma}\rangle}{E}-1\right]f_{lj}=0, \tag{2.29}$$

where

$$\langle\mathbf{l}.\boldsymbol{\sigma}\rangle=j(j+1)-l(l+1)-\frac{3}{4}. \tag{2.30}$$

At large distances the system contains only Coulomb interaction (2.17).

2.3 S-Matrix

In this section we will investigate the so-called scattering states, i.e. real solutions of the Schrödinger equation with positive energy. The formalism presented below is common for boson and fermion cases. In order to simplify notations we will use boson channel notation l. For the fermion emission this index should be replaced by $l \to (lj)$.

2.3.1 Scattering States

In the external region, where the nuclear interaction vanishes, the solution is a combination of the Coulomb functions. The standard form that one adopts is the following

$$\begin{aligned} f_l^{(\mathrm{ext})}(E,r) &\sim G_l(\chi,\rho)\sin\delta_l(E)+F_l(\chi,\rho)\cos\delta_l(E) \\ &= \frac{i}{2}e^{-i\delta_l(E)}\left[H_l^{(-)}(\chi,\rho)-\mathscr{S}_l(E)H_l^{(+)}(\chi,\rho)\right]. \end{aligned} \tag{2.31}$$

Notice that this form is valid for spherical emitters, where in each spherical channel, corresponding to a given angular momentum l, there is an incoming wave $H_l^{(-)}$. Later on, in the Section devoted to the R-matrix method, we will give a more general expression corresponding to deformed emitters. We will show below that the phase shift is a real number, so that the S-matrix, defined by

$$\mathscr{S}_l(E)=e^{2i\delta_l(E)}, \tag{2.32}$$

satisfies the unitarity condition

$$\mathscr{S}_l(E)\mathscr{S}_l^{\dagger}(E)=1. \tag{2.33}$$

To evaluate the phase shift, and therefore the S-matrix, one requires the continuity of the external and internal wave functions and of the corresponding derivatives at the point $r = R$. The internal wave function should be regular in origin

$$f_l^{(\text{int})}(r \to 0) \to r^{l+1}. \tag{2.34}$$

This is achieved through the matching of the internal and external logarithmic derivatives of the wave function $f_l(r)$, i.e.

$$\beta_l^{(\text{int})}(R) \equiv \frac{1}{f_l^{(\text{int})}(R)} \frac{df_l^{(\text{int})}(R)}{dr} = \beta_l^{(\text{ext})}(R) \equiv \frac{1}{f_l^{(\text{ext})}(R)} \frac{df_l^{(\text{ext})}(R)}{dr}. \tag{2.35}$$

From Eq. 2.31, writing the trigonometric functions in terms of exponentials, one gets

$$\mathscr{S}_l(E) = e^{2i\bar{\delta}_l(E,R)} \frac{\beta_l^{(int)}(E,R) - D_l(E,R) + iP_l(E,R)}{\beta_l^{(int)}(E,R) - D_l(E,R) - iP_l(E,R)}. \tag{2.36}$$

where we have defined

$$\begin{aligned}
D_l(E,R) &\equiv kR \frac{F_l'(\chi,\kappa R) + iG_l'(\chi,\kappa R)}{F_l^2(\chi,\kappa R) + G_l^2(\chi,\kappa R)} = i\kappa R \frac{\left[H_l^{(-)}(\chi,\kappa R)\right]'}{\left|H_l^{(+)}(\chi,\kappa R)\right|^2}, \\
P_l(E,R) &\equiv \frac{\kappa R}{F_l^2(\chi,\kappa R) + G_l^2(\chi,\kappa R)} = \frac{\kappa R}{\left|H_l^{(+)}(\chi,\kappa R)\right|^2}, \\
e^{2i\bar{\delta}_l(E,R)} &\equiv \frac{F_l(\chi,\kappa R) + iG_l(\chi,\kappa R)}{F_l(\chi,\kappa R) - iG_l(\chi,\kappa R)} = \frac{H_l^{(-)}(\chi,\kappa R)}{H_l^{(+)}(\chi,\kappa R)},
\end{aligned} \tag{2.37}$$

with, e.g. $F_l'(\chi,\kappa R) \equiv dF_l(\chi,\rho)/d\rho|_{\rho=\kappa R}$. We have also used the property that the Wronskian for Coulomb functions is unity, i. e. $F_l'(\chi,\rho)G_l(\chi,\rho) - G_l'(\chi,\rho)F_l(\chi,\rho) = 1$ for all values of ρ [5]. One notices that $\bar{\delta}_l$ is real, since the Coulomb functions are real. Moreover, it vanishes inside the barrier for narrow resonant states due to very small values of the regular Coulomb function, i.e. in this region it is $e^{2i\bar{\delta}_l(E,R)} \approx 1$.

2.3.2 Resonances

Close to the resonant energy E_n one can expand the logarithmic derivative as follows

$$\beta_l^{(\text{int})}(E,R) \approx \beta_l(E_n,R) + \beta_l'(E_n,R)(E - E_n), \tag{2.38}$$

and the S-matrix becomes

$$\mathscr{S}_{nl}(E,R) = e^{2i\bar{\delta}_l(E,R)} \frac{E - E_n - \Delta_{nl}(E,R) - \frac{i}{2}\Gamma_{nl}(E,R)}{E - E_n - \Delta_{nl}(E,R) + \frac{i}{2}\Gamma_{nl}(E,R)} \tag{2.39}$$

where we introduced the energy shift and decay width as

$$\begin{aligned} \Delta_{nl}(E,R) &\equiv \frac{D_l(E,R) - \beta_l(E_n,R)}{\beta_l'(E_n,R)}, \\ \Gamma_{nl}(E,R) &\equiv -\frac{2P_l(E,R)}{\beta_l'(E_n,R)}. \end{aligned} \tag{2.40}$$

In order to estimate the derivative of the logarithmic derivative with respect to the energy we consider two internal solutions of Eq. 2.65, namely $f_1(r) \equiv f_l(E_1, r)$ and $f_2(r) \equiv f_l(E_2, r)$. Since they obey the Schrödinger equation one gets

$$f_2(r)\frac{d^2f_1(r)}{dr^2} - f_1(r)\frac{d^2f_2(r)}{dr^2} = \frac{2\mu}{\hbar^2}(E_2 - E_1)f_1(r)f_2(r). \tag{2.41}$$

By integrating both sides from 0 to R, dividing by $f_1(R)f_2(R)$ and using the fact that the internal solution should be regular in origin, i.e. $f_l(E, r = 0) = 0$, one obtains

$$\begin{aligned} \frac{R}{f_1(R)}\frac{df_1(R)}{dr} - \frac{R}{f_2(R)}\frac{df_2(R)}{dr} &= \beta(E_1,R) - \beta(E_2,R) \\ = (E_2 - E_1)\frac{2\mu R}{\hbar^2 f_1(R)f_2(R)} \int_0^R f_1(r)f_2(r)dr. \end{aligned} \tag{2.42}$$

In the limit $E_2 \to E_1 = E$ and normalizing to unity the internal wave function one gets

$$-\frac{d}{dE}\beta_l(E,R) = \frac{2\mu R}{\hbar^2 f_l^2(E,R)} \equiv \gamma_l^{-2}(E,R), \tag{2.43}$$

where we introduced the reduced width γ_l. Thus, the decay width in Eq. 2.40 acquires the standard form, as a product between the penetrability and reduced width squared, i.e.

$$\Gamma_{nl}(E,R) = 2P_l(E,R)\gamma_l^2(E_n,R), \tag{2.44}$$

which is a real positive number. We will see later that the same factorization of the decay width is also given by the R-matrix theory [6].

According to Eq. 2.39 the maximum value of the S-matrix occurs when the denominator is a minimun, i.e. for the energy $E = E_{nl}^{(0)}(E,R) = E_n + \Delta_l(E,R)$. In terms of this energy the S-matrix reads

$$\mathscr{S}_{nl}(E,R) = e^{2i\bar{\delta}_l(E,R)} \frac{E - E_{nl}^{(0)}(E,R) - \frac{i}{2}\Gamma_{nl}(E,R)}{E - E_{nl}^{(0)}(E,R) + \frac{i}{2}\Gamma_{nl}(E,R)}. \tag{2.45}$$

Physically, this equation is interpreted as the S-matrix corresponding to a resonance with energy $E_{nl}^{(0)}(E,R)$ and width $\Gamma_{nl}(E, R)$.

So far we have been careful to show the dependence upon E and R of the parameters entering the S-matrix, i.e. $E_{nl}^{(0)}(E,R)$ and $\Gamma_{nl}(E, R)$. However it has to be stressed that the dependence upon R is artificial, since the theory should not depend upon the matching radius if this point is properly chosen, i.e. beyond the range of the nuclear force. The dependence upon the energy, based in the approximation (2.38), is a more serious point. Wigner realized that this dependence is irrelevant in the analysis of observable resonances, since in that case the width is so small that in the energy range of the resonance Γ can be considered a constant. In the rest of this section we will only consider narrow resonances. That is the S-matrix has the form

$$\mathscr{S}_{nl}(E) = e^{2i\bar{\delta}_l} \frac{E - E_{nl}^{(0)} - \frac{i}{2}\Gamma_{nl}}{E - E_{nl}^{(0)} + \frac{i}{2}\Gamma_{nl}} = e^{2i\bar{\delta}_l}\left[1 - \frac{i\Gamma_{nl}}{E - E_{nl}^{(0)} + \frac{i}{2}\Gamma_{nl}}\right]. \tag{2.46}$$

where $E_{nl}^{(0)}$ and Γ_{nl} are real positive numbers.

It is important to keep in mind that Eq. 2.46 was obtained by assuming that Γ_{nl} is small and therefore it is only valid for narrow (and therefore isolated) resonances.

2.3.3 *Poles of the S-Matrix*

The poles of the S-matrix characterize the type of the resonant state. From Eq. 2.46 one sees that the energies and widths of narrow resonances can be obtained by calculating the complex poles of the S-matrix. But this is a very difficult task, because Γ_l can be many orders of magnitude smaller than $E_l^{(0)}$ and the computation of the complex energy $E_l^{(0)} + i\Gamma_l/2$ would require a very high precision. We will come back to this problem in Sect.2.4. But still one can compute the energy of the resonance as the pole of the S-matrix and then the width as the corresponding residues. To see this, we first notice that the number $\bar{\delta}_l$ (called "hard sphere phase shift") represents the contribution of the continuum background to the S-matrix and its value is negligible close to a narrow resonance. Therefore, the residues of the S-matrix at the pole n is

$$\mathrm{Res}[\mathscr{S}_{nl}(E)] = \lim_{E \to E_{nl}^{(0)} - \frac{i}{2}\Gamma_{nl}} \left(E - E_{nl}^{(0)} + \frac{i}{2}\Gamma_{nl}\right)\mathscr{S}_{nl}(E) = -i\Gamma_{nl}, \tag{2.47}$$

which is an important result since it shows that if the resonance is isolated then the residues of the S-matrix is a pure imaginary number. On the contrary, for resonances that are not isolated the residua are generally complex quantities [7].

Let us mention that by exchanging $H_l^{(+)}$ and $H_l^{(-)}$ between them in Eq. 2.31 one obtains a symmetric pole. Thus, by writing

$$\kappa_{nl} = \kappa_{nl}^{(0)} - i\lambda_{nl}, \tag{2.48}$$

where $\lambda_{nl} = \Gamma_{nl}/2$ the poles of the S-matrix are [8, 9]:

(a) decay states (Gamow resonances) with $\kappa_{nl}^{(0)} > 0$, $\lambda_{nl} > 0$ (dark squares in Fig. 2.2);
(b) capture states with $\kappa_{nl}^{(0)} < 0$, $\lambda_{nl} > 0$ (open squares in Fig. 2.2).

Let us mention that for negative energies the S-matrix has only imaginary poles, corresponding in (2.25) to the following asymptotics

$$f_l(\rho) \to exp[\mp(\rho + \frac{1}{2}l\pi)], \tag{2.49}$$

These are

(c) bound states, for which $\kappa_{nl}^{(0)} = 0$ and $\lambda_{nl} < 0$ (dark circles in Fig. 2.2);
(d) antibound states with $\kappa_{nl}^{(0)} = 0$, $\lambda_{nl} > 0$ (open circles in Fig. 2.2).

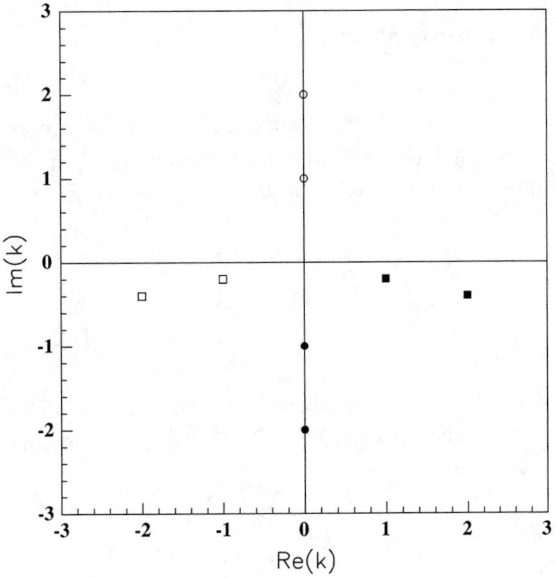

Fig. 2.2 Poles of the S-matrix: bound states (*dark circles*), antibound states (*open circles*), decay states (*dark squares*), capture states (*open squares*)

From Eq. 2.46 one also obtains

$$\delta_{nl} = \arctan\frac{\Gamma_{nl}/2}{E_{nl}^{(0)} - E}, \tag{2.50}$$

which shows that a resonance appears when the phase shift increases as it approaches the value $\delta_{nl} = \pi/2$ as a function of the energy assuming, also here, that the hard sphere phase shift is negligible. This criterion is often used to determine the energies of resonances.

Equation 2.50 allows one to evaluate the decay width by using still another expression, namely

$$\Gamma_{nl} = -2\left[\frac{\partial ctg\,\delta_{nl}}{\partial E}\right]^{-1}_{E=E_{nl}^{0}}. \tag{2.51}$$

Finally, it is worthwhile to mention that from Eq. 2.46 the cross section corresponding to the scattering of the particle at the energy of the resonance acquires the form

$$\sigma_{nl}(E) = (2l+1)\frac{\pi}{k^2}\frac{\Gamma_{nl}^2}{(E - E_{nl}^{(0)})^2 + (\Gamma_{nl}/2)^2}. \tag{2.52}$$

This formula was derived by G. Breit and E. P. Wigner [10] to explain the capture of slow neutrons. It is one of the most successful expressions written in quantum physics, as shown by its extensive use in the study of resonances ever since. It was by comparing with the experiment Wigner interpreted the number Γ as the width of the resonance. Since the imaginary part of the S-matrix pole is $-\ \Gamma/2$ (Eq. 2.46), this interpretation coincided with the Gamow interpretation of the width.

2.4 Gamow States

The states with complex energies $E = E_l^{(0)} - \frac{i}{2}\Gamma_l$, corresponding to the poles of the type (a) and (b), are called Gamow outgoing/ingoing states. According to the representation of the S-matrix (2.45) the scattering state (2.31) is given by

$$\begin{aligned} f_l^{(\mathrm{ext})}(r) \sim & \left(E - E_l^{(0)} + \frac{i}{2}\Gamma_l\right)H_l^{(\mp)}(\chi,\rho) \\ & - \left(E - E_l^{(0)} - \frac{i}{2}\Gamma_l\right)H_l^{(\pm)}(\chi,\rho), \end{aligned} \tag{2.53}$$

and the first term vanishes. Thus, the Gamow states are eigenstates of the stationary system of equations (2.14) with the following asymptotics

$$f_l^{(\mathrm{ext})}(r) = N_l H_l^{(\pm)}(\chi,\rho). \tag{2.54}$$

We can now formulate the complex eigenvalue problem for the Gamow states. The internal component of the relative wave function $f_l^{(\mathrm{int})}(r)$ should be regular in origin (2.34). The continuity of logarithmic derivatives for the internal and external wavefunction component of the resonant type (2.54) at some large radius R, where the interaction is given by the spherical Coulomb potential, can be fulfilled only for a discrete set of complex values of the wave number κ_{nl} (2.48), given by the above cases (a) and (b).

The angular momentum representation can be generalized for deformed nuclei, where both emitted fragments have structure and they can be left in some excited states. The dynamics of the decaying system is described by the following stationary Schrödinger equation

$$\mathbf{H}\Psi_{J_iM_i}(\mathbf{x}_1,\mathbf{x}_2,\mathbf{r}) = E\Psi_{J_iM_i}(\mathbf{x}_1,\mathbf{x}_2,\mathbf{r}). \tag{2.55}$$

We consider that the Hamiltonian describing binary emission is given by the following general ansatz

$$\mathbf{H} = -\frac{\hbar^2}{2\mu}\Delta_r + \mathbf{H}_1(\mathbf{x}_1) + \mathbf{H}_2(\mathbf{x}_2) + V(\mathbf{x}_1,\mathbf{x}_2,\mathbf{r}), \tag{2.56}$$

where $\mathbf{x}_k$ denote the internal coordinates of fragments and $\mathbf{r}$ the distance between them. We denote by V the inter-fragment potential and by $\mathbf{H}_k$ the Hamiltonians describing the internal motion of emitted fragments, i.e.

$$\mathbf{H}_k\Phi_{J_kM_k}(\mathbf{x}_k) = E_k\Phi_{J_kM_k}(\mathbf{x}_k), = 1,2, \tag{2.57}$$

where E_k are the excitation energies of emitted fragments and $\Phi_{J_kM_k}(\mathbf{x}_k)$ their eigenstates, satisfying the orthonormality condition

$$\langle \Phi_{J_kM_k}|\Phi_{J'_kM'_k}\rangle = \delta_{J_kJ'_k}\delta_{M_kM'_k}. \tag{2.58}$$

The external solution can be written as a superposition of different outgoing channels $c \equiv (J_1, J_2, J_c, j_c)$, describing all possible binary splittings, i.e.

$$\Psi_{J_iM_i}(\mathbf{x}_1,\mathbf{x}_2,\mathbf{r}) = \sum_c \Psi^{(c)}_{J_iM_i}(\mathbf{x}_1,\mathbf{x}_2,\mathbf{r}) = \sum_c \frac{f_c(r)}{r}\mathcal{Y}^{(c)}_{J_iM_i}(\mathbf{x}_1,\mathbf{x}_2,\hat{r}), \tag{2.59}$$

where, with $\hat{r} \equiv (\phi,\theta)$, we introduced the core-angular harmonics

$$\langle \mathbf{x}_1,\mathbf{x}_2,\hat{r}|c\rangle \equiv \mathcal{Y}^{(c)}_{J_iM_i}(\mathbf{x}_1,\mathbf{x}_2,\hat{r}) = \left\{[\Phi_{J_1}(\mathbf{x}_1)\otimes\Phi_{J_2}(\mathbf{x}_2)]_{J_c}\otimes\mathcal{Y}_{j_c}(\hat{r})\right\}_{J_iM_i}. \tag{2.60}$$

As usually, the symbol $[...\otimes...]_{JM}$ denotes the angular momentum coupling. Thus, the total spin is decomposed in each channel as follows

$$\begin{aligned}\mathbf{J}_i &= \mathbf{J}_c + \mathbf{j}_c\\ \mathbf{J}_c &= \mathbf{J}_1 + \mathbf{J}_2.\end{aligned} \tag{2.61}$$

These functions are also introduced within the R-matrix theory [6, 11, 12], were they are called surface functions. Obviously when one of the emitted fragments is structureless, like for instance in proton or α emission, one has $\Phi_{J_2}(\mathbf{x}_2) = 1$. The harmonics $\mathcal{Y}_{j_c m_c}(\hat{r})$ describe the angular relative motion and they coincide with usual spherical harmonics (2.10) in the case both fragments are bosons. Here j_c is an integer number. They coincide with spin-orbital harmonics (2.28) for the fermion (proton or neutron) emission, where j_c is half integer.

Due to the orthonormality of their components, the core-angular harmonics are mutually orthonormal, i.e.

$$\langle c|c'\rangle = \langle \mathcal{Y}^{(c)}_{J_iM_i}|\mathcal{Y}^{(c')}_{J_iM_i}\rangle = \delta_{cc'}. \tag{2.62}$$

We split the inter-fragment potential into a spherical and a deformed component, i.e. $V = V_0 + V_d$. By projecting out a given channel c and taking into account the orthonormality condition (2.62) one obtains the coupled channels system of equations for the radial wave functions

$$\frac{d^2f_c(r)}{dr^2} = \left\{\frac{l_c(l_c+1)}{r^2} + \frac{2\mu}{\hbar^2}[V_0(r) - E_c]\right\}f_c(r) + \frac{2\mu}{\hbar^2}\sum_{c'} V_d^{(cc')}(r)f_{c'}(r), \tag{2.63}$$

where the channel energy is defined as $E_c = E - E_1 - E_2$ and

$$V_d^{(cc')}(r) = \langle \mathcal{Y}^{(c)}_{J_iM_i}|V_d|\mathcal{Y}^{(c')}_{J_iM_i}\rangle. \tag{2.64}$$

At large distances only the spherical component is dominant and therefore the above system becomes decoupled

$$\left[-\frac{d^2}{d\rho_c^2} + \frac{l_c(l_c+1)}{\rho_c^2} + \frac{V_0(r)}{E_c} - 1\right]f_c(r) = 0, \tag{2.65}$$

in terms of the reduced radius for a given channel "c" $\rho_c = \kappa_c r$, where momentum is defined by $\kappa_c = \sqrt{2\mu E_c}/\hbar$. Here one has only the spherical Coulomb interaction (2.17) and the system (2.65) has a similar to (2.20) form for each angular momentum in the channel c, i.e.

$$\left[-\frac{d^2}{d\rho_c^2} + \frac{l_c(l_c+1)}{\rho_c^2} + \frac{\chi_c}{\rho_c} - 1\right]f_c(\chi_c, \rho_c) = 0, \tag{2.66}$$

where the channel Coulomb parameter is given by

$$\chi_c = 2\frac{Z_1Z_2e^2}{\hbar v_c}, \tag{2.67}$$

with the asymptotic channel velocity defined as follows

$$v_c = \sqrt{\frac{2E_c}{\mu}} = \frac{\hbar\kappa_c}{\mu}. \tag{2.68}$$

The outgoing/ingoing solutions of Eq. 2.66 are the standard Coulomb–Hankel waves (2.24). Thus, the general solution is an eigenstate of the stationary system of equations (2.63) with the following asymptotics

$$f_c(r) \to_{r\to\infty} f_c^{(\mathrm{ext})}(r), \tag{2.69}$$

where $f_c^{(\mathrm{ext})}(r)$ satisfies Eq. 2.66, i.e. it is a linear combination of (2.24). As in the spherical case, by denoting with $f_c^{(\mathrm{int})}(r)$ the internal components of the relative wave function regular in the origin, the continuity of the logarithmic derivatives of the wavefunction components at some large radius R, where the interaction becomes spherical, is given by a similar to Eq. 2.35 condition for each channel, i.e.

$$\beta_c^{(\mathrm{int})}(R) \equiv \frac{1}{f_c^{(\mathrm{int})}(R)} \frac{df_c^{(\mathrm{int})}(R)}{dr} = \beta_c^{(\mathrm{ext})}(R) \equiv \frac{1}{f_c^{(\mathrm{ext})}(R)} \frac{df_c^{(\mathrm{ext})}(R)}{dr}. \tag{2.70}$$

It can be fulfilled only for a discrete set of complex values of the wave number κ_n for the above (a)–(d) cases.

As in the spherical case, the cases (a) and (b) are respectively satisfied by the following asymptotics

$$f_c^{(\mathrm{ext})}(r) = N_c H_{l_c}^{(\pm)}(\chi_c, \rho_c), \tag{2.71}$$

where N_c are the scattering amplitudes in the channel c.

2.5 Decay Width and Half Life

For Gamow states (a) the imaginary part of κ according to (2.7) is negative and the modulus of the outgoing wave increases at large distances. By considering Eqs. 2.6 and 2.7, one obtains that the matter density decreases according to the following rule

$$|\Phi(t,\mathbf{r})|^2 = |\Psi(\mathbf{r})|^2 e^{-\Gamma t/\hbar}, \tag{2.72}$$

which is nothing else but the well-known exponential decay law, giving the number of nuclei at a certain moment

$$N(t) = N(0)e^{-\lambda t}, \tag{2.73}$$

where the decay constant is given by $\lambda = \Gamma/\hbar$. The half life is defined as the interval of time satisfying the condition $N(T) = N(0)/2$, i.e.

$$T_{1/2} = \frac{\hbar \ln 2}{\Gamma} = \frac{4.56 \times 10^{-22}}{\Gamma}, \tag{2.74}$$

where Γ is in MeV and T in seconds.

The decay width can be determined, in principle, by solving the coupled channels system (2.63) with the matching conditions (2.70) in the complex energy plane. The evaluation of the S-matrix poles can be performed by using the same procedure as the one used to evaluate bound states. The difference is that, now, one has to introduce the outgoing boundary condition given by Eq. 2.71. There are standard computer codes to do this, e.g. the codes of Refs. [13, 14]. These codes evaluate the energies corresponding to all poles of the S-matrix. The real energies define either the bound, or the antibound states. The complex energies, which are close to the real energy axis correspond to narrow resonances and therefore they accept the interpretation given above to such energies. That is, the real part is the position of the decaying resonance and minus twice the imaginary part is the corresponding width. However, in observable emission processes, the value of the imaginary part is usually much smaller, in absolute value, than the corresponding real part and its calculation is a difficult numerical task. But even if this calculation is possible, we want to stress that not always does the imaginary part of the energies correspond to the width of a resonance.

There is also an equivalent way to determine the width. Let us consider the stationary Schrödinger equation and its complex conjugate

$$\begin{aligned}\left(E-\frac{i}{2}\Gamma\right)\Psi &= \left(-\frac{\hbar^2}{2\mu}\nabla^2+V\right)\Psi\\ \left(E+\frac{i}{2}\Gamma\right)\Psi^* &= \left(-\frac{\hbar^2}{2\mu}\nabla^2+V\right)\Psi^*.\end{aligned} \tag{2.75}$$

One multiplies to the left the first relation by Ψ^* and the second one by Ψ. By substracting the two equalities one obtains

$$\Gamma\Psi^*\Psi = \frac{\hbar^2}{2\mu i}\left(\Psi^*\nabla^2\Psi - \Psi\nabla^2\Psi^*\right). \tag{2.76}$$

Here we considered that the potential operator V is Hermitian and therefore the corresponding difference vanishes after the volume integration. We then integrate this relation over internal variables $\mathbf{x}_1$, $\mathbf{x}_2$ and the relative coordinate $\mathbf{r}$ inside a sphere with a large radius. By transforming the right hand side term into a surface integral one obtains the following expression of the decay width

$$\Gamma = \frac{\hbar\oint \mathcal{J}(\hat{r})d\hat{r}}{\int\mathcal{P}(\mathbf{r})d\mathbf{r}}. \tag{2.77}$$

Here we introduced the internal probability

$$\mathcal{P}(\mathbf{r}) = \int d\mathbf{x}_1\int d\mathbf{x}_2|\Psi|^2, \tag{2.78}$$

and the probability flux

$$\mathscr{J}(\hat{r}) = \frac{\hbar}{2\mu i} \int d\mathbf{x}_1 \int d\mathbf{x}_2 (\Psi^* \nabla \Psi - \Psi \nabla \Psi^*) r^2. \tag{2.79}$$

We consider that the wave function is normalized to unity inside the considered sphere. In this way we suppose that the two fragments exist with the unity probability inside this volume. This statement is in an aparent contradiction with the emission process, leading to a decrease of the internal probability. Anyway, due to the very small value of the decay width compared with the emission energy one can use this condition, for a relative large time interval compared with the characteristic nuclear time [15].

On the surface of the sphere the gradient operator acts only on the radial direction $\nabla \to \mathbf{e}_r \frac{\partial}{\partial r}$, i.e.

$$\mathscr{J}(\hat{r}) \to \frac{\hbar}{2\mu i} \int d\mathbf{x}_1 \int d\mathbf{x}_2 \left(\Psi^* \frac{\partial \Psi}{\partial r} - \Psi \frac{\partial \Psi^*}{\partial r} \right) r^2. \tag{2.80}$$

Thus, by using the channel expansion (2.59) and

$$\frac{\partial \Psi^{(c)}}{\partial r} \to i\kappa_c \Psi^{(c)}, \tag{2.81}$$

the angular distribution becomes

$$\Gamma(\hat{r}) = \hbar \mathscr{J}(\hat{r}) = \hbar \sum_{cc'} v_c \lim_{r \to \infty} r^2 \int d\mathbf{x}_1 \int d\mathbf{x}_2 \Psi^{(c)*} \Psi^{(c')}. \tag{2.82}$$

By using the orthogonality of the core-angular harmonics (2.62) the decay width, proportional to the total probability flux through the surface of this sphere, becomes

$$\Gamma = \oint \Gamma(\hat{r}) d\hat{r} = \hbar \sum_c v_c \lim_{r \to \infty} \oint r^2 d\hat{r} \int d\mathbf{x}_1 \int d\mathbf{x}_2 |\Psi^{(c)}|^2. \tag{2.83}$$

By using the asymptotic relation (2.71) and the fact that the modulus of the outgoing Coulomb–Hankel wave function is unity, as seen from Eq. 2.24, one obtains

$$\Gamma = \sum_c \hbar v_c |N_c|^2 \equiv \sum_c \Gamma_c. \tag{2.84}$$

Thus, the total decay width can be written as a sum of partial decay widths corresponding to the considered channels. The equality between internal and external radial wave functions together with Eq. 2.71, i.e.

$$f_c^{(\mathrm{int})}(E_c, R) = f_c^{(\mathrm{ext})}(E_c, R) = N_c H_{l_c}^{(+)}(\chi_c, \kappa_c R), \tag{2.85}$$

determines the scattering amplitude

$$N_c = \frac{f_c^{(\mathrm{int})}(E_c, R)}{H_{l_c}^{(+)}(\chi_c, \kappa_c R)}. \tag{2.86}$$

Notice that N_c does not depend upon R, since both internal and external components satisfy the same Schrödinger equation. By inserting this value in the expression of the decay width (2.83) one obtains the following relation

$$\Gamma_c = \hbar v_c \left| \frac{f_c^{(\mathrm{int})}(E_c, R)}{H_{l_c}^{(+)}(\chi_c, \kappa_c R)} \right|^2 = 2P_{l_c}(E_c, R)\gamma_c^2(E_c, R), \tag{2.87}$$

where χ_c is the Coulomb parameter corresponding to the resonant complex energy. Here we introduced the standard penetrability and reduced width squared [6]

$$\begin{aligned} P_{l_c}(E_c, R) &= \frac{\kappa_c R}{\left| H_{l_c}^{(+)}(\chi_c, \kappa_c R) \right|^2}, \\ \gamma_c^2(E_c, R) &= \frac{\hbar^2}{2\mu R} |f_c^{(\mathrm{int})}(E_c, R)|^2. \end{aligned} \tag{2.88}$$

The form of the above decay width at the energy $E = E_c$, in terms of the penetrability and reduced width, is the same as in Eq. 2.44.

2.6 Decay Rules for the Half Life

According to the factorization of the decay width (2.87) and the above relation for the penetrability (2.88) the half life (2.74) is proportional to the modulus squared of the Coulomb–Hankel function inside the barrier. In this region it practically coincides with the irregular Coulomb function and has a very simple WKB ansatz, given in Appendix (14.2) by

$$\begin{aligned} H_l^{(+)}(\chi, \rho) &\approx (ctg\,\alpha)^{1/2} exp[\chi(\alpha - \sin\alpha\cos\alpha)] C_l \\ &= \left(\frac{1}{x} - 1 \right)^{-1/4} exp\left[\chi\left(\arccos\sqrt{x} - \sqrt{x(1-x)} \right) \right] C_l \\ &\equiv H_0^{(+)}(\chi, \rho) C_l \end{aligned} \tag{2.89}$$

where, with the external turning point $R_b = Z_1 Z_2 e^2/E$ and barrier energy $V_0 = Z_1 Z_2 e^2/R$, we introduced the following notations

$$\cos^2\alpha = x = \frac{\rho}{\chi} = \frac{R}{R_b} = \frac{E}{V_0}$$
$$C_l = exp\left[\frac{l(l+1)}{\chi}\sqrt{\frac{\chi}{\rho}-1}\right]. \tag{2.90}$$

Thus, the logarithm of the half life, corrected by the exponential centrifugal factor C_l defined by the second line of this relation, should be proportional to the Coulomb parameter, i.e.

$$\log_{10}T_{\rm red} = a_0\chi + b_0, \tag{2.91}$$

where we defined the reduced half life by

$$T_{\rm red} = \frac{T_{1/2}}{C_l^2} = \frac{\ln 2}{v_l}\left|\frac{H_0^{(+)}(\chi,\rho)}{f_l^{(\rm int)}(R)}\right|^2. \tag{2.92}$$

This relation is also called Geiger–Nuttall law, discovered in 1911 [16, 17] for α-decay between ground states (where the angular momentum carried by the α-particle is $l = 0$) and it can be written as follows

$$\log_{10}T_{1/2} = a\frac{Z}{\sqrt{Q_\alpha}} + b, \tag{2.93}$$

where Z is the charge of the left daughter nucleus and Q_α the Q-value of the α-particle. As we pointed out the explanation of this law was given by G. Gamow in 1928 [2], in terms of the quantum-mechanical penetration of the Coulomb barrier, given by the first line of Eq. 2.88. This quantity is characterized by the Coulomb parameter, which is proportional to the ratio $Z/\sqrt{Q_\alpha}$.

A special situation occurs in the case of proton emission, when the angular momentum of the emitted proton in general is different from zero. The half lives systematics for known proton emitters [18, 19, 20] is given in Fig. 2.3a. The picture becomes much simpler for the reduced half life (2.92) in Fig. 2.3b.

The Geiger–Nuttall law for proton emitters can be reproduced by the formula

$$\begin{aligned} log_{10}T_{red}^{(k)} &= a_k(\chi - 20) + b_k, \\ a_1 &= 1.31, \quad b_1 = -2.44, \quad Z<68 \\ a_2 &= 1.25, \quad b_2 = -4.71, \quad Z>68, \end{aligned} \tag{2.94}$$

where $k = 1$ corresponds to the upper line in Fig. 2.3b. The standard errors are $\sigma_1 = 0.26$ and $\sigma_2 = 0.23$, corresponding to a mean factor less than two. Here we considered the geometrical radius, i.e. $R = 1.2(A_D^{1/3} + 1)$. We will give in the next Section the explanation for this specific form of the Geiger–Nuttall law for proton emitters.

The two-proton emission was predicted in 1960 [21], but at the moment, the experimental material consists of few cases only. In Fig. 2.4, it is shown the

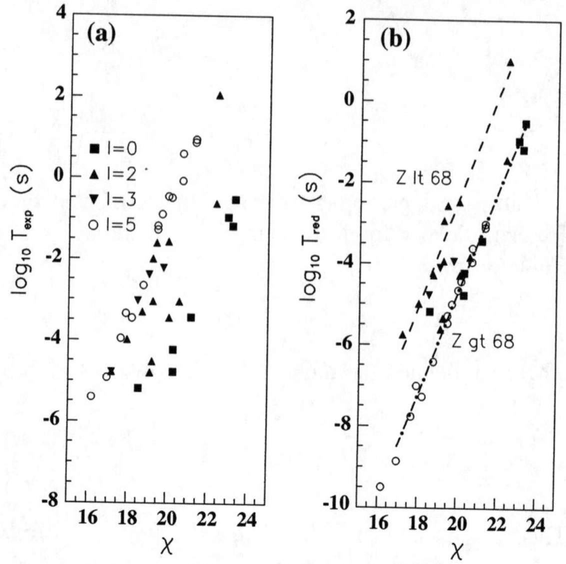

Fig. 2.3 **a** Logarithm of half lives for proton emitters versus the Coulomb parameter (2.67). *Different symbols* denote angular momenta carried by the emitted proton. **b** Logarithm of reduced half lives (2.92) for proton emitters versus the Coulomb parameter

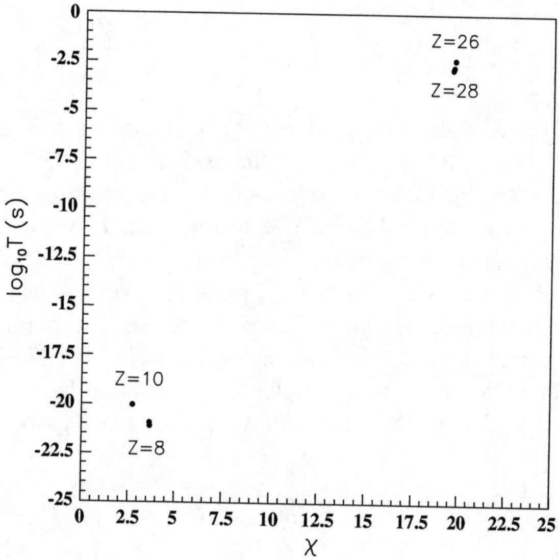

Fig. 2.4 Logarithm of half lives for two-proton emitters versus the Coulomb parameter (2.67). The charge of the mother nucleus is indicated

logarithm of the half life versus the Coulomb parameter, by considering a di-proton emission.

The α-decays between ground states are characterized by a remarkable regularity, especially for transitions between ground states of even–even nuclei. The logarithm of half lives along various isotopic chains lie on separate lines, as it is shown in Fig. 2.5. This feature is known as the Viola–Seaborg rule [22], i.e.

Fig. 2.5 Logarithm of half lives for α-decays from even–even nuclei versus Coulomb parameter (2.67). *Different lines* connect decays from nuclei with the same charge number

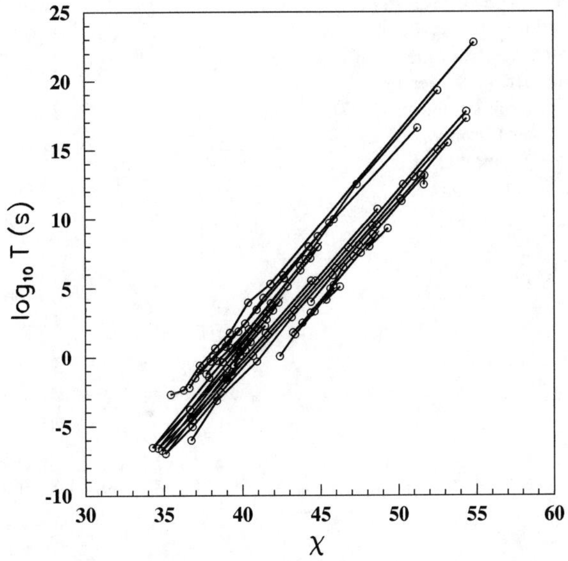

$$\log_{10}T_{1/2} = \frac{a_1 Z + a_2}{\sqrt{Q_\alpha}} + b_1 Z + b_2, \tag{2.95}$$

and it is connected with different α-particle reduced widths, multiplying the penetrability in (2.87), as will be shown latter. Still in doing systematics along neutron chains, there are important deviations with respect to this rule, as for instance in α-decay from odd mass nuclei, and this feature is strongly connected with nuclear structure details.

By using Eq. 2.89 a semi-empirical α-decay universal law for even–even emitters can be written as follows

$$\begin{aligned} \log_{10}T_{1/2} &= -\log_{10}P_\alpha - 20.446 + C(Z,N) \\ -\log_{10}P_\alpha &= A(Z,N)\sqrt{\frac{A_D}{A_P Q_\alpha}}[\arccos\sqrt{X} - \sqrt{X(1-X)}] \\ X = \frac{R_t}{R_b}, \; R_t &= 1.2249(A_D^{1/3} + 4^{1/3}), \; R_b = \frac{2Z_D e^2}{Q_\alpha}, \end{aligned} \tag{2.96}$$

where the functions *A*(*Z*, *N*) and *C*(*Z*, *N*) in terms of the parent proton and neutron numbers are given in Ref. [23].

In Ref. [24] a simpler universal law for even–even α-emitters is given by

$$\log_{10}T_{1/2} = 9.54\frac{Z_D^{0.6}}{\sqrt{Q_\alpha}} - 51.37. \tag{2.97}$$

Recently in Ref. [25] another type of law, taking into account all relevant dependencies, was proposed

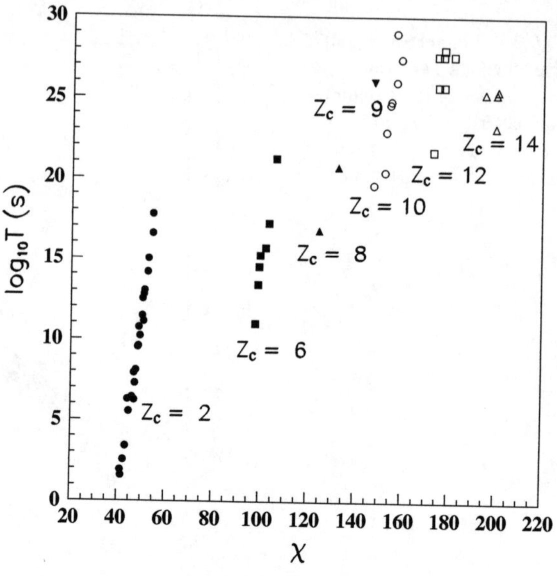

Fig. 2.6 Logarithm of half lives for heavy cluster decays and the corresponding α-decays from the same mother nuclei versus Coulomb parameter (2.67). *Different symbols* denote charge number of the emitted cluster

$$\log_{10}T_{1/2} = a + b\frac{A^{1/6}Z^{1/2}}{\mu} + c\frac{Z}{\sqrt{Q_\alpha}} + d\frac{A^{1/6}\sqrt{l(l+1)}}{Q_\alpha} + e[(-)^j - 1], \quad (2.98)$$

where A and Z are the parent mass and charge numbers respectively, with different sets of parameters for even–even, odd–odd, even–odd and odd–even α-emitters.

Viola–Seaborg rule can be generalized for heavy-cluster decays [26], as it is shown in Fig. 2.6. Here, the angular momenta carried by emitted fragments are zero. Thus, a similar to Eq. 2.96 universal law for the heavy cluster emission $P \to D + C$ is given by the following ansatz

$$\begin{aligned} &\log_{10}T_{1/2} = -\log_{10}P_C - 22.169 + 0.598(A_C - 1) \\ &-\log_{10}P_C = 0.22873\sqrt{\mu Z_D Z_C R_b}[\arccos\sqrt{Y} - \sqrt{Y(1-Y)}], \quad \mu = \frac{A_D A_C}{A_P} \\ &Y = \frac{R_t}{R_b}, \quad R_t = 1.2249(A_D^{1/3} + A_C^{1/3}), \quad R_b = 1.43998\frac{Z_D Z_C}{Q}. \end{aligned} \quad (2.99)$$

By using the expansion in power series of $\cos\alpha$ in Eq. 2.89, a simplified version of the above law was recently proposed in Ref. [27], namely

$$\log_{10}T_{1/2} = a\chi' + b\rho' + c, \quad (2.100)$$

in terms of the following two variables

$$\begin{aligned} \chi' &= Z_D Z_C \sqrt{A/Q} \\ \rho' &= \sqrt{Z_D Z_C A(A_D^{1/3} + A_C^{1/3})}, \quad A = \frac{A_D A_C}{A_P}. \end{aligned} \quad (2.101)$$

In Ref. [28] it was proposed the generalization of the Viola–Seaborg rule for the heavy cluster emission

$$\log_{10} T_{1/2} = a_1 \frac{Z_D Z_C}{\sqrt{Q}} + a_2 Z_D Z_C + b_2 + c_2, \tag{2.102}$$

with the following set of parameters

$$a_1 = 1.517,\ a_2 = 0.053,\ b_2 = -92.911,\ c_2 = 1.402,$$

where c_2 is the blocking parameter for odd-mass nuclei.

Finally, let us mention that it is not possible to derive a Viola–Seaborg rule for cold fission fragments. This feature was evidenced in Ref. [29] by using the Two Center Shell Model described in Sect.12.2. Here, the energy surface of the fissioning system is computed within the liquid drop model plus shell corrections in terms of several coordinates: the distance between emitted fragments, mass asymmetry, deformations and neck coordinate. The double magicity of ^{4}He and ^{208}Pb leads to very pronounced valleys of the total energy surface with a constant mass asymmetry during the whole α-decay of cluster emission process, respectively. Thus, the penetration has a simple expression given by Eq. (2.88), leading to the Viola–Seaborg rule, as we will show in the next section. On the contrary, the cold fission path proceeds through a sadle point of the potential energy surface, close to the double magic nucleus ^{132}Sn, but the mass asymmetry changes during the fission process after this point and the maximum yield corresponds to a different partition. The penetration given by the semiclassical approach (5.2), is computed along the fission path and obviously the half life has a more complex structure which does not satisfy the Viola–Seaborg rule.

Thus, one concludes that all cold emission processes have a common physical root, namely *the cold rearrangement of nucleons during the splitting process to a more stable di-nuclear configuration in which one of fragments has a double-magic structure*.

2.7 Decay Rule for the Reduced Width

The Viola–Seaborg rule (2.95) has a simple explanation in terms of the following schematic cluster-daugher potential [30]

$$\begin{aligned} V(r) &= \hbar\omega \frac{\beta(r - r_0)^2}{2} + v_0,\ r \leq r_B \\ &= \frac{Z_D Z_C e^2}{r},\ r > r_B, \end{aligned} \tag{2.103}$$

plotted in Fig. 2.7 for a particular value $r_0 = 0$.

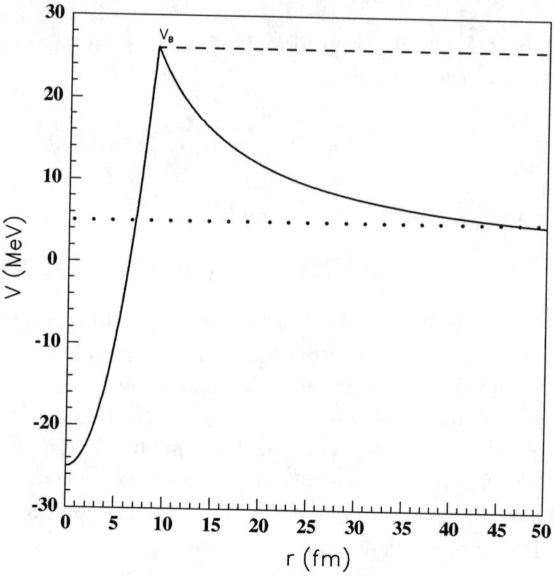

Fig. 2.7 The α-core potential (2.103) with $r_0 = 0$, $v_0 = -25$ MeV (*solid line*) and its barrier value (*dashed line*). Q-value is denoted by a *dotted line*

By considering Q-value as the first eigenstate in a spherical shifted harmonic oscillator well (see Appendix (14.8)), together with the continuity condition at the top of the barrier r_B, i.e.

$$
\begin{aligned}
Q - v_0 &= \frac{1}{2}\hbar\omega \\
\hbar\omega\frac{\beta(r_B - r_0)^2}{2} + v_0 &= \frac{Z_D Z_C e^2}{r_B},
\end{aligned}
\tag{2.104}
$$

one obtains the following relation

$$
\hbar\omega\frac{\beta(r_B - r_0)^2}{2} = V_{\text{frag}}(r_B) + \frac{1}{2}\hbar\omega,
\tag{2.105}
$$

where we introduced the so called fragmentation (or driving) potential as the difference between the Coulomb barrier and Q-value

$$
V_{\text{frag}}(r_B) = \frac{Z_D Z_C e^2}{r_B} - Q.
\tag{2.106}
$$

Let us stress on the fact that the above driving potential is a rough estimate of the interaction responsible for the emission process within the two potential model described in Sect. 4.6 in Chap. 4 by Eq. 4.51, i.e. the difference between the dashed and solid lines in Fig. 2.7. As we will show in that Section, this interaction connects the initial bound state to the final state in continuum.

According to Eq. 2.87 the logarithm of the decay width is a sum of two components

$$\log_{10}\Gamma = \log_{10} 2P(r_B) + log_{10}\gamma^2(r_B). \qquad (2.107)$$

The first component, in the above relation, contains the logarithm of the Coulomb–Hankel function inside the Coulomb barrier which, according to Eq. 2.89, is proportional to the Coulomb parameter χ. The second part contains the reduced width squared which, according to Eq. 2.88, is proportional to the modulus of the internal wave function squared. For a shifted harmonic oscillator well, one obtains (see Appendix (14.8)) $|f_{\rm int}|^2 = A^2 exp[-\beta(r - r_0)^2]$ and

$$\log_{10}\gamma^2(r_B) = -\frac{log_{10}e^2}{\hbar\omega}V_{\rm frag}(r_B) + \log_{10}\frac{\hbar^2 A^2}{2\mu e r_B}. \qquad (2.108)$$

Let us stress on the fact that the above relation does not depend upon the radius r_0.

In Fig. 2.8 we plotted the logarithm of the experimental reduced width squared by using Eq. (2.108) versus the neutron number by different open symbols, corresponding to five different regions of even–even α emitters, namely

$$\begin{array}{lll} (1) & Z<82,\ 50\leq N<82 & \text{(stars)}, \\ (2) & Z<82,\ 82\leq N<126 & \text{(open crosses)}, \\ (3) & Z\geq 82,\ 82\leq N<126 & \text{(open circles)}, \\ (4) & Z\geq 82,\ 126\leq N<152 & \text{(open squares)}, \\ (5) & Z\geq 82,\ N\geq 152 & \text{(open triangles)}. \end{array} \qquad (2.109)$$

In our calculations we used the value of the touching radius, i.e. $r_B = 1.2(A_D^{1/3} + R_C^{1/3})$.

By dark circles in Fig. 2.8, it is given the linear fit of the logarithm of the reduced width squared, in terms of the fragmentation potential (2.106), separately

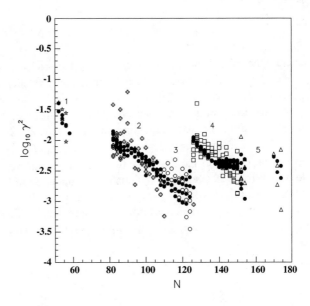

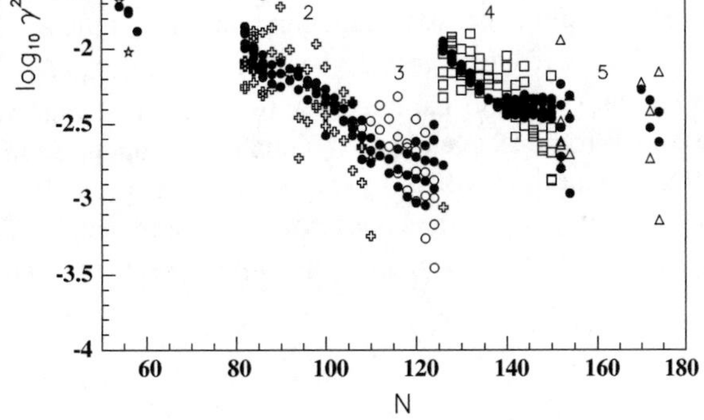

Fig. 2.8 The logarithm of the reduced width squared versus neutron number for five regions of the nuclear chart described by (2.109) (*open symbols*). By *dark circles* are given the results of the linear fit in terms of the fragmentation potential (2.106), separately for each region

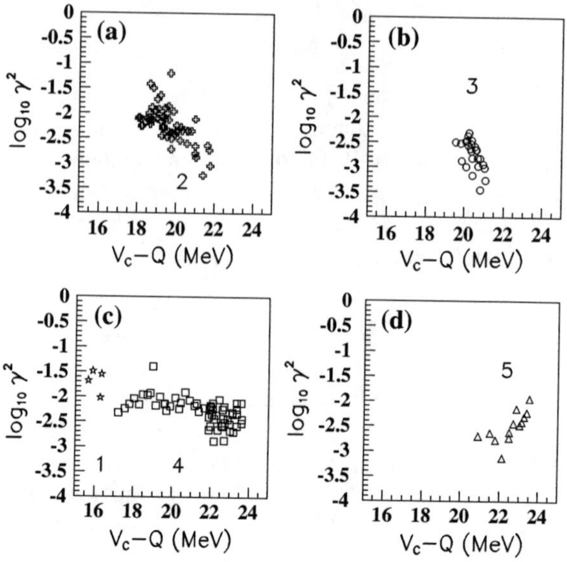

Fig. 2.9 The logarithm of the reduced width squared versus the fragmentation potential (2.106) for five regions of the nuclear chart described by (2.109)

for each region. In order to see this dependence better, in Fig. 2.9 it is given the logarithm of the experimental reduced width versus the fragmentation potential (2.106). Indeed, one sees a nice linear dependence for the regions 1–4, because they contain long isotopic chains, while in the last region 5 one has not more than two isotopes/chain. This is the reason why, except for the last region 5, the reduced width decreases with respect to the fragmentation potential, according to the theoretical prediction given by Eq. 2.108.

In this way one obtains that indeed the logarithm of the half life is of the Viola–Seaborg type

$$\log_{10} T_{1/2} = c_1(r_B)\chi + c_2 V_{\text{frag}}(r_B) + c_3(r_B, A^2), \qquad (2.110)$$

because the fragmentation potential contains the product $Z_D Z_C$. Let us mention that in the above relation we neglected for the weaker dependence on the second argument $\rho_B = \kappa r_B = \sqrt{2\mu Q} r_B/\hbar$ of the Coulomb–Hankel function (2.89) [27]. Obviously, the sum does not depend upon the radius and the free term depends on the logarithm of the wave function amplitude squared.

In Fig. 2.10 we plotted the difference $\log_{10} T_{1/2} - c_2 V_{\text{frag}}(r_B) - c_3(r_B, A^2)$ versus the Coulomb parameter χ, by using the same five symbols for the above described regions. Amazingly enough, we obtained three lines, corresponding to different amplitudes of the wave function at the radius r_B. The regions 1 and 4, corresponding to emitters above double magic nuclei ^{50}Sn and ^{208}Pb, respectively, have practically the same internal amplitudes A. The same is true for the regions 3 and 5.

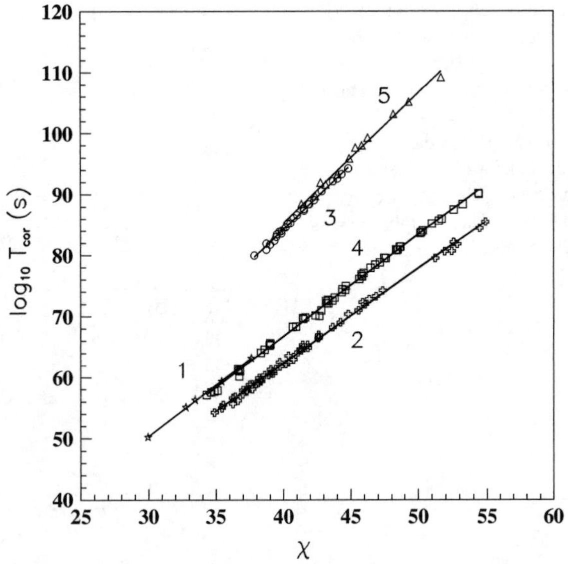

Fig. 2.10 The difference $\log_{10} T_{1/2} - c_2 V_{\rm frag}(r_B) - c_3(r_B, A^2)$ versus the Coulomb parameter χ for five different regions described by (2.109). The *straight lines* are the corresponding linear fits

Finally we mention that the linear dependence of $\log_{10}\gamma^2$ versus the fragmentation potential (2.108) remains valid for any kind of cluster emission. This fact is nicely confirmed by α and cluster emission processes in Fig. 2.11a, where we plotted the dependence between the corresponding experimental values for the same decays in Fig. 2.6. The straight line is the linear fit for cluster emission processes, except α-decay

$$\log_{10}\gamma^2 = -0.586(V_C - Q) + 15.399. \tag{2.111}$$

As already mentioned, for α-decays the fit gives several parallel lines corresponding to regions (2.109) in Fig. 2.8.

The above value of the slope $-\log_{10}e^2/\hbar\omega$ in Eq. 2.108 leads to $\hbar\omega \approx 1.5\,\text{MeV}$, with the same order of magnitude as in the α-decay case. The relative large scattering of experimental data around the straight line in Fig. 2.11a can be explained by the simplicity of the used cluster-core potential emission processes, (2.103).

Let us mention that a relation expressing the spectroscopic factor (proportional to the integral of the reduced width squared) for cluster emission processes was derived in Ref. [31]

$$S = S_\alpha^{(A_C-1)/3}, \tag{2.112}$$

where A_C is the mass of the emitted light cluster and $S_\alpha \sim 10^{-2}$. As can be seen from Fig. 2.11b, between A_C and $V_{\rm frag}$ there exists a rather good linear dependence and therefore the above scaling law can be easily understood in terms of the fragmentation potential.

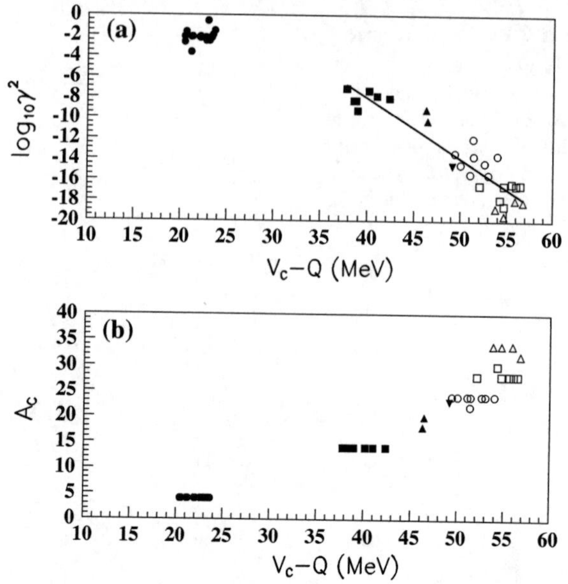

Fig. 2.11 **a** The logarithm of the reduced width squared versus the fragmentation potential (2.106). *Different symbols* correspond to cluster decays in Fig. 2.6. The *straight line* is the linear fit (2.111) for cluster emission processes, except α-decay. **b** Cluster mass number versus the fragmentation potential

Concerning the reduced widths of proton emitters, in Refs. [20, 32] it was pointed out the correlation between the reduced width and the quadrupole deformation. This fact can be seen in Fig. 2.12a, where the region with $Z < 68$ corresponds to $\beta > 0.1$ (open circles), while the other one with $Z > 68$ to $\beta < 0.1$ (dark circles). The two linear fits have obviously different slopes. This dependence is induced by the propagator matrix, entering the definition of the

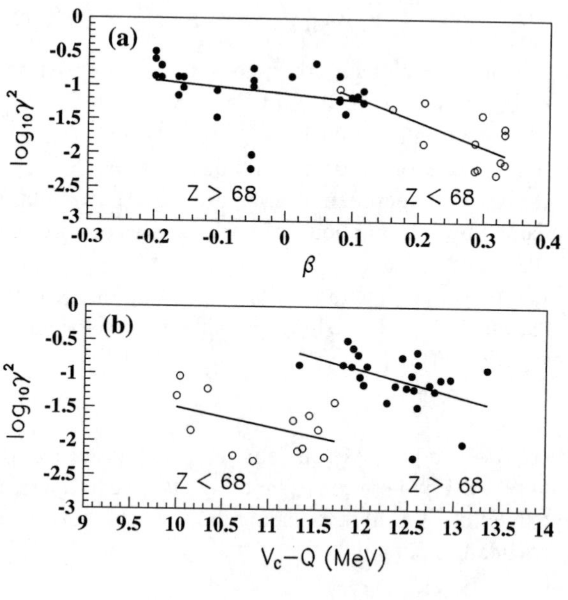

Fig. 2.12 **a** The logarithm of the reduced width squared versus the quadrupole deformation. By *open circles* are given proton emitters with $Z < 68$, while by *dark circles* those with $Z > 68$. The two regression lines fit the corresponding data. **b** The logarithm of the reduced width squared versus the fragmentation potential (2.106). The symbols are the same as in **a**

deformed reduced width, given by Eq. 4.17. Notice that the two dark circles with the smallest reduced widths correspond to the heaviest emitters with $Z > 80$.

At the same time one sees from Fig. 2.12b that the same data are clustered into two regions, which can be directly related with the fragmentation potential (2.106). Here the two linear fits in terms of the fragmentation potential, corresponding to the two regions of charge numbers, have roughly the same slopes, but different values in origin, i.e.

$$\begin{aligned} \log_{10}\gamma^2 &= -0.283(V_C - Q) + 1.329, \quad Z < 68 \\ \log_{10}\gamma^2 &= -0.365(V_C - Q) + 3.440, \quad Z > 68. \end{aligned} \tag{2.113}$$

The ho energy is $\hbar\omega \approx 1.5\,\text{MeV}$ for proton emission, i.e. the same order as for heavy cluster radioactivity and α-decay.

Thus, the two different lines in Fig. 2.3b can be directly connected with similar lines in Fig. 2.12b. They correspond to different orders of magnitude of the fragmentation potential, giving different orders of magnitude to wave functions and therefore to reduced widths.

2.8 Inter-Fragment Potential

As it was already mentioned, we can describe various emission processes, from the proton emission up to the cold fission, within the stationary coupled channels formalism. We suppose that the emitted fragments are already born and their motion is fully described by the Schrödinger equation with a two-body potential, defined for all inter-fragment distances. Obviously, such a description is strictly valid only for the particle (proton/neutron) emission. In the general case when both emitted fragments have structure, like for instance in α-decay, this potential picture is an idealization. Anyway, the emitted fragments are already formed in the region around the geometrical touching point, i.e. at the nuclear surface and only here one can determine a two-body potential. In the overlapping region the Pauli principle acts and the two fragments loose their identity. The equivalent potential becomes non-local and for a correct treatment, it is necessary the antisymmetrization of the wave function within the so-called Resonating Group Method (RGM), as it is described in the review [33], devoted to the microscopic description of cluster emission. Unfortunately, this method is adequate to describe the emission of relative light particle from nuclei close to a double magic nucleus, as it is the α-decay from ^{212}Po.

Anyway, a reasonable way to simulate the Pauli principle is the introduction of a repulsive core. As many calculations showed, the shape of this potential is not important, its only role consists in adjusting the energy of the resonant state in the resulting pocket-like potential to the experimental Q-value. The reason for this is that only the external part of the potential is important, in order to determine the asymptotics of the wave function and therefore the physical observables, like channel decay widths.

2.9 Double Folding Potential

The most general method to estimate the interaction potential between two composite fragments is the double folding procedure. We will suppose that both emitted fragments can be excited during the decay process. We separate the rotational degrees of freedom from other internal coordinates i.e. $\mathbf{x_k} = (\alpha_k, \omega_k)$, where ω_k are the Euler rotational coordinates. A good approximation of the Hamiltonian, describing the binary emission process, is given by the following ansatz

$$\mathbf{H} = -\frac{\hbar^2}{2\mu}\Delta_r + \mathbf{H}_1(\alpha_1) + \mathbf{H}_2(\alpha_2) + \mathbf{T}_1(\omega_1) + \mathbf{T}_2(\omega_2) + V(\alpha_1, \alpha_2, \omega_1, \omega_2, \mathbf{r}), \tag{2.114}$$

where $\mathbf{H}_k(\alpha_k)$ is connected with the internal dynamics of fragments, while $\mathbf{T}_k(\omega_k)$ with their rotational motion. This Hamiltonian describes a large variety of situations, i.e. proton/neutron emission, α-decay, heavy cluster emission and fission.

The double folding procedure is described in many text-books and review papers, e.g. [34], and consists of the following six dimensional integral

$$\begin{aligned} V(\alpha_1, \alpha_2, \omega_1, \omega_2, \mathbf{r}) &= \int d\mathbf{r}_1 \int d\mathbf{r}_2 \rho_1(\alpha_1, \mathbf{r}_1)\rho_2(\alpha_2, \mathbf{r}_2)v(\mathbf{r}_{12}) \\ \mathbf{r}_{12} &\equiv \mathbf{r} + \mathbf{r}_2 - \mathbf{r}_1, \end{aligned} \tag{2.115}$$

where $\mathbf{r}_k$ is the radius giving the nucleon position inside the k-th nucleus, as seen in Fig. 2.13, ρ_k is the nuclear density of the k-th fragment and $v(\mathbf{r}_{12})$ is the nucleon–nucleon force. The most popular two-body interaction, used to describe heavy ion scattering, is given by a superposition of Yukawa potentials, simulating the exchange of different mesons, called also M3Y interaction [35]. It is given by the following relation

$$v(\mathbf{r}_{12}) = v_{00}(r_{12}) + \hat{J}_{00}\delta(\mathbf{r}_{12}) + v_{01}(r_{12})\boldsymbol{\tau}_1 \cdot \boldsymbol{\tau}_2 + \frac{e^2}{r_{12}}, \tag{2.116}$$

where the central and isospin parts have respectively the following expressions

$$\begin{aligned} v_{00}(r) &= \left[7999\frac{e^{-4r}}{4r} - 2134\frac{e^{-2.5r}}{2.5r}\right] \text{MeV} \\ v_{01}(r) &= \left[-4885.5\frac{e^{-4r}}{4r} + 1175.5\frac{e^{-2.5r}}{2.5r}\right] \text{MeV}. \end{aligned} \tag{2.117}$$

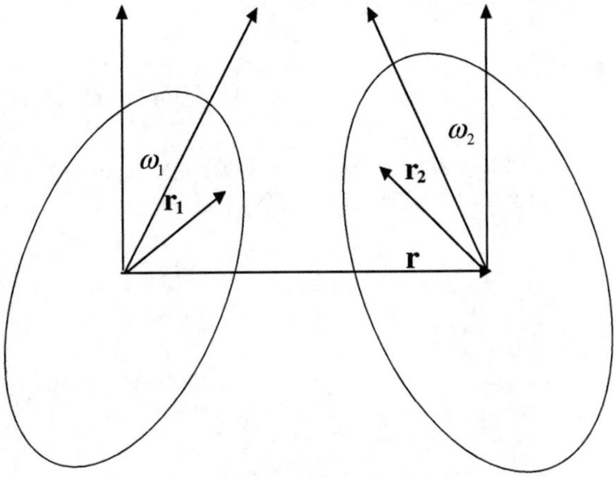

Fig. 2.13 Geometry of the double folding interaction

The second term in Eq. 2.116 approximates the single-nucleon exchange effects through a zero-range pseudopotential with the strength $\hat{J}_{00} = -262\,\mathrm{MeV\,fm}^3$.

Let us consider for simplicity that both nuclei are axially symmetric, but the generalization to the triaxial case is straightforward. The radial components of the nuclear densities are given by the standard multipole expansion, which can be written in both intrinsic and laboratory systems of coordinates as follows

$$\begin{aligned}\rho(\alpha_k, \mathbf{r_k}) &= \sum_{\lambda} \rho_\lambda(\alpha_k, r_k) Y_{\lambda 0}(\hat{r}'_k) \\ &= \sum_{\lambda\mu} \rho_\lambda(\alpha_k, r_k) D^\lambda_{\mu 0}(\omega_k) Y_{\lambda\mu}(\hat{r_k}), \quad k = 1, 2.\end{aligned} \tag{2.118}$$

We then expand the two-body interaction in Fourier components

$$v(\mathbf{r} + \mathbf{r}_2 - \mathbf{r}_1) = \int q^2 dq d\hat{q} \tilde{v}(q) e^{i\mathbf{qr}} e^{i\mathbf{qr}_2} e^{-i\mathbf{qr}_1}, \tag{2.119}$$

where one has for the Yukawa-like interaction

$$\tilde{v}(q) = \frac{1}{(2\pi)^3} \int v_0 \frac{e^{ikr}}{r} e^{i\mathbf{q}.\mathbf{r}} d\mathbf{r} = \frac{1}{(2\pi)^3} \frac{4\pi v_0}{q^2 + k^2}. \tag{2.120}$$

By using the multipole representation of the plane wave

$$e^{i\mathbf{q}.\mathbf{r}} = 4\pi \sum_l i^l j_l(qr) \sum_m Y_{lm}(\hat{q}) Y^*_{lm}(\hat{r}), \tag{2.121}$$

we obtain

$$V(\alpha_1, \alpha_2, \omega_1, \omega_2, \mathbf{r}) = V_0(\alpha_1, \alpha_2, r) + V_d(\alpha_1, \alpha_2, \omega_1, \omega_2, \mathbf{r}), \tag{2.122}$$

where the deformed part of the potential is given by

$$V_d(\alpha_1,\alpha_2,\omega_1,\omega_2,\mathbf{r}) = \sum_{\lambda_1\lambda_2\lambda_3} V_{\lambda_1\lambda_2\lambda_3}(\alpha_1,\alpha_2,r) \times \left\{ \left[D_0^{\lambda_1}(\omega_1) \otimes D_0^{\lambda_2}(\omega_2) \right]_{\lambda_3} \otimes Y_{\lambda_3}(\hat{r}) \right\}_0 . \tag{2.123}$$

Here the term $(\lambda_1\lambda_2\lambda_3) = (000)$ is excluded from summation. The radial formfactor in (2.123) is given by the integration over the angular coordinate $\hat{q}$

$$V_{\lambda_1\lambda_2\lambda_3}(\alpha_1,\alpha_2,r) = i^{\lambda_1-\lambda_2+\lambda_3}\frac{(4\pi)^3}{\sqrt{4\pi}}\hat{\lambda}_1\hat{\lambda}_2\langle\lambda_1,0;\lambda_2,0|\lambda_3,0\rangle \times \int_0^\infty q^2 dq\tilde{v}(q)\tilde{\rho}_{\lambda_1}(q)\tilde{\rho}_{\lambda_2}(q)j_{\lambda_3}(qr). \tag{2.124}$$

where we introduced the radial Fourier transform of densities

$$\tilde{\rho}_{\lambda_k}(q) = \int_0^\infty r_k^2 dr_k \rho_{\lambda_k}(r_k) j_{\lambda_k}(qr_k). \tag{2.125}$$

The spherical part of the potential can be written as a particular case of the above double folding potential, i.e.

$$V_0(\alpha_1,\alpha_2,r) = \frac{1}{\sqrt{4\pi}} V_{000}(\alpha_1,\alpha_2,r). \tag{2.126}$$

In the following we will describe various particular cases of the inter-fragment potential.

2.9.1 Boson Emission

Let us consider that both fragments are even–even nuclei with internal structure, like for instance in fission processes. They are left in ground or excited state and this kind of spectroscopy is characterized by a double fine structure.

Experimental measurements by electron scattering and microscopic calculations showed that the density distribution of fragments can be approximated by a Fermi-like shape, i.e.

$$\rho(\alpha_k,\mathbf{r}'_k) = \frac{\rho_k^{(0)}}{1+e^{[r_k-R(\hat{r}'_k)]/a_k}}, \quad k=1,2, \tag{2.127}$$

where the radius of the nuclear surface is given by

$$R(\hat{r}'_k) = R_0\left[1+\sum_{\lambda>0}\sum_{\nu}\alpha_{\lambda\nu}Y_{\lambda\nu}(\hat{r}'_k)\right], \tag{2.128}$$

and the central densities are normalized by the total number of protons and neutrons separately.

The case when $\lambda_2 = 0$, $\lambda_1 = \lambda_3 = \lambda$ describes the emission of a structureless particle "2" in the field of the daughter nucleus "1". This kind of spectroscopy is characterized by a single fine structure. The typical case is the α-decay, but the emission of spherical heavy clusters is also described by this particular ansatz. In all these cases the proton density of the light cluster is given by a Gaussian-like distribution, i.e.

$$\rho(\mathbf{r}_2) = \frac{Z_2}{b^3\pi^{3/2}} e^{-(r_2/b)^2}, \tag{2.129}$$

and a similar expression for the neutron density. The parameter of the distribution width is $b = 1.19$ fm for an α-particle, $b = 1.58$ fm for ^{12}C, $b = 1.74$ fm for ^{14}C [36]. The expansion (2.123) becomes the usual multipole–multipole interaction

$$V(\alpha, \omega, \mathbf{r}) = V_0(\alpha, r) + \sum_{\lambda > 0} V_\lambda(\alpha, r) \sum_{\mu} D^{\lambda}_{\mu 0}(\omega) Y_{\lambda\mu}(\hat{r}). \tag{2.130}$$

The last sum over μ represents nothing else but the rotation (by Euler angles ω) in the intrinsic system of coordinates

$$V(\alpha, \mathbf{r}') = V_0(\alpha, r) + \sum_{\lambda > 0} V_\lambda(\alpha, r) Y_{\lambda 0}(\hat{r}'). \tag{2.131}$$

2.9.2 Fermion Emission

Let us suppose that the light particle is a fermion (proton or neutron) and the daughter nucleus has axial symmetry. The interaction potential is given by the following ansatz

$$V(\mathbf{x}, \mathbf{r}, \mathbf{s}) = V_0(r, \mathbf{s}) + V_d(\mathbf{x}, \mathbf{r}, \mathbf{s}), \tag{2.132}$$

where $V_0(r, \mathbf{s})$ is the spherical component of the interaction

$$V_0(r, \mathbf{s}) \equiv V_0(r) + V_{so}(r)\mathbf{l}\cdot\boldsymbol{\sigma}. \tag{2.133}$$

Here $V_0(r)$ is the central interaction between the left nucleus and the fermion, while $V_{so}(r)$ is the spin-orbit interaction ($\sigma = 2s$). The central potential includes the nuclear part $V_N(r)$ and the Coulomb interaction $V_C(r)$. The folding procedure gives as a result the nuclear potential with a Woods–Saxon shape, like it is given in Fig. 2.1.

The deformed part can be written in a similar way as (2.130), i.e.

$$V_d(\mathbf{x}, \mathbf{r}, \mathbf{s}) \equiv \sum_{\lambda>0} V_\lambda(r, \mathbf{s}) Q_\lambda(\mathbf{x}) \cdot T_\lambda(\hat{r}). \tag{2.134}$$

The scalar product is, as usually, $Q_\lambda(\mathbf{x}) \cdot T_\lambda(\hat{r}) = \sum_\mu Q^*_{\lambda\mu}(\mathbf{x}) T_{\lambda\mu}(\hat{r})$where$Q_{\lambda\mu}(\mathbf{x})$ is the λ-pole operator which depends upon the collective coordinate $\mathbf{x}$. For example, in the case of collective rotations this coordinate corresponds to the Euler rotation angles $\omega \equiv (\phi, \theta, \psi)$, while for vibrational modes it corresponds to the quadrupole coordinates $\alpha_{2\mu}$.

The multipole operator $T_{\lambda\mu}$ has a more complex structure, as it is given in Appendix (14.1), i.e.

$$\begin{aligned} V_\lambda(r, \mathbf{s}) Q_\lambda(\mathbf{x}) \cdot T_\lambda(\hat{r}) &\to V_{c,\lambda}(r) Q_\lambda(\mathbf{x}) \cdot Y_\lambda(\hat{r}) \\ &+ V^{(\lambda-1)}_{\mathrm{so},\lambda}(r) Q_\lambda(\mathbf{x}) \cdot T^{(\lambda-1)}_\lambda(\hat{r}, \sigma) + V^{(\lambda+1)}_{\mathrm{so},\lambda}(r) Q_\lambda(\mathbf{x}) \cdot T^{(\lambda+1)}_\lambda(\hat{r}, \sigma), \end{aligned} \tag{2.135}$$

where the differential operator $T^{(\lambda\pm1)}_{\lambda\mu}$ is defined in Eq. 14.33 and describes the deformed spin-orbit part of the mean field.

2.9.3 Vibrational Nuclei

Let us consider an emission process where the light particle is structureless (a proton or an α-particle) and the heavy fragment is left in a vibrational state. Let us also suppose that the deformed interaction (2.131) has equipotential surfaces given by the ansatz (2.128). This means that the potential an be written as follows

$$V(\alpha, \mathbf{r}') = V\left(\frac{r}{1 + \sum_{\lambda>0} \sum_\nu \alpha_{\lambda\nu} Y_{\lambda\nu}(\hat{r}')}\right). \tag{2.136}$$

By expanding this function around the spherical shape one obtains the general form of the interaction between a vibrational core and the emitted light particle

$$\begin{aligned} V(\alpha, \mathbf{r}') &\approx V(0, r) + \sum_{\lambda>0} \sum_\nu \frac{\partial V(0, r)}{\partial \alpha_{\lambda\nu}} \alpha_{\lambda\nu} \\ &= V(0, r) - r\frac{\partial V(0, r)}{\partial r} \sum_{\lambda>0} \sum_\nu \alpha_{\lambda\nu} Y_{\lambda\nu}(\hat{r}'). \end{aligned} \tag{2.137}$$

In the laboratory system it becomes

$$V(\alpha, \omega, \mathbf{r}') = V_0(0, r) - r\frac{\partial V_\lambda(0, r)}{\partial r} \sum_{\lambda>0} \sum_{\mu\nu} \alpha_{\lambda\nu} D^\lambda_{\mu\nu}(\omega) Y_{\lambda\mu}(\hat{r}). \tag{2.138}$$

2.9.4 Triaxial Nuclei

The above potential is a particular case of a general triaxial interaction

$$\begin{aligned} V(\omega, \mathbf{r}, \mathbf{s}) &= V_0(r, \mathbf{s}) + \sum_{\lambda > 0} V_\lambda(r, \mathbf{s}) \cdot D^\lambda(\omega) \cdot T_\lambda(\hat{r}), \\ &\equiv V_0(r, \mathbf{s}) + \sum_{\lambda > 0, \mu} V_{\lambda\mu}(r, \mathbf{s}) \sum_\nu D^\lambda_{\nu\mu}(\omega) T_{\lambda\nu}(\hat{r}), \end{aligned} \tag{2.139}$$

where the multipole operators $T_{\lambda\mu}$ correspond to the usual spherical harmonics for the central part of the potential, while for the spin-orbit interaction (which includes derivatives) $T_{\lambda\mu}$ are the differential operators given by Eq. 14.33.

Therefore, one obtains in the intrinsic system the expression

$$V(\omega, \mathbf{r}', \mathbf{s}) = V_0(r, \mathbf{s}) + \sum_{\lambda > 0, \mu} V_{\lambda\mu}(r, \mathbf{s}) T_{\lambda\mu}(\hat{r}'). \tag{2.140}$$

2.10 Spectroscopic Factor

The above described double folding procedure supposes that the two fragments are already born. The parameters of the M3Y interaction (2.117) correspond to a double folding potential describing the heavy ion scattering. Yet, the wave function describing the emission process $P \to D + C$, where $P(D)$ is the parent (daughter) nucleus, has a cluster-like ansatz, i.e. it is a superposition of different mutually orthogonal channel components, similar with Eq. 2.59

$$\Psi_{J_i M_i}(\mathbf{x}_P, \mathbf{r}) \to \sum_c \mathscr{F}_c(r) \mathscr{Y}^{(c)}_{J_i M_i}(\mathbf{x}_D, \mathbf{x}_C, \hat{r}), \tag{2.141}$$

where $\mathbf{x}$ indicates the internal coordinate. We did not write the equality sign because, in general, the wave function of the initial configuration, given by the left hand side, does not contain a 100% cluster-like representation, as it is written by the right hand side.

Indeed, the ratio between the computed half life, by using the phenomenological double folding potential, and the experimental value is less than unity. It is called phenomenological spectroscopic factor

$$S = \frac{T_{\text{phen}}}{T_{\text{exp}}}. \tag{2.142}$$

This is due to the fact that actually the emitted fragments do not exist during the decay process, but they are born with certain probability. In deriving the expression of the decay width (2.83) we divided the outgoing flux to the volume integral of the wave function squared over the internal volume, by considering its value unity. Actually this volume integral gives the creation probability of emitted fragments.

The amplitude of the cluster-like ansatz contained in the initial wave function (2.141) is the overlap between the initial wave function and the product between the internal wave functions of the emitted fragments

$$\mathcal{F}_c(r) = \langle \mathcal{Y}^{(c)}_{J_iM_i} | \Psi_{J_iM_i} \rangle. \tag{2.143}$$

It is also called preformation amplitude and will be extensively analyzed in the second part of the book. Here, we only give the main ideas connected with this concept.

2.10.1 Particle Emission

For a proton emission from an odd–even emitter, connecting deformed nuclei in their ground states, the channel is given by the spin projection, i.e. $c \equiv K$. Let us denote by $a^{\dagger}_{\pi K}$ the *particle* creation operator ($\pi =$ proton). The initial wave function in the pairing approach is a superposition of different proton excitations of the parent Bardeen–Cooper–Schriffer (BCS) vacuum $\alpha^{\dagger}_{\pi K}|\text{BCS}_{\pi P}\rangle$. Since

$$a^{\dagger}_{\pi K}|\text{BCS}_{\pi D}\rangle = \left[u^{(D)}_{\pi K} \alpha^{\dagger}_{\pi K} + v^{(D)}_{\pi K} \alpha_{\pi \overline{K}} \right] |\text{BCS}_{\pi D}\rangle. \tag{2.144}$$

The preformation amplitude becomes

$$\mathcal{F}_K \equiv \langle \text{BCS}_{\pi D} | a_{\pi K} \alpha^{\dagger}_{\pi K} | \text{BCS}_{\pi P} \rangle = u^{(D)}_{\pi K} \langle \text{BCS}_{\pi D} | \text{BCS}_{\pi P} \rangle \approx u^{(D)}_{\pi K}. \tag{2.145}$$

where the last approximation reflects the blocking effect of the odd proton. It is important to point out that in proton emission the spectroscopic amplitude $\mathcal{F}_K$ is a constant, corresponding to the BCS amplitude around the Fermi surface, i.e. $u_K \sim \sqrt{0.5}$. It multiplies the scattering amplitude N_K.

2.10.2 Cluster Emission

The situation changes for cluster emission. The preformation amplitude (2.143) is a function of the radius between emitted fragments and it plays the role of the "internal" wave function. It should satisfy the matching condition with respect to the corresponding "external" channel radial component at certain radius R, i.e.

$$\mathscr{F}_c(R) = \frac{f_c(R)}{R}$$
$$\mathscr{F}'_c(R) = \left[\frac{f_c(r)}{r}\right]'_{r=R}. \tag{2.146}$$

In the second part, devoted to microscopic approaches, we will analyze in detail the properties of the preformation amplitude. Here we give only some preliminary details.

Let us illustrate how to estimate the overlap integral (2.143) in the case of α-decays involving transitions between ground states. The main idea is to find an α-like four body creation operator connecting daughter with parent nuclei, i.e.

$$|\Psi_P\rangle = P^\dagger_\alpha|\Psi_D\rangle. \tag{2.147}$$

If both parent and daughter are deformed nuclei, described within the pairing approach, than one has the following factorization

$$P^\dagger_\alpha = P^\dagger_\pi P^\dagger_\nu, \tag{2.148}$$

in terms of proton and neutron two body creation operators

$$P^\dagger_\tau = \sum_{K>0} X_{\tau K} a^\dagger_{\tau K} a^\dagger_{\tau\overline{K}}, \quad \tau = \pi, \nu. \tag{2.149}$$

The expansion coefficients are given by the following ansatz

$$X_{\tau K} = \langle \mathrm{BCS}_{\tau P}|a^\dagger_{\tau K} a^\dagger_{\tau\overline{K}}|\mathrm{BCS}_{\tau D}\rangle \approx u^{(D)}_{\tau K} v^{(D)}_{\tau K}. \tag{2.150}$$

Thus the overlap integral becomes

$$\mathscr{F}_\alpha(\mathbf{r}) = \langle \Psi_D \Psi_\alpha|P^\dagger_\alpha|\Psi_D\rangle = \sum_{KK'>0} X_{\pi K} X_{\nu K'} \langle \Psi_\alpha|a^\dagger_{\pi K} a^\dagger_{\pi\overline{K}} a^\dagger_{\nu K'} a^\dagger_{\nu\overline{K}'}|0\rangle, \tag{2.151}$$

where Ψ_α is the α-particle wave function, written as a product of three Gaussians in relative proton–neutron coordinates [37–39]. This four-body overlap integral will be computed in the second part of the book, by using the standard recoupling of two proton and two neutron single particle states, from absolute to the relative and center of mass coordinates [40].

In the case of the cluster emission, one obtains a similar representation [41]. For instance in ^{14}C emission a good approximation of the parent wave function, at distances where Pauli principle is less important, is given by

$$|\Psi_P\rangle \approx P^\dagger_{\alpha_1} P^\dagger_{\alpha_2} P^\dagger_{\alpha_3} P^\dagger_\nu |\Psi_D\rangle, \tag{2.152}$$

and the preformation factor is given by a similar expression, i.e.

$$\mathscr{F}_{^{14}C}(\mathbf{r}) = \langle \Psi_D \Psi_{^{14}C} | P^{\dagger}_{\alpha_1} P^{\dagger}_{\alpha_2} P^{\dagger}_{\alpha_3} P^{\dagger}_{\nu} | \Psi_D \rangle, \tag{2.153}$$

where $\Psi_{^{14}C}$ is the ^{14}C wave function, written as a product of several Gaussians in relative coordinates.

One defines the microscopic spectroscopic factor for transitions connecting ground states by the following integral

$$S_{gs} = \sum_{c} \int_{0}^{\infty} |r\mathscr{F}_c(r)|^2 dr. \tag{2.154}$$

It gives the order of magnitude of the cluster content inside the parent wave function. In principle it should have the same order of magnitude as the spectroscopic factor defined by Eq. 2.142. Actually they are quite different and the ratio S/S_{gs} defines the amount of the additional clustering with respect to the microscopic estimate, given by the preformation amplitude (2.143).

References

1. Unger, H.-J.: On the factorisation of the wave function and the green function in the region of isolated poles of the S-function. Nucl. Phys. A **104**, 564–576 (1967)
2. Gamow, G.: Zur Quantentheorie des Atomkernes. Z. Phys. **51**, 204–212 (1928)
3. Condon, E.U., Gurwey, R.W.: Wave mechanics and radioactive disintegration. Nature **22**, 439 (1928)
4. Civitarese, O., Gadella, M.: Physical and mathematical aspects of Gamow States. Phys. Rep. **396**, 41–113 (2004)
5. Abramowitz, M., Stegun, I.A. (eds.): Handbook of Mathematical Functions. Dover Publications Inc., New York (1983)
6. Lane, A.M., Thomas, R.G.: R-Matrix theory of nuclear reactions. Rev. Mod. Phys. **30**, 257–353 (1958)
7. Vertse, T., Liotta, R.J., Maglione, E.: Exact and approximate calculation of giant resonances. Nucl. Phys. A **584**, 13–34 (1995)
8. Berggren,T.: The use of resonant states in Eigenfunction expansions of scattering and reaction amplitudes. Nucl. Phys. A **109**, 265–287 (1968)
9. Berggren, T.: On the interpretation of cross sections for production of resonant final states. Phys. Lett. B **73**, 389–392 (1978)
10. Breit, G., Wigner, E.P.: Capture of slow neutrons. Phys. Rev. **49**, 519–531 (1936)
11. Teichmann, T., Wigner, E.P.: Sum rules in the dispersion theory of nuclear reactions. Phys. Rev. **87**, 123–135 (1952)
12. Thomas, R.G.: A formulation of the theory of alpha-particle decay from time-independent equations. Prog. Theor. Phys. **12**, 253–264 (1954)
13. Ixaru, L., Rizea, M., Vertse, T.: Piecewiese perturbation methods for calculating Eigensolutions of complex optical potential. Comput. Phys. Commun. **85**, 217–230 (1995)
14. Vertse, T., Pál, K.F., Balogh, A.: GAMOW, a program for calculating the resonant state solution of the Radial Schrödinger Equation in an arbitrary optical potential. Comput. Phys. Commun. **27**, 309–322 (1982)

15. Taylor, J.R.: Scattering Theory. Wiley, New York (1972)
16. Geiger, H., Nuttall, J.M.: The ranges of the α particles from various substances and a relation between range and period of transformation. Philos. Mag. **22**, 613–621 (1911)
17. Geiger, H.: Reichweitemessungen an α-Strahlen. Z. Phys. **8**, 45–57 (1922)
18. Sonzogni, A.A.: Proton radioactivity in $Z > 50$ nuclides. Nucl. Data Sheets **95**, 1–48 (2002)
19. Delion, D.S., Liotta, R.J., Wyss, R.: Theories of proton emission. Phys. Rep. **424**, 113–174 (2006)
20. Delion, D.S., Liotta, R.J., Wyss, R.: Systematics of proton emission. Phys. Rev. Lett. **96**, 072501/1–4 (2006)
21. Goldansky, V.I.: On neutron deficient isotopes of light nuclei and the phenomena of proton and two-proton radioactivity. Nucl. Phys. **19**, 482–495 (1960)
22. Viola, V.E., Seaborg, G.T.: Nuclear systematics of the heavy elements-II. J. Inorg. Nucl. Chem. **28**, 741–761 (1966)
23. Hatsukawa, Y., Nakahara, H., Hoffman, D.C.: Systematics of alpha decay half lives. Phys. Rev. C **42**, 674–682 (1990)
24. Brown, B.A.: Simple relation for alpha decay half-lives. Phys. Rev. C **46**, 811–814 (1992)
25. Denisov, V.Y., Khudenko, A.A.: α-Decay half-lives: empirical relations. Phys. Rev. C **79**, 054614/1–5 (2009)
26. Poenaru, D.N., Nagame, Y., Gherghescu, R.A., Greiner, W.: Systematics of cluster decay modes. Phys. Rev. C **65**, 054308/1–6 (2002)
27. Qi, C., Xu, F.R., Liotta, R.J., Wyss, R.: Universal decay law in charged-particle emission and exotic cluster radioactivity. Phys. Rev. Lett. **103**, 072501/1–4 (2009)
28. Ren, Z., Xu, C., Wang, Z.: New perspective on complex cluster radioactivity of heavy nuclei. Phys. Rev. C **70**, 034304/1–8 (2004)
29. Mirea, M., Delion, D.S., Săndulescu, A.: Microscopic cold fission yields of ^{252}Cf. Phys. Rev. C **81**, 044317/1–4 (2010)
30. Delion, D.S.: Universal decay rule for reduced widths. Phys. Rev. C **80**, 024310/1–7 (2009)
31. Blendowske, R., Fliessbach, T., Walliser, H.: From α-decay to exotic decays—a unified model. Z. Phys. A **339**, 121–128 (1991)
32. Medeiros, E.L., Rodrigues, M.M.N., Duarte, S.B., Tavares, O.A.P.: Systematics of half-lives for proton radioactivity. Eur. J. Phys. A **34**, 417–427 (2007)
33. Lovas, R.G., Liotta, R.J., Insolia, A., Varga, K., Delion, D.S.: Microscopic theory of cluster radioactivity. Phys. Rep. **294**, 265–362 (1998)
34. Carstoiu, F., Lombard, R.J.: A new method of evaluating folding type integrals. Ann. Phys. (NY) **217**, 279–303 (1992)
35. Bertsch, G., Borysowicz, J., McManus, H., Love, W.G.: Interactions for inelastic scattering derived from realistic potentials. Nucl. Phys. A **284**, 399–419 (1977)
36. Dao Khoa, T.: α-Nucleus potential in the double-folding model. Phys. Rev. C **63**, 034007/1–15 (2001)
37. Mang, H.J., Rasmussen, J.O.: Mat. Fys. Skr. Dan. Vid. Selsk. **2**(3), (1962)
38. Mang, H.J.: Alpha decay. Ann. Rev. Nucl. Sci. **14**, 1–28 (1964)
39. Poggenburg, J.K., Mang, H.J., Rasmussen, J.O.: Theoretical alpha-decay rates for the Actinide region. Phys. Rev. **181**, 1697–1719 (1969)
40. Delion, D.S., Insolia, A., Liotta, R.J.: New single particle basis for microscopic description of decay processes. Phys. Rev. C **54**, 292–301 (1996)
41. Delion, D.S., Insolia, A., Liotta, R.J.: Pairing correlations and quadrupole deformation effects on the ^{14}C decay. Phys. Rev. Lett. **78**, 4549–4552 (1997)

Chapter 3
Core-Angular Harmonics

3.1 Definition

The system of differential equations (2.63) describing emission processes can be written in two different ways, depending upon the structure of the emitted light cluster, namely

(a) boson emission and
(b) fermion emission.

First of all, let us introduce the wave function describing collective rotational and vibrational excitations of fragments within the geometric Bohr–Mottelson model [1]. A rotational band in even–even axially symmetric nuclei, with a given intrinsic angular momentum projection K, is defined as follows

$$\overline{\mathscr{D}}^{J*}_{MK}(\omega) = \frac{1}{\sqrt{2(1+\delta_{K,0})}}\left[\mathscr{D}^{J*}_{MK}(\omega) + (-)^{J+K}\mathscr{D}^{J*}_{M,-K}(\omega)\right]. \tag{3.1}$$

The ground rotational band with $K = 0$ is given by the normalized Wigner function $\mathscr{D}^{J*}_{M0}$ defined in Appendix (14.3). The rotational bands in triaxial nuclei are given by a superposition of states (3.1), i.e.

$$\Phi^{(\alpha)}_{JM}(\omega) = \sum_{K\geq 0} c(\alpha, J, K)\overline{\mathscr{D}}^{J*}_{MK}(\omega). \tag{3.2}$$

The rotational band can be also built on top of a single particle or a vibrational state. A rotational band in the even (proton)–odd (neutron) nucleus is defined as follows

$$\Psi^{J}_{MK}(\omega, \mathbf{r}'_n) = \frac{1}{\sqrt{2(1+\delta_{K,0})}}[\mathscr{D}^{J*}_{MK}(\omega)\Phi_K(\mathbf{r}'_n) + (-)^{J+K}\mathscr{D}^{J*}_{M,-K}(\omega)\Phi_{\tilde{K}}(\mathbf{r}'_n)]. \tag{3.3}$$

D. S. Delion, *Theory of Particle and Cluster Emission*, Lecture Notes in Physics, 819,
DOI: 10.1007/978-3-642-14406-6_3, © Springer-Verlag Berlin Heidelberg 2010

A similar ansatz has a rotational band built on a vibrational state

$$\Psi^J_{MK}(\omega,\alpha) = \frac{1}{\sqrt{2(1+\delta_{K0})}}[\mathscr{D}^{J*}_{M,K}(\omega)\Phi_K(\alpha) + (-)^{J+K}\mathscr{D}^{J*}_{M,-K}(\omega)\Phi_{\tilde{K}}(\alpha)]. \quad (3.4)$$

Here the wave function Φ_K describes an internal excitation: single particle motion in (3.3) or surface vibrations in (3.4).

It is also possible to describe collective excitations of even–even nuclei in an unified way, by using the Coherent State Model (CSM) [2, 3]. The wave function describing the ground band is given by

$$|\varphi^{(g)}_{JM}\rangle = N^{(g)}_J P^J_{M0} \exp[d(b^\dagger_{20} - b_{20})]|0\rangle, \quad (3.5)$$

where $b^\dagger_{2\mu}$ is the quadrupole creation boson operator, and the angular momentum projector is expressed in terms of the rotation operator $\hat{R}(\omega)$ as follows

$$P^J_{MK} = \frac{2J+1}{8\pi^2}\int D^J_{MK}(\omega)\hat{R}(\omega)d\omega. \quad (3.6)$$

It turns out that the expectation values of an ho Hamiltonian, i.e.

$$\mathscr{H}_J(d) = \langle\varphi^{(g)}_{JM}|\sum_\mu b^\dagger_{2\mu}b_{2\mu}|\varphi^{(g)}_{JM}\rangle, \quad (3.7)$$

have a vibrational sequence for small values of the deformation parameter $d < 1$ and a rotational shape for large values $d > 3$ [2, 3]. In this way it becomes possible to describe the structure of vibrational, transitional and rotational even–even nuclei by a smooth variation of the deformation parameter d.

A state in an odd–even nucleus is built as a coupling of the even–even core with the proton wave function $\psi^{(\lambda)}_{jm}(\mathbf{r}_p,\mathbf{s}_p) = u_l(r_p)\mathscr{Y}^{(\lambda\frac{1}{2})}_{jm}(\hat{r}_p,\mathbf{s}_p)$, where the spin-angular part is defined below by (3.22) and the radial wave function is an eigenstate of the spherical Woods-Saxon sp mean field corresponding to the even–even core. The coefficients of the resulting eigenstates

$$\Psi^{(n)}_{IM}(\mathbf{x},\mathbf{r}_p,\mathbf{s}_p) = \sum_{J\lambda j} X_I(J\lambda j;n)\left[\varphi^{(g)}_J(\mathbf{x}) \otimes \psi^{(\lambda)}_j(\mathbf{r}_p,\mathbf{s}_p)\right]_{IM}, \quad (3.8)$$

are determined by diagonalizing the particle–core interaction. The above wave function is the Nilsson counterpart in the laboratory system of coordinates, taking also into account the core degrees of freedom. For large deformations one obtains an expression

$$\Psi^{(n)}_{IM}(\omega,\mathbf{r}_p,\mathbf{s}_p) = \sum_{J\lambda j} X_I(J\lambda j;n)\left[\mathscr{D}^{J*}_0(\omega) \otimes \psi^{(\lambda)}_j(\mathbf{r}_p,\mathbf{s}_p)\right]_{IM}, \quad (3.9)$$

which is similar with the standard Nilsson wave function in the laboratory system of coordinates.

3.2 Boson Emission

In this section we will consider that both emitted fragments are bosons. Let us first analyze the most general case of the binary cold fission. Most of binary cold fission processes refer to the initial ground state of an even–even nucleus with $J_i = 0$. The relative motion of emitted fragments is described by the standard angular harmonics $Y_{lm}(\hat{r})$. Various channels are defined by $c \equiv (J_1 J_2 l)$, where J_k are the spins of the fragments, while l is the relative angular momentum. Thus, one has $j_c = l$ in the definition of channel core-angular harmonics (2.60), i.e.

$$\mathcal{Y}_{J_1 J_2 l}(\mathbf{x}_1, \mathbf{x}_2, \hat{r}) = \left\{ [\Phi_{J_1}(\mathbf{x}_1) \otimes \Phi_{J_2}(\mathbf{x}_2)]_l \otimes Y_l(\hat{r}) \right\}_0. \tag{3.10}$$

As a particular case, for the α-decay or heavy cluster emission process

$$P(J_i M_i) \rightarrow D(J_f M_f) + C(l), \tag{3.11}$$

connecting the parent (P) and daughter (D) nuclei, the light fragment (C) is a structureless boson with $J_2 = 0$ and therefore one has $\Phi_{J_2}(\mathbf{x}_2) = 1$.

The most general form of core-angular harmonics is given by

$$\mathcal{Y}^{(J_f l)}_{J_i M_i}(\mathbf{x}, \hat{r}) = \left[\Phi_{J_f}(\mathbf{x}) \otimes Y_l(\hat{r}) \right]_{J_i M_i}. \tag{3.12}$$

If the initial spin is zero, the wave function describing transitions to excited state in the daughter nucleus has a simpler ansatz

$$\mathcal{Y}_J(\omega, \hat{r}) = [\Phi_J(\omega) \otimes Y_J(\hat{r})]_0. \tag{3.13}$$

The α-decay from an odd–even nucleus is described by the following core-angular harmonics

$$\mathcal{Y}^{(nIl)}_{J_i M_i}(\mathbf{x}, \mathbf{r}_p, \mathbf{s}_p, \hat{r}) = \left[\Psi^{(n)}_{IM}(\mathbf{x}, \mathbf{r}_p, \mathbf{s}_p) \otimes Y_l(\hat{r}) \right]_{J_i M_i}. \tag{3.14}$$

The coupled system of equations gets a simpler form and it can be written in terms of the channel reduced radius as follows

$$\frac{d^2 f_c(r)}{d\rho_c^2} = \sum_{c'} \mathcal{A}_{cc'}(r) f_{c'}(r), \tag{3.15}$$

where the coupling matrix is given by

$$\mathcal{A}_{cc'}(r) = \left[\frac{l_c(l_c+1)}{\rho_c^2} + \frac{V_0(r)}{E_c} - 1 \right] \delta_{cc'} + \frac{1}{E_c} \left\langle \mathcal{Y}^{(c)}_{J_i M_i} | V_d(r) | \mathcal{Y}^{(c')}_{J_i M_i} \right\rangle. \tag{3.16}$$

Table 3.1 Core-angular harmonics describing boson emission processes

No.	Harmonics	Definition	Parent nucleus Daughter state
1.	$Y_{lm}(\hat{r})$	$\Theta_{lm}(\theta)\Phi_m(\phi)$	Even–even Adiabatic core
2.	$\mathcal{Y}^{(J_f l)}_{J_i M_i}(\omega, \hat{r})$	$\left[\mathcal{D}^{J_f *}_0(\omega) \otimes Y_l(\hat{r})\right]_{J_i M_i}$	Even–even Ground rotational band
3.	$\mathcal{Y}_J(\omega, \hat{r})$	$\left[\mathcal{D}^{J*}_0(\omega) \otimes Y_J(\hat{r})\right]_0$	Even–even ground Ground rotational band
4.	$\mathcal{Y}^{(J_f K_f l)}_{J_i M_i}(\omega, \hat{r})$	$\left[\mathcal{D}^{J_f *}_{K_f}(\omega) \otimes Y_l(\hat{r})\right]_{J_i M_i}$	Even–even K_f rotational band
5.	$\mathcal{Y}^{(\alpha J_f l)}_{J_i M_i}(\omega, \hat{r})$	$\left[\Phi^{(\alpha)}_{J_f}(\omega) \otimes Y_l(\hat{r})\right]_{J_i M_i}$	Triaxial even–even Triaxial rotational band
6.	$\mathcal{Y}^{(\lambda l)}_{J_i M_i}(\alpha_\lambda, \hat{r})$	$\left[\Phi_\lambda(\alpha_\lambda) \otimes Y_l(\hat{r})\right]_{J_i M_i}$	Even–even Vibrational state
7.	$\mathcal{Y}^{(J_f K_f l)}_{J_i M_i}(\omega, \alpha, \hat{r})$	$\left[\Psi^{J_f}_{K_f}(\omega, \alpha) \otimes Y_l(\hat{r})\right]_{J_i M_i}$	Even–even Rotational − vibrational state
8.	$\mathcal{Y}_{J_1 J_2 l}(\omega_1, \omega_2, \hat{r})$	$\left\{\left[\mathcal{D}^{J_1 *}_0(\omega_1) \otimes \mathcal{D}^{J_2 *}_0(\omega_1)\right]_l \otimes Y_l(\hat{r})\right\}_0$	Even–even Rotational + rotational state
9.	$\mathcal{Y}^{(nIl)}_{J_i M_i}(\mathbf{x}, \mathbf{r}_p, \mathbf{s}_p \hat{r})$	$\left[\Psi^{(n)}_{IM}(\mathbf{x}, \mathbf{r}_p, \mathbf{s}_p) \otimes Y_l(\hat{r})\right]_{J_i M_i}$	Odd–even Even–even

The matrix element of the deformed potential is given by the following standard expression

$$V^{(cc')}_d(r) \equiv \left\langle \mathcal{Y}^{(c)}_{J_i M_i} | V_d(r) | \mathcal{Y}^{(c')}_{J_i M_i} \right\rangle = \sum_{\lambda > 0} V_\lambda(r) \left\langle \mathcal{Y}^{(c)}_{J_i M_i} | Q_\lambda(\omega) \cdot Y_\lambda(r) | \mathcal{Y}^{(c)}_{J_i M_i} \right\rangle$$
$$= \sum_{\lambda > 0} (-)^{\lambda + J_f - J'_f} \begin{Bmatrix} J_f & J'_f & \lambda \\ l' & l & J_i \end{Bmatrix} \langle \Phi_{J_f} || Q_\lambda || \Phi_{J'_f} \rangle V_\lambda(r) \langle Y_l || Y_\lambda || Y_{l'} \rangle. \tag{3.17}$$

Various core-angular harmonics involved in boson emission are given in Table 3.1.

3.3 Fermion Emission

In this section we will consider that at least one of the emitted fragments is a fermion. As a typical example of the fermion emission is proton emission from proton rich nuclei. Neutron/proton delayed emission from giant resonant states or beta delayed particle emission can also be considered in this class. All these processes can be described within the same formalism. Thus, let us consider a general fermion emission process

$$P(J_iM_i) \rightarrow D(J_fM_f) + p(lj). \tag{3.18}$$

The wave function of the decaying system has the following form

$$\Psi_{J_iM_i}(\mathbf{x}, \mathbf{r}, \mathbf{s}) = \sum_{J_flj} \frac{f_{J_flj}(r)}{r} \mathcal{Y}^{(J_flj)}_{J_iM_i}(\mathbf{x}, \hat{r}, \mathbf{s}). \tag{3.19}$$

where $J_i(J_f)$ is the spin of the mother (daughter) nucleus but, as mentioned above, this index also labels all other quantum numbers, while (l, j) are the angular momentum and spin of the outgoing fermion. These angular momenta should satisfy the triangular relation, i.e.

$$|J_i - J_f| \le j \le |J_i + J_f|. \tag{3.20}$$

If the emitter is an odd–even nucleus and the daughter nucleus is left in its ground state, one has $J_f = 0$ and the fermion carries the spin $j = J_i$.

In the expansion (3.19) we introduced the core-spin-orbit harmonics defined by the following tensor

$$\mathcal{Y}^{(J_flj)}_{J_iM_i}(\mathbf{x}, \hat{r}, \mathbf{s}) = \left[\Phi_{J_f}(\mathbf{x}) \otimes \mathcal{Y}^{(l\frac{1}{2})}_{j}(\hat{r}, \mathbf{s})\right]_{J_iM_i}, \tag{3.21}$$

where the angular part, describing fermion motion, is given by

$$\mathcal{Y}^{(l\frac{1}{2})}_{jm}(\hat{r}, \mathbf{s}) = \left[i^l Y_l(\hat{r}) \otimes \chi_{\frac{1}{2}}(\mathbf{s})\right]_{jm}. \tag{3.22}$$

Here $\chi_{\frac{1}{2}}(\mathbf{s})$ denotes the spin function. The core-angular harmonics involved in fermion emission are given in Table 3.2.

Table 3.2 Core-angular harmonics describing fermion emission processes

No.	Harmonics	Definition	Parent nucleus Daughter state
1.	$\mathcal{Y}^{(l\frac{1}{2})}_{jm}(\hat{r}, \mathbf{s})$	$\left[i^l Y_l(\hat{r}) \otimes \chi_{\frac{1}{2}}(\mathbf{s})\right]_{jm}$	Odd–even Adiabatic core
2.	$\mathcal{Y}^{(J_flj)}_{J_iM_i}(\omega, \hat{r}, \mathbf{s})$	$\left[\mathcal{D}^{J_f*}_{0}(\omega) \otimes \mathcal{Y}^{(l\frac{1}{2})}_{j}(\hat{r}, \mathbf{s})\right]_{J_iM_i}$	Odd–even Ground rotational band
3.	$\mathcal{Y}^{(J_fK_flj)}_{J_iM_i}(\omega, \hat{r}, \mathbf{s})$	$\left[\overline{\mathcal{D}}^{J_f*}_{K_f}(\omega) \otimes \mathcal{Y}^{(l\frac{1}{2})}_{j}(\hat{r}, \mathbf{s})\right]_{J_iM_i}$	Odd–even K_f rotational band
4.	$\mathcal{Y}^{(\alpha J_flj)}_{J_iM_i}(\omega, \hat{r}, \mathbf{s})$	$\left[\Phi^{(\alpha)}_{J_f}(\omega) \otimes \mathcal{Y}^{(l\frac{1}{2})}_{j}(\hat{r}, \mathbf{s})\right]_{J_iM_i}$	Triaxial odd–even Triaxial rotational band
5.	$\mathcal{Y}^{(\lambda lj)}_{J_iM_i}(\alpha_\lambda, \hat{r}, \mathbf{s})$	$\left[\Phi_\lambda(\alpha_\lambda) \otimes \mathcal{Y}^{(l\frac{1}{2})}_{j}(\hat{r}, \mathbf{s})\right]_{J_iM_i}$	Odd–even Vibrational state
6.	$\mathcal{Y}^{(J_fK_flj)}_{J_iM_i}(\omega, \hat{r}_p, \mathbf{r}'_n)$	$\left[\Psi^{J_f*}_{K_f}(\omega, \mathbf{r}'_n) \otimes \mathcal{Y}^{(l\frac{1}{2})}_{j}(\hat{r}_p, \mathbf{s}_p)\right]_{J_iM_i}$	Odd–odd neutron K_f rotational band

According to Appendix (14.4), the matrix elements of the deformed interaction, entering the system of differential equations (2.63), are given by

$$\begin{aligned}\frac{2\mu}{\hbar^2}V_d^{(cc')}(r,\mathbf{s})f_{c'}(r) &= \frac{2\mu}{\hbar^2}\sum_{\lambda>0}V_\lambda(r,\mathbf{s})\left\langle \mathcal{Y}^{(c)}_{J_iM_i}|Q_\lambda(\omega).T_\lambda(r)|\mathcal{Y}^{(c)}_{J_iM_i}\right\rangle f_{c'}(r)\\ &= \frac{2\mu}{\hbar^2}\sum_{\lambda>0}(-)^{\lambda+J_f-J_f'}\begin{Bmatrix} J_f & J_f' & \lambda\\ j' & j & J_i\end{Bmatrix}\langle J_f||Q_\lambda||J_f'\rangle\\ &\times\left[\left(V_{d\lambda}(r)\langle lj||Y_\lambda||l'j'\rangle + U^{(0)}_{\lambda 0}(l,j,l',j')\right)f_{c'}(r)\right.\\ &\left.+\,U^{(1)}_{\lambda 0}(l,j,l',j')\frac{df_{c'}(r)}{dr}\right]\\ &\equiv \mathcal{C}_{cc'}(r,\mathbf{s})f_{c'}(r)+\mathcal{D}_{cc'}(r,\mathbf{s})\frac{df_{c'}(r)}{dr},\end{aligned} \tag{3.23}$$

where the functions $U^{(k)}_{\lambda\mu}$ are defined in Appendix (14.4). Thus, the system of differential equations (2.63) can be written as follows

$$\frac{d^2f_c(r)}{dr^2}=\sum_{c'}\left\{[\mathcal{V}_c(r,\mathbf{s})\delta_{cc'}+\mathcal{C}_{cc'}(r,\mathbf{s})]f_{c'}(r)+\mathcal{D}_{cc'}(r,\mathbf{s})\frac{df_{c'}(r)}{dr}\right\}, \tag{3.24}$$

where $\mathcal{V}_c$ is the spherical part. Therefore one obtains

$$\frac{dY}{dr}=\mathcal{A}Y,\quad Y=\begin{bmatrix}\mathbf{f}\\ \mathbf{f}'\end{bmatrix}, \tag{3.25}$$

where we used a column vector to denote the wave function components $\mathbf{f}$. The matrix $\mathcal{A}$ is given by

$$\mathcal{A}=\begin{bmatrix}0 & \mathcal{I}\\ \mathcal{V}\mathcal{I}+\mathcal{C} & \mathcal{D}\end{bmatrix}, \tag{3.26}$$

where $\mathcal{I}$ denotes the unity matrix.

3.4 Angular Distribution

The most sensitive tool to probe the wave function of the relative motion is given by the angular distribution.

3.4.1 Fermion Emission

Let us first estimate angular distribution in the fermion emission. The typical example is given by proton emission [4]. In order to estimate the decay probability per unit of solid angle in a given direction $\hat{r}$, given by Eq. 2.82, one has to integrate the expression of the decay width over the core variable, i.e.

$$\Gamma_{J_f}(\hat{r}) = \hbar v \lim_{r\to\infty} \sum_{lj} | \sum_{M_f} \langle J_f, M_f; j, M_i - M_f | J_i, M_i \rangle \mathcal{Y}^{(l\frac{1}{2})}_{j,M_i-M_f}(\hat{r}) f_{J_f lj}(r)|^2, \quad (3.27)$$

where for simplicity we considered the same velocity (i.e. the same Q-value) for all considered channels. By using the asymptotic form of the radial components (2.71) and the Coulomb–Hankel asymptotic form given in Appendix (14.2) one gets

$$\begin{aligned}\Gamma_{J_f}(\hat{r}) = \hbar v \sum_{lj} | \sum_{M_f} \langle J_f, M_f; j, M_i - M_f | J_i, M_i \rangle \mathcal{Y}^{(l\frac{1}{2})}_{j,M_i-M_f}(\hat{r}) \\ \times N_{J_f lj}[\cos(\phi_0 - \phi_l) - i \sin(\phi_0 - \phi_l)]|^2,\end{aligned} \quad (3.28)$$

where the Coulomb angle ϕ_l is defined in the Appendix (14.2). After some straightforward algebra one obtains

$$\begin{aligned}\Gamma_{J_f}(\hat{r}) = \hbar v \sum_{lj} \sum_{l'j'} \sum_{M_f} \langle J_f, M_f; j, M_i - M_f | J_i, M_i \rangle \langle J_f, M_f; j', M_i - M_f | J_i, M_i \rangle \\ \times \mathcal{Y}^{(l\frac{1}{2})*}_{j,M_i-M_f}(\hat{r}) \mathcal{Y}^{(l'\frac{1}{2})}_{j',M_i-M_f}(\hat{r}) N_{J_f lj} N_{J_f l'j'} \cos(\phi_l - \phi_{l'}).\end{aligned} \quad (3.29)$$

By using Eq. 14.73, expressing the product of spin-orbital harmonics, one gets

$$\begin{aligned}\Gamma(\hat{r}) &= \frac{\Gamma}{4\pi} W^{(F)}(\theta), \\ W^{(F)}(\theta) &= 1 + \sum_{L>0} A_L^{(F)} P_L(\cos\theta),\end{aligned} \quad (3.30)$$

where the coefficients of Legendre polynomials are given by

$$\begin{aligned}A_L^{(F)} = \left(\sum_{lj} N^2_{J_f lj} \right)^{-1} (-)^{J_i + \frac{1}{2} - J_f} \frac{\hat{L}}{\hat{J}_i} \langle J_i, M_i; L, 0 | J_i, M_i \rangle \\ \times \sum_{ljl'j'} (-)^{\frac{-L+j-j'-l+l'}{2}} Z^{(F)}(j, J_i, j', J_i; J_f, L) N_{J_f lj} N_{J_f l'j'} \cos(\phi_l - \phi_{l'}),\end{aligned} \quad (3.31)$$

in terms of the following $Z^{(F)}$-functions, depending upon Racah W-symbols and Clebsch–Gordan coefficients

$$Z^{(F)}(a,b,c,d;e,f) = (-)^{\frac{f-a+c}{2}} \hat{a}\hat{b}\hat{c}\hat{d} W(a,b,c,d;e,f) \left\langle a, \frac{1}{2}; c, -\frac{1}{2} | f, 0 \right\rangle. \quad (3.32)$$

The orientation of the mother nucleus is determined by the spin projection M_i. In experimental situations the value of the initial projection spin M_i is randomly distributed. Therefore in the above relation one should consider an average value of the Clebsch–Gordan coefficient $\langle J_i, M_i; L, 0|J_i, M_i \rangle_{av}$ [5].

It is important to point out that in transitions connecting ground states the normalization factors $N_{J_f lj}$ cancel out and therefore the angular distribution becomes an universal function, independent of nuclear structure details, i.e.

$$A_L^{(F)} = (-)^{M_i+1/2}(2J_i + 1)\langle J_i, -M_i; J_i, M_i|L, 0\rangle \left\langle J_i, \frac{1}{2}; J_i, -\frac{1}{2}|L, 0 \right\rangle. \tag{3.33}$$

In Ref. [4] the angular distribution of emitted protons from ^{131}Eu was evaluated by using a formalism equivalent to that presented here. It was concluded that indeed the transition to the ground state has an universal behaviour, while that to an excited state can be an important tool for the spin-parity assignment of the emitted proton.

3.4.2 Boson Emission

This feature is at variance with the boson emission, as for instance in α-decay connecting ground states ($J_i = J_f$) of odd-mass nuclei, where the anisotropy is very sensitive to the nuclear deformation.

Let us consider that the odd-particle is not active during the decay process. We use a similar to (3.27) expression of the angular distribution, i.e integrated over the core coordinates. The only difference in (3.29) is that we use Eq. 14.69, expressing the product of two spherical harmonics. The resulting sum over projection M_f for the product of three Clebsch–Gordan coefficients can be recoupled by using the Racah W-coefficient. The final result is similar to the fermion case (3.30), i.e.

$$\begin{aligned} \Gamma(\hat{r}) &= \frac{\Gamma}{4\pi} W^{(B)}(\theta), \\ W^{(B)}(\theta) &= 1 + \sum_{L \geq 1} A_L^{(B)} P_L(\cos\theta), \end{aligned} \tag{3.34}$$

and the coefficients $A_L^{(B)}$ have a similar structure

$$\begin{aligned} A_L^{(B)} &= \left(\sum_l N_{J_f l}^2\right)^{-1} (-)^{J_i - J_f} \frac{\hat{L}}{\hat{J}_i} \langle J_i, M_i; L, 0|J_i, M_i\rangle \\ &\times \sum_{ll'} (-)^{\frac{L-l+l'}{2}} Z^{(B)}(l, J_i, l', J_i; J_f, L) N_{J_f l} N_{J_f l'} \cos *(\phi_\lambda - \phi_{\lambda'}), \end{aligned} \tag{3.35}$$

where we introduced the $Z^{(B)}$ symbols

$$Z^{(B)}(a, b, c, d; e, f) = (-)^{\frac{f-a+c}{2}} \hat{a}\hat{b}\hat{c}\hat{d} W(a, b, c, d; e, f)\langle a, 0; c, 0|f, 0\rangle. \tag{3.36}$$

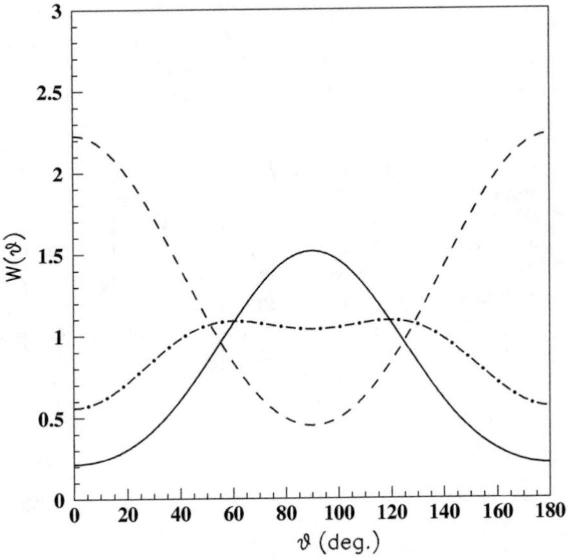

Fig. 3.1 Angular distribution corresponding to initial angular momentum $M_i = 1/2$ (*solid line*), 3/2 (*dash-dotted line*) and 5/2 (*dashed line*) for the quadrupole deformation $\beta_2 = 0.2$ [6, 7]. The transition is $^{241}\text{Am} \rightarrow ^{237}\text{Np} + \alpha$, where $K_i = K_f = J_i = \frac{5}{2}^-$

Therefore, the emitted α-particle carries many angular momenta $|J_i - J_f| \leq l \leq J_i + J_f$, while the spin of the emitted proton in (3.33) has only the value $j = J_i$.

It turns out that the angular distribution is very sensitive to the quadrupole deformation. In Refs. [6, 7], it was obtained a good agreement with the experiment, concerning the anisotropy $\Gamma(\pi/2)/\Gamma(0)$ for the following transition

$$^{241}\text{Am} \rightarrow ^{237}\text{Np} + \alpha, \tag{3.37}$$

in which $K_i = K_f = J_i = \frac{5}{2}^-$. This result is very relevant, taking into account that it was considered a realistic value of the deformation and the angular distribution is strongly dependent upon the initial spin projection M_i, as it is shown in Fig. 3.1.

In Refs. [8, 9] it was developed a similar formalism for the well deformed nuclei and the anisotropies in α-decays from ^{221}Fr, ^{227}Pa, ^{241}Am, ^{253}Es, were estimated by obtaining a good agreement with experimental data.

In Ref. [10] the same formalism was used, but within the quadrupole–quadrupole weak coupling α-core scheme, in order to analyze the anisotropic emission in ^{205,207,219}Rn, ^{221}Fr, ^{227,229}Pa, ^{229}U.

References

1. Bohr, A., Mottelson, B.: Nuclear Structure. Benjamin, New York (1975)
2. Raduta, A.A., Ceausescu, V., Gheorghe, A., Dreizler, R.M.: Boson description of ^{190}Pt and ^{192}Pt. Phys. Lett. B **99**, 444–448 (1981)

3. Raduta, A.A., Ceausescu, V., Gheorghe, A., Dreizler, R.M.: Phenomenological description of three interacting collective bands. Nucl. Phys. A **381**, 253–276 (1982)
4. Kadmensky, S.G., Sonzogni, A.A.: Proton angular distributions from oriented proton-emitting nuclei. Phys. Rev. C **62**, 054601/1–5 (2000)
5. Rose, M.E.: Elementary Theory of Angular Momentum. Wiley, New York (1957)
6. Delion, D.S., Insolia, A., Liotta, R.J.: Anisotropy in alpha decay of odd-mass deformed nuclei. Phys. Rev. C **46**, 884–888 (1992)
7. Delion, D.S., Insolia, A., Liotta, R.J.: Alpha widths in deformed nuclei: microscopic approach. Phys. Rev. C **46**, 1346–1354 (1992)
8. Stewart, T.L., Kermode, M.W., Beachey, D.J., Rowley, N., Grant, I.S., Kruppa, A.T.: Alpha decay of deformed actinide nuclei. Phys. Rev. Lett. **77**, 36–39 (1996)
9. Stewart, T.L., Kermode, M.W., Beachey, D.J., Rowley, N., Grant, I.S., Kruppa, A.T.: α-Particle decay through a deformed barrier. Nucl. Phys. A **611**, 332–354 (1996)
10. Berggren, T.: Anisotropic alpha decay from oriented odd-mass isotopes of some Light Actinides. Phys. Rev. C **50**, 2494–2507 (1994)

Chapter 4
Coupled Channels Methods

4.1 Numerical Integration

As in the previous chapters, we will use the short-hand notation "c" to denote quantum numbers of a given channel, labelling the wave function, e.g. (J_1, J_2, J_c, j_c). By l_c we will understand the angular momentum of the channel "c".

In this section, we will present the numerical integration procedure to find resonant solutions of the system (3.25). The solution of the coupled channels equations should be regular at origin. This condition provides a set of N linear independent vector functions satisfying the following conditions

$$\mathscr{R}_{ca}(r) \rightarrow_{r\rightarrow 0} \delta_{ca} r^{l_c+1}, \tag{4.1}$$

if the potential has a Woods–Saxon shape. The first index "c" of this radial wave function denotes the basis, while the second one "a" the eigenvalue.

In a matrix notation this relation can be written as follows

$$\begin{bmatrix} \mathscr{R}_{11} & \mathscr{R}_{21} & \dots & \mathscr{R}_{N1} \\ \mathscr{R}_{12} & \mathscr{R}_{22} & \dots & \mathscr{R}_{N2} \\ \dots & \dots & \dots & \dots \\ \mathscr{R}_{1N} & \mathscr{R}_{2N} & \dots & \mathscr{R}_{NN} \end{bmatrix} \rightarrow_{r\rightarrow 0} \begin{bmatrix} r^{l_1+1} & 0 & \dots & 0 \\ 0 & r^{l_2+1} & \dots & 0 \\ 0 & 0 & \dots & r^{l_N+1} \end{bmatrix}. \tag{4.2}$$

Each left-hand side column is found by using a forward integration with the initial conditions in the corresponding right-hand side column.

If the interaction has a strong repulsive core in the origin, the initial integration condition at a certain radius r_0 inside the core is given by a different condition, namely

$$\mathscr{R}_{ca}(r) \rightarrow_{r\rightarrow r_0} \delta_{ca}\varepsilon, \tag{4.3}$$

where ε is an arbitrary small number.

It is very important to point out on the following: it is necessary to integrate the system of equations only once by considering any of the initial conditions. This procedure provides a matrix connecting the initial wave function with the final

D. S. Delion, *Theory of Particle and Cluster Emission*, Lecture Notes in Physics, 819, DOI: 10.1007/978-3-642-14406-6_4, © Springer-Verlag Berlin Heidelberg 2010

one, the so-called propagator matrix. The method is described in Appendix (14.6) for the Runge–Kutta propagator. The fundamental system of solutions at the radius r is generated by a multiplication between the propagator matrix and the matrix of initial conditions (4.2). If the interaction does not contain the first derivative, one can use the Numerov method described in Appendix (14.5).

In general the forward integration is stable until some point R where the nuclear potential vanishes and total the potential reaches a maximum. The internal solution can be written as a superposition of these fundamental solutions with some unknown coefficients

$$f_c^{(\mathrm{int})}(r) = \sum_a \mathscr{R}_{ca}(r) M_a. \tag{4.4}$$

In matrix notation the vector of solutions is given as the product of the internal matrix of fundamental solutions times the matrix of unknown coefficients, i.e. $\mathbf{f}^{(\mathrm{int})}(r) = \mathscr{R}(r)\mathbf{M}$.

Beyond R, the forward integration procedure is not convenient, because the regular increasing solution “contaminates” the irregular decreasing one. Nevertheless, this integration can be performed from large distances where the interaction becomes spherical. Here we define the $N \times N$ matrix of outgoing fundamental solutions (Gamow matrix) with the following asymptotics

$$\begin{aligned} \mathscr{H}_{ca}^{(+)}(r) &\equiv \mathscr{G}_{ca}(r) + i\mathscr{F}_{ca}(r) \to_{r\to\infty} \delta_{ca} H_{l_c}^{(+)}(\chi_c, \kappa_c r) \\ &\equiv \delta_{ca}[G_{l_c}(\chi_c, \kappa_c r) + iF_{l_c}(\chi_c, \kappa_c r)], \end{aligned} \tag{4.5}$$

which, as in (4.2), can be written in matrix notation

$$\begin{bmatrix} \mathscr{H}_{11}^{(+)} & \mathscr{H}_{21}^{(+)} & \dots & \mathscr{H}_{N1}^{(+)} \\ \mathscr{H}_{12}^{(+)} & \mathscr{H}_{22}^{(+)} & \dots & \mathscr{H}_{N2}^{(+)} \\ \dots & \dots & \dots & \dots \\ \mathscr{H}_{1N}^{(+)} & \mathscr{H}_{2N}^{(+)} & \dots & \mathscr{H}_{NN}^{(+)} \end{bmatrix} \to_{r\to\infty} \begin{bmatrix} H_{l_1}^{(+)} & 0 & \dots & 0 \\ 0 & H_{l_2}^{(+)} & \dots & 0 \\ 0 & 0 & \dots & H_{l_N}^{(+)} \end{bmatrix}. \tag{4.6}$$

The matrix $\mathscr{H}$ is found by backward integration starting from a large distance. The cth component of the external solution is built as a linear superposition of the column-vectors in $\mathscr{H}$, i.e.

$$f_c^{(\mathrm{ext})}(r) = \sum_a \mathscr{H}_{ca}^{(+)}(r) N_a \to_{r\to\infty} N_c H_{l_c}^{(+)}(\chi_c, \kappa_c r), \tag{4.7}$$

or in matrix notation $\mathbf{f}^{(\mathrm{ext})}(r) = \mathscr{H}^{(+)}(r)\mathbf{N}$.

The matching constants are found using the continuity conditions at a certain radius R. The corresponding system of equations is given by

$$\begin{aligned} f_c^{(\mathrm{int})}(R) &= f_c^{(\mathrm{ext})}(R), \\ \frac{d}{dr} f_c^{(\mathrm{int})}(R) &= \frac{d}{dr} f_c^{(\mathrm{ext})}(R). \end{aligned} \tag{4.8}$$

In order to find a non-trivial solution the following secular equation has to be fulfilled

$$\begin{vmatrix} \mathscr{R}(R) & \mathscr{H}^{(+)}(R) \\ \frac{d}{dr}\mathscr{R}(R) & \frac{d}{dr}\mathscr{H}^{(+)}(R) \end{vmatrix} \approx \begin{vmatrix} \mathscr{R}(R) & \mathscr{G}(R) \\ \frac{d}{dr}\mathscr{R}(R) & \frac{d}{dr}\mathscr{G}(R) \end{vmatrix} = 0. \tag{4.9}$$

The solutions corresponding to the first determinant provide complex energies, corresponding to the S-matrix poles. They define the deformed Gamow decaying states. For a large Coulomb barrier the resonance is very narrow and therefore the regular Coulomb functions have vanishing values inside the barrier. Thus, the use of real irregular Coulomb functions in the second equality (4.9) is a very good approximation. Let us also mention that at large distances, but still inside the Coulomb barrier, the system (4.9) becomes decoupled, i.e. one obtains the usual condition for logarithmic derivatives (2.70) corresponding to each channel

$$\begin{aligned} \frac{1}{\mathscr{R}_{cc}(R)}\frac{d\mathscr{R}_{cc}(R)}{dr} &= \frac{1}{\mathscr{H}^{(+)}_{cc}(\chi_c,\kappa_c R)}\frac{d\mathscr{H}^{(+)}_{cc}(\chi_c,\kappa_c R)}{dr} \\ &\approx \frac{1}{\mathscr{G}_{cc}(\chi_c,\kappa_c R)}\frac{d\mathscr{G}_{cc}(\chi_c,\kappa_c R)}{dr}. \end{aligned} \tag{4.10}$$

The external matching constants can be obtained by inverting (4.7) together with the first condition (4.8), i.e.

$$N_a = \sum_c \left[\mathscr{H}^{(+)}_{ac}(R)\right]^{-1} f^{(\mathrm{int})}_c(R), \tag{4.11}$$

where the wave function is normalized in the internal region, i.e.

$$\int_0^{R_e} \sum_c |f_c(r)|^2 dr = 1. \tag{4.12}$$

Here R_e is the external turning point, where $V(R_e) = E$. These external normalization constants, which are called scattering amplitudes, can be written in terms of the propagator operator as follows

$$N_c = \frac{1}{H^{(+)}_{l_c}(\kappa_c R)} \sum_{c'} \mathscr{K}_{cc'}(R) f^{(\mathrm{int})}_{c'}(R), \tag{4.13}$$

where the propagator matrix describes the transition from the initial state $\mathscr{H}^{(+)}$ to the final solution $H^{(+)}$, i.e.

$$\mathscr{K}_{cc'}(R) = H^{(+)}_{l_c}(\chi_c,\kappa_c R)\left[\mathscr{H}^{(+)}_{cc'}(R)\right]^{-1} = \delta_{cc'} + \Delta\mathscr{K}_{cc'}(R). \tag{4.14}$$

It is defined in such a way, that for a spherical Coulomb field it becomes the unity matrix and, therefore, $\Delta\mathscr{K}_{cc'}(R) = 0$. It turns out that even for large quadrupole deformations, e.g. $\beta_2 = 0.4$, the correcting matrix is rather small outside the nuclear surface, i.e. $\max[\Delta\mathscr{K}_{cc'}(R)] < 0.1$. In Sect. 4.9 we give this matrix in the case of the very deformed proton emitter ^{131}Eu. It is worthwhile to point out that this operator is the exact counterpart of the the so-called WKB Fröman matrix, introduced in the theory of the α-decay from deformed nuclei in Ref. [1]. For a spherical potential this matrix is obviously diagonal and it is given by the ratio between the outgoing Coulomb–Hankel function $H_{l_c}^{(+)}$ and the integrated solution $\mathscr{H}_{cc}^{(+)}(R)$.

The relation (4.13) leads to a factorized form of the partial decay width which is similar to the spherical relation, i.e.

$$\Gamma_c = 2P_{l_c}(R)\left|\gamma_c^{(d)}(R)\right|^2, \tag{4.15}$$

where the penetrability has the standard form for spherical emitters, i.e.

$$P_{l_c}(E_c,R) \equiv \frac{\kappa_c R}{F_{l_c}^2(\chi_c,\kappa_c R) + G_{l_c}^2(\chi_c,\kappa_c R)} = \frac{\kappa_c R}{\left|H_{l_c}^{(+)}(\chi_c,\kappa_c R)\right|^2}, \tag{4.16}$$

while the deformed reduced width is given by

$$\begin{aligned} \gamma_c^{(d)}(R) &= \sqrt{\frac{\hbar^2}{2\mu R}} s_c(R), \\ s_c(R) &\equiv \sum_{c'} \mathscr{K}_{cc'}(R) f_{c'}^{(\text{int})}(R). \end{aligned} \tag{4.17}$$

It is also possible to have an alternative matrix factorization of the following form

$$\Gamma_c = 2\sum_{c'}\left|P_{cc'}^{1/2}(R)\gamma_{c'}(R)\right|^2, \tag{4.18}$$

where now we introduced the matrix elements of the deformed penetrability

$$P_{cc'}^{1/2}(R) = P_{l_c}^{1/2}(E_c,R)\mathscr{K}_{cc'}(R) = \sqrt{\kappa_c R}\left[\mathscr{H}_{cc'}^{(+)}(R)\right]^{-1}, \tag{4.19}$$

but with the standard spherical reduced width

$$\gamma_c(R) = \sqrt{\frac{\hbar^2}{2\mu R}} f_c^{(\text{int})}(R). \tag{4.20}$$

For large radii, where the interaction becomes spherical, one obtains obviously the diagonal relation (2.87) in both cases.

By inverting Eq. 4.11, one obtains the components of the wave function in terms of partial channels widths

$$f_c^{(\text{int})}(r) = \sum_a \mathscr{H}_{ca}^{(+)}(r)\sqrt{\frac{\Gamma_a}{\hbar v_a}}. \tag{4.21}$$

Thus, in order to determine the radial components of the wave function, it is necessary to know all relevant partial decay widths. Obviously the wave function components depend upon the concrete form of the used potential, determining the fundamental matrix of Gamow solutions. For this reason, instead of the so-called hindrance factors, which are model dependent

$$\mathrm{HF}_c = \left| \frac{f_0^{(\mathrm{int})}(r)}{f_c^{(\mathrm{int})}(r)} \right|^2, \tag{4.22}$$

where by "0" we denoted the ground state channel, it is more appropriate to use the intensity ratios

$$I_c = \log_{10} \frac{\Gamma_0}{\Gamma_c}. \tag{4.23}$$

4.2 Integration Procedures

If the numerical integration is performed in the external region on the real axis, this method gives reliable solutions for the irregular Coulomb matrix $\mathscr{G}$. Likewise the approximation $\mathscr{H}^{(+)}(R) \approx \mathscr{G}(R)$ is very good to estimate the scattering amplitudes (and therefore the decay widths) by using Eq. 4.13. On the other hand, the regular Coulomb matrix $\mathscr{F}$ is necessary only to obtain the very small imaginary part of the energy, that is the solutions of Eq. 4.9, which is minus twice the decay width. The difficult task of evaluating the functions $\mathscr{F}$ and $\mathscr{G}$ simultaneously can be accomplished by using the complex scaling method [2]. In this method the integration is performed in the complex radial plane, thus avoiding the drastic decrease of the regular solution.

Historically the first generally available computer code to evaluate Gamow states in spherical symmetric potentials, is the code GAMOW [3]. It computes all poles of the S-matrix, including wide resonances and antibound states. Later, improved numerical techniques were used to evaluate those poles with high accuracy [4]. This was necessary since, as already mentioned, measurable half lives correspond to imaginary parts of the energy which are, in absolute value, many orders of magnitude smaller than the corresponding real part.

In Ref. [5] a different approach is proposed. In order to avoid drastic variations of the solutions inside the barrier, the coupled system of equations is solved by using an equivalent system of equations for the matrix of the logarithmic derivative of the wave function, i.e.

$$\frac{d\phi_{cc'}}{dr} + \sum_a [\phi_{ca} - \mathscr{D}_{ca}]\phi_{ac'} - \mathscr{V}_c \delta_{cc'} - \mathscr{C}_{cc'} = 0. \tag{4.24}$$

where $\mathscr{V}, \mathscr{C}, \mathscr{D}$ denote the matrices introduced in Eq. 3.24. The logarithmic derivative matrix is

$$\phi \equiv \frac{d\mathscr{R}}{dr}[\mathscr{R}]^{-1}. \tag{4.25}$$

This procedure can be used for a non-singular matrix of solutions $\mathscr{R}$, i. e. inside the Coulomb barrier, and is stable for both the regular and irregular parts of the Gamow matrix $\mathscr{H}^{(+)}$. On the other hand, the system (4.24) is nonlinear and the matrix $\mathscr{P}(r, r_0)$, connecting the points r_0 and r depends not only upon the potential, but also on the solution.

4.3 Diagonalization Method

In emission processes, the Coulomb barrier is much higher than the energy of the emitted fragments. As a result, the ratio between the irregular and regular Coulomb waves at the matching radius $r = R$ is

$$\frac{G_l(\chi, kR)}{F_l(\chi, kR)} > 10^{10}. \tag{4.26}$$

The resonant state in the internal region $[0, R]$ is practically real and it behaves like a bound state due to the exponential decreasing character of the irregular Coulomb wave for $r > R$. This property implies that one can obtain the wave function in the internal region by using a harmonic oscillator (ho) representation. To do this in our coupled channels case we rewrite the coupled system of equations (2.63) in the following form

$$\begin{aligned} H_{l_c}^{(\beta_0)} f_c(r) &\equiv -\frac{\hbar\omega}{2\beta_0}\left[\frac{d^2}{r^2} - \frac{l_c(l_c+1)}{r^2}\right] f_c(r) \\ &= [E - E_c - V_0(r)] f_c(r) - \sum_{c'} V_d^{(cc')}(r) f_{c'}(r), \end{aligned} \tag{4.27}$$

where we introduced the ho parameter

$$\beta_0 = \frac{\mu\omega}{\hbar}. \tag{4.28}$$

Expanding the internal radial wave function in this basis one gets

$$f_c^{(\text{int})}(r) = \sum_n^{n_{\max}} d_{nc} \overline{\mathscr{R}}_{nl_c}^{(\beta)}(r), \tag{4.29}$$

where the new ho parameter β gives the best fit of the spherical interaction $V_0(r)$ with an ho potential. The eigenvalue system of equations is readily obtained from (4.27). Notice that the ho equation

$$\begin{aligned} H_l^{(\beta_0)}\overline{\mathscr{R}}_{nl}^{(\beta_0)}(r) &= -\frac{\hbar\omega}{2\beta_0}\left[\frac{d^2}{r^2} - \frac{l(l+1)}{r^2}\right]\overline{\mathscr{R}}_{nl}^{(\beta_0)}(r) \\ &= \hbar\omega\left(2n + l + \frac{3}{2} - \frac{\beta_0 r^2}{2}\right)\overline{\mathscr{R}}_{nl}^{(\beta_0)}(r) \end{aligned} \tag{4.30}$$

has an analytical solution

$$\begin{aligned} \overline{\mathscr{R}}_{nl}^{(\beta_0)}(r) &\equiv \langle r|\beta_0 nl\rangle = r\mathscr{R}_{nl}^{(\beta_0)}(r) \\ &= (-)^n\left[\frac{2\beta_0^{l+3/2}n!}{\Gamma(n+l+3/2)}\right]^{1/2} r^{l+1}e^{-\frac{\beta_0 r^2}{2}}L_n^{l+1/2}(\beta_0 r^2), \end{aligned} \tag{4.31}$$

where $\mathscr{R}_{nl}^{(\beta_0)}(r)$ is the standard radial ho wave function, defined in Appendix (14.8), depending upon the Laguerre polynomial $L_n^{\alpha}(x)$. These functions obey the orthonormality condition, which reads

$$\langle\beta_0 nl|\beta_0 n'l\rangle = \int_0^{\infty}\overline{\mathscr{R}}_{nl}^{(\beta_0)}(r)\overline{\mathscr{R}}_{n'l}^{(\beta_0)}(r)dr = \delta_{nn'}. \tag{4.32}$$

One finally gets

$$\sum_{n'c'} H_{nc,n'c'}^{(\beta)}d_{n'c'} = Ed_{nc}, \tag{4.33}$$

where the Hamiltonian matrix is

$$\begin{aligned} H_{nc,n'c'}^{(\beta)} = &\left\{\left[f\hbar\omega\left(2n + l_c + \frac{3}{2}\right) + E_c\right]\delta_{nn'}\right. \\ &\left.+\left[\langle\beta nl_c|V_0(r)|\beta n'l_c\rangle - \frac{f\hbar\omega}{2}\langle\beta nl_c|\beta r^2|\beta n'l_c\rangle\right]\right\}\delta_{cc'} \\ &+ \langle\beta nl_c|V_d^{(cc')}(r)|\beta n'l_{c'}\rangle. \end{aligned} \tag{4.34}$$

The new parameter $\beta = f\beta_0$ is chosen to minimize the contribution of the difference between the spherical and ho potential.

On concludes that any emission process is characterized by two parameters:

(1) the ho parameter β, defining the size of the internal part of the interaction and
(2) the Coulomb parameter χ (2.67), describing the external dynamics of emitted fragments.

These two parameters, β and χ are apparently independent. Actually they are connected and we will discuss this point in Sect. 10.4 in Chap. 10.

As a rule, the dimension of the algebraic system of equations (4.33) is by one order of magnitude larger than the dimension of the differential equations system (4.27), but this drawback is compensated by the convenience of having a standard eigenvalue problem. A good approximation of the matching constant, and therefore of the decay width, is given by Eq. 4.11 using the internal wave function provided by the diagonalization procedure and the irregular Coulomb matrix $\mathcal{G}$ defined by (4.5).

4.4 Analytical Continuation Method

One can exploit the intuitive idea that a resonant Gamow state can be obtained from a bound state by changing the strength of the potential. Recently unbound states in exotic nuclei were studied by using this idea [6–8].

In this method one replaces the interaction V by λV, where the strength $\lambda < 1$ is a well-behaved parameter. One can prove that the wave number $\kappa = \sqrt{2\mu E}/\hbar$ is an analytical function of the strength λ. Near the threshold, where $\kappa(\lambda_0) = 0$, one has [9]

$$\kappa(\lambda) \sim i\sqrt{\lambda - \lambda_0}, \tag{4.35}$$

for any angular momentum. In Ref. [10] it is proposed to use the Padé approximant of the second kind, as a tool to get an analytical continuation of the wave number in the complex λ plane, i. e. to extend the wave number from a bound state to a resonant region. Therefore one writes

$$\begin{aligned} \kappa \approx \kappa^{(N,M)}(x) &= i\frac{c_0 + c_1 x + c_2 x^2 + \cdots + c_M x^M}{1 + d_1 x + d_2 x^2 + \cdots + d_N x^N}, \\ x &\equiv \sqrt{\lambda - \lambda_0}. \end{aligned} \tag{4.36}$$

After determining the threshold value λ_0 one evaluates the coefficients of the Padé approximant $\kappa^{(N,M)}(\lambda_i)$ in (4.36) by solving the Schrödinger equation, corresponding to the bound states of the potential $\lambda_i V$ for several values of $\lambda_i < 1$. Finally, the analytical continuation, given by $\kappa = \kappa^{(N,M)}(\lambda = 1)$, is used to extrapolate the bound state wave function $\psi_l^B(\kappa r)$ into the scattering region.

As it was shown in Ref. [10], this method gives an excellent agreement with the solution found by a direct integration of the Schrödinger equation both in spherical and deformed nuclei.

4.5 Distorted Wave Approach (DWA)

We have seen that inside the Coulomb barrier practically the Gamow state is real and coincides with the irregular Coulomb wave. In other words the particle spends a long time inside the nucleus and, therefore, the wavefunction in this region is like that corresponding to a bound state. Thus, in this region, the resonant state can be obtained by using a harmonic oscillator representation. At large distances, the difference between the Gamow state and the irregular wave increases with increasing radius. Yet, it is possible to restore the Gamow wave function starting with the internal solution, by using the distorted wave approach (DWA).

The Green function for deformed nuclei with given quantum numbers J_iM_i, can be written as

$$G^{(+)}(\mathbf{x},\mathbf{r};\mathbf{x}',\mathbf{r}') = -\frac{2\mu}{\hbar^2\kappa}\sum_c \frac{H_{l_c}^{(+)}(\chi,\kappa r)}{r}|c;J_iM_i\rangle\langle c;J_iM_i|\frac{F_{l_c}(\chi,\kappa r')}{r'}, \qquad (4.37)$$

where $|c;JM\rangle$ is as in (3.21) with $c\equiv(J_f lj)$. The outgoing wave function at large distances is

$$\begin{aligned}\Psi_{J_iM_i}^{(+)}(\mathbf{x},\mathbf{r}) = &-\frac{2\mu}{\hbar^2\kappa}\sum_c \frac{H_{l_c}^{(+)}(\chi,\kappa r)}{r}|c;J_iM_i\rangle \\ &\times \int r'^2 dr' \langle c;J_iM_i|\frac{F_{l_c}(\chi,\kappa r')}{r'}\Big|V(\mathbf{x}',\mathbf{r}') - V_C^{(0)}(r')\Big|\Psi_{J_iM_i}^{(\mathrm{res})}\rangle,\end{aligned} \qquad (4.38)$$

where the resonant wave function $\Psi_{J_iM_i}^{(\mathrm{res})}$ is evaluated in the internal region [0, R] and is normalized to unity here. Its evaluation is performed by using a diagonalization procedure as in the previous section. Here we will use a different notation, namely $f_c^{(\mathrm{res})}(r)\equiv f_{nc}(r)$. This outgoing wave function has the standard form, i.e.

$$\Psi_{J_iM_i}^{(+)}(\mathbf{x},\mathbf{r}) = \sum_c N_c^{\mathrm{DW}}\frac{H_{l_c}^{(+)}(\chi,\kappa r)}{r}\mathcal{Y}_{J_iM_i}^{(c)}(\mathbf{x},\hat{r}), \qquad (4.39)$$

where the distorted wave (DW) channel scattering amplitude, determining the partial decay width, is given by

$$\begin{aligned}N_c^{\mathrm{DW}} = &-\frac{2\mu}{\hbar^2\kappa}\int_0^R r dr F_{l_c}(\chi,\kappa r) \\ &\times \left\langle \mathcal{Y}_{J_iM_i}^{(c)}(\mathbf{x},\hat{r})\left[V(\mathbf{x},\mathbf{r}) - V_C^{(0)}(r)\right]\Psi_{J_iM_i}^{(\mathrm{res})}(\mathbf{x},\mathbf{r})\right\rangle.\end{aligned} \qquad (4.40)$$

By inserting here the expansion of the resonant wave function (3.19), one obtains

$$\begin{aligned}N_c^{\mathrm{DW}} = &-\frac{2\mu}{\hbar^2\kappa}\sum_{c'}\int_0^R dr F_{l_c}(\chi,\kappa r) f_{c'}^{(\mathrm{res})}(r) \\ &\times \left\langle \mathcal{Y}_{J_iM_i}^{(c)}(\mathbf{x},\hat{r})\left[V(\mathbf{x},\mathbf{r}) - V_C^{(0)}(r)\right]\mathcal{Y}_{J_iM_i}^{(c')}(\mathbf{x},\hat{r})\right\rangle.\end{aligned} \qquad (4.41)$$

The expansion of the interaction in partial waves, as in Eq. 2.132, leads to

$$
\begin{aligned}
N_c^{\mathrm{DW}} &= -\frac{2\mu}{\hbar^2\kappa}\int_0^R drF_{l_c}(\chi,\kappa r)\left[V_0(r)-V_C^{(0)}(r)\right]f_c^{(\mathrm{res})}(r) \\
&-\frac{2\mu}{\hbar^2\kappa}\sum_{\lambda>0}\sum_{c'}\int_0^R drF_{l_c}(\chi,\kappa r)V_\lambda(r)f_{c'}^{(\mathrm{res})}(r)\left\langle \mathcal{Y}_{J_iM_i}^{(c)}|Q_\lambda(\omega).T_\lambda(\hat{r})|\mathcal{Y}_{J_iM_i}^{(c')}\right\rangle,
\end{aligned}
\tag{4.42}
$$

which contains the same matrix elements as in the coupled system of equations (2.63). This relation is very useful to estimate the matching constant numerically and therefore the decay width. When $R\rightarrow\infty$ the above relation becomes exact and one can show that $N_c^{\mathrm{DW}}=N_c$ [11], as we will describe below for spherical emitters.

In the spherical case of boson emission, where $c\equiv l$, one obtains a very simple expression. The angular part of the matrix element is diagonal in the lm quantum numbers and therefore one obtains

$$
\psi_{lm}^{(+)}(\mathbf{r},\mathbf{s}) = N_l^{\mathrm{DW}}\frac{H_l^{(+)}(\chi,\kappa r)}{r}Y_{lm}(\hat{r}),
\tag{4.43}
$$

where the normalization factor is proportional to the Gell-Mann-Goldberger matrix element [12], i.e.

$$
N_l^{\mathrm{DW}} = -\frac{2\mu}{\hbar^2\kappa}\int_0^\infty F_l(\chi,\kappa r)\left[V_0(r)-V_C^{(0)}(r)\right]f_l^{(\mathrm{res})}(r)dr.
\tag{4.44}
$$

For fermion emission one replaces $Y_{lm}(\hat{r})$ by spin-orbital harmonics $\mathcal{Y}_{jm}^{(l)}(\hat{r},s)$.

By making the substitution

$$
V_0(r)-V_C^{(0)}(r)\rightarrow\left[H+\frac{\hbar^2}{2\mu}\frac{d^2}{dr^2}\right]_{\mathrm{right}}-\left[H_0+\frac{\hbar^2}{2\mu}\frac{d^2}{dr^2}\right]_{\mathrm{left}},
\tag{4.45}
$$

one obtains for the normalization factor (4.44) the following useful expression

$$
\begin{aligned}
N_l^{\mathrm{DW}} &= -\frac{1}{\kappa}\int_0^\infty\frac{1}{dr}\left[F_l(\chi,\kappa r)\frac{df_l(r)}{dr}-f_l\frac{dF_l(\chi,\kappa r)}{dr}\right]dr \\
&= -\frac{1}{\kappa}\left[F_l(\chi,\kappa r)\frac{df_l(r)}{dr}-f_l\frac{dF_l(\chi,\kappa r)}{dr}\right]_0^\infty.
\end{aligned}
\tag{4.46}
$$

We use the property that at the lower limit the integrand vanishes (both functions are regular), while at the upper limit the relation (2.71) is applied. By taking into account the value of the Wronskian for these Coulomb functions, one obtains the scattering amplitude

$$N_l^{\rm DW} = N_l. \tag{4.47}$$

The expression (4.44) gives a very useful estimate of the decay width, when the integration is performed over a finite interval [0, R], which, as we saw above, is a very good approximation for narrow resonances. The regular radial wave function $f_l(r)$ can be found by a diagonalization procedure using a spherical harmonic oscillator basis. Thus, one obtains values for the width which do not differ appreciably from the exact result (4.44) [11].

We also stress on the fact that, according to (4.46), the scattering amplitude can be expressed in two equivalent ways, namely

(1) as a volume integral on the difference between distorted and initial interactions, or
(2) as a surface integral on the distorted and initial wave functions.

One of the first calculations of half lives in proton emission processes was performed in Ref. [13], and later in [14] within the DWA formalism, which was previously used by the same authors in α-decay studies [15]. The distorted wave formula (4.42) was used for the matching constant, considering a large radial interval for the resonant wave function. It was assumed that the wave function is a superposition of a shell model, an intermediate and a cluster component, i.e.

$$\Psi^{\rm (res)}_{J_iM_i}(\mathbf{r}) = \Psi^{\rm (sh)}_{J_iM_i}(\mathbf{r}) + \Psi^{\rm (int)}_{J_iM_i}(\mathbf{r}) + \Psi^{\rm (cl)}_{J_iM_i}(\mathbf{r}). \tag{4.48}$$

The proton shell model component is defined in the internal region, that is from the origin until the distance where the nuclear density is 10% of the central value. It satisfies the Schrödinger equation, corresponding to a central proton-daughter Woods–Saxon potential with the Becchetti–Greenlees parametrisation [16]. The wave function is multiplied by the spectroscopic BCS amplitude u_{K_i}. The intermediate region is the region where the density decreases between 10% and 1% from the central value. The cluster component is defined beyond the radius where the nuclear density is 1% with respect to the central value. It is the solution of the coupled channels system (2.63), within the formalism developed for rotational odd–even emitters. They estimated the half lives of the spherical proton emitters

$$^{147}\mathrm{Tm},\ ^{151}\mathrm{Lu}$$

by using different potentials. The calculated values thus obtained agree with each other within a factor of three. They showed that the half lives are very sensitive to the quantum numbers assigned to the states. This sensitivity led the authors to conclude that in both nuclei the emitter decay from a $1h_{11/2}$ proton orbital (see Table I of Ref. [13]). Predictions of quantum numbers for several possible emitters were also made. The following spherical emitters were studied

$$^{109}\mathrm{I},\ ^{113}\mathrm{Cs}.$$

Due to the fact that the energies of the excited 2^+ states are rather high (720 keV for ^{109}I and 400 keV for ^{113}Cs), the two functions in the intermediate region are close to each other. Therefore the resulting wave function was computed here by using an interpolation procedure. The dependence of the half life upon deformation was investigated. It turns out that the cluster component gives not more than 20% contribution. It was found a good agreement with experimental data, corresponding to realistic deformations.

In Ref. [17] the authors investigated the deformed emitters

$$^{147,147*}\mathrm{Tm},\ ^{150,151}\mathrm{Lu}$$

within the same formalism. The star denotes an isomeric state. They were able to reproduce experimental data for the deformation parameters predicted by the systematics.

4.6 Two Potential Method

A particular case of the distorted wave formalism is given by the so-called two potential approach of Gurwitz and Kalbermann [18, 19]. The initial potential is splitted into two parts with respect to a radius r_B, where the barrier has the maximal value, i.e.

$$V(r) = U(r) + W(r). \tag{4.49}$$

The first part describes the proton motion in a bound potential

$$\begin{aligned} U(r) &= V(r), \quad r \le r_B, \\ &= V_B \equiv V(r_B), \quad r > r_B, \end{aligned} \tag{4.50}$$

while the second one describes the transition to the continuum

$$\begin{aligned} W(r) &= 0, \quad r < r_B, \\ &= V(r) - V_B, \quad r \ge r_B. \end{aligned} \tag{4.51}$$

In Fig. 2.7 we plotted by the solid line a simple potential, describing the main features of the emission process, namely

$$\begin{aligned} V(r) &= \left(\frac{C}{r_B} - v_0\right)\frac{r^2}{r_B^2} + v_0, \quad r \le r_B \\ &= \frac{C}{r}, \quad r > r_B, \end{aligned} \tag{4.52}$$

where $C = Z_1 Z_2 e^2$ and v_0 is the depth fixing the eigenvalue to the experimental Q-value. The difference between this potential and the maximal value (dashed line) is the "transition operator" W(r) triggering the emission. Indeed, the initial wave function satisfy the Schrödinger equation

$$\mathbf{H}_0\Phi_0 = (T + U)\Phi_0 = E_0\Phi_0. \tag{4.53}$$

The perturbation W changes the state of the emitted light fragment from its initial to its final value. Since $W(r)$ does not vanish at infinity one solves the problem in a shifted potential $\tilde{W}(r) = W(r) + V_B$. The partial decay width is given by the distorted wave formula [12], i.e.

$$\Gamma_l = \frac{4\mu}{\hbar^2 k}\left|\int_{r_B}^{\infty} \phi_l(r)W(r)\chi_l(r)dr\right|^2, \tag{4.54}$$

where $\phi_l(r)$ is the radial part of Φ_0 and $\chi_l(r)$ is the regular wave function of the Hamiltonian $T + \tilde{W}$. This method was used to describe α-decays between ground states [20, 21] and the anisotropic α-particle emission in odd–mass nuclei [22].

4.7 Intrinsic System of Coordinates

For well deformed nuclei the so-called strong interaction scheme is very suitable, i.e. one can consider the motion of the light fragment in the intrinsic frame of the deformed daughter nucleus. In this section we will investigate emission processes by using the intrinsic system of coordinates for both fermion and boson light fragments, by using a common ansatz for the relative angular wave function, namely

$$\mathcal{Y}^{(ls)}_{jm}(\hat{r}, \mathbf{s}) \rightarrow \left\{ \begin{array}{ll} \mathcal{Y}^{(l\frac{1}{2})}_{jm}(\hat{r}, \mathbf{s}), & s = \frac{1}{2} \\ Y_{jm}(\hat{r}), & s = 0 \end{array} \right\}. \tag{4.55}$$

In the last case one obviously has $l = j$.

For the fermion case we consider proton emission from odd–even rotational nuclei. Boson case corresponds to α or cluster decays between excited states of even–even rotational nuclei. First we will consider the ground rotational band of the daughter nucleus. In this case the collective coordinates ω are the Euler angles, which define the position of the major axes in the even–even daughter nucleus. We will investigate axially symmetric nuclei in this section.

Let us consider proton emission from odd–even nuclei. We will assume that the odd–even parent nucleus in the ground state has spin J_i and that the final state of the even–even core belongs to the ground band. The wave functions of this band are given by the normalized Wigner functions, i.e.

$$\Phi_{J_f M_f}(\omega) = \mathcal{D}^{J_f *}_{M_f 0}(\omega) \equiv \sqrt{\frac{2J_f + 1}{8\pi^2}} D^{J_f *}_{M_f 0}(\omega) = \frac{1}{\sqrt{2\pi}} Y_{J_f M_f}(\theta, \phi). \tag{4.56}$$

We use the Rose convention to define Wigner functions, as described in the Appendix (14.3). The core-spin-orbit harmonics (3.21) become in this case

$$\mathscr{Y}^{(J_f lj)}_{J_i M_i}(\omega, \hat{r}, \mathbf{s}) \equiv \langle \omega, \hat{r}, \mathbf{s} | J_f lj; J_i M_i \rangle = \left[\mathscr{D}^{J_f *}_0(\omega) \otimes \mathscr{Y}^{(ls)}_j(\hat{r}, \mathbf{s}) \right]_{J_i M_i}. \tag{4.57}$$

The multipole operator for axially deformed potentials is $Q_{\lambda\mu} = D^{\lambda*}_{u_0}$ and the core-proton interaction is

$$\begin{aligned} V(\omega, \mathbf{r}, \mathbf{s}) &= V_0(r, \mathbf{s}) + \sum_{\lambda > 0} V_\lambda(r, \mathbf{s}) \sum_\mu D^\lambda_{\mu 0}(\omega) T_{\lambda\mu}(\hat{r}) \\ &= V_0(r, \mathbf{s}) + \sum_\lambda \sqrt{\frac{4\pi}{2\lambda + 1}} V_\lambda(r, \mathbf{s}) \sum_\mu Y^*_{\lambda\mu}(\omega) T_{\lambda\mu}(\hat{r}). \end{aligned} \tag{4.58}$$

where the multipole operators $T_{\lambda\mu}$ are as in (14.33).

Within the strong coupling scheme the wave function in the intrinsic system of coordinates $\mathbf{r}'$, or K-system, is determined by the Euler angles ω. By changing the coordinates to this system, the multipole operators transform according to

$$\hat{R}(\omega) T_{JM}(\hat{r}) = \sum_{M'} D^J_{M'M}(\omega) T_{JM'}(\hat{r}) = T_{JM}(\hat{r}'). \tag{4.59}$$

and the potential becomes

$$V(\omega, \mathbf{r}, \mathbf{s}) = V_0(r, \mathbf{s}) + \sum_{\lambda > 0} V_\lambda(r, \mathbf{s}) T_{\lambda 0}(\hat{r}'). \tag{4.60}$$

The energies of the rotational eigenstates (4.56) are

$$E_{J_f} = \frac{\hbar^2}{2\mathscr{I}} J_f (J_f + 1), \tag{4.61}$$

where $\mathscr{I}$ denotes the moment of inertia.

The angular harmonics (4.55) transform according to

$$\mathscr{Y}^{(ls)}_{jM}(\hat{r}, \mathbf{s}) = \sum_{K_i} D^{j*}_{MK_i}(\omega) \mathscr{Y}^{(ls)}_{jK_i}(\hat{r}', \mathbf{s}). \tag{4.62}$$

and one obtains for the core-orbital harmonics the value

$$\mathscr{Y}^{(J_f lj)}_{J_i M_i}(\omega, \hat{r}, \mathbf{s}) = \sum_{K_i > 0} A^{J_i K_i}_{j J_f} \mathscr{Z}^{(ljK_i)}_{J_i M_i}(\omega, \hat{r}', \mathbf{s}), \tag{4.63}$$

where we introduced the intrinsic core-orbital harmonics as follows

$$\begin{aligned} \mathscr{Z}^{(ljK_i)}_{J_i M_i}(\omega, \hat{r}', \mathbf{s}) = \frac{1}{\sqrt{2}} [&\mathscr{D}^{J_i *}_{M_i K_i}(\omega) \mathscr{Y}^{(ls)}_{jK_i}(\hat{r}', \mathbf{s}) \\ &+ (-)^{J_i + K_i} \mathscr{D}^{J_i *}_{M_i -K_i}(\omega) \mathscr{Y}^{(ls)}_{j\bar{K}_i}(\hat{r}', \mathbf{s})], \end{aligned} \tag{4.64}$$

while for the transformation constants, which virtually are Clebsh–Gordan coefficients, we used the notation of Ref. [23], i.e.

$$A^{J_iK_i}_{jJ_f} \equiv (-)^{J_i-J_f-K_i}\sqrt{1+(-)^{J_f}}\langle J_i,K_i;j,-K_i|J_f,0\rangle. \tag{4.65}$$

For transitions to the ground state $J_f = 0$ and this coefficient has a simple form, namely

$$A^{J_iK_i}_{j0} = \sqrt{\frac{2}{2j+1}}. \tag{4.66}$$

The intrinsic system of functions (4.64) is orthonormal and the A-coefficients give the orthonormal transformation between the laboratory $(J_f lj)$ and intrinsic (ljK_i) representations, i.e.

$$\begin{aligned}\sum_{K_i>0} A^{J_iK_i}_{jJ_f}A^{J_iK_i}_{jJ'_f} &= \delta_{J_fJ'_f},\\ \sum_{J_f} A^{J_iK_i}_{jJ_f}A^{J_iK'_i}_{jJ_f} &= \delta_{K_iK'_i},\end{aligned} \tag{4.67}$$

and therefore the relation (4.64) can be inverted

$$\mathscr{L}^{(ljK_i)}_{J_iM_i}(\omega,\hat{r}',\mathbf{s}) = \sum_{J_f} A^{J_iK_i}_{jJ_f}\mathscr{Y}^{(J_flj)}_{J_iM_i}(\omega,\hat{r},\mathbf{s}). \tag{4.68}$$

The wave function (3.19) has a similar form in the intrinsic system, i.e.

$$\Psi_{J_iM_i}(\omega,\mathbf{r}',\mathbf{s}) = \sum_{ljK_i>0}\frac{g_{ljK_i}(r)}{r}\mathscr{L}^{(ljK_i)}_{J_iM_i}(\omega,\hat{r}',\mathbf{s}), \tag{4.69}$$

where the intrinsic radial wave function is connected with its laboratory counterpart by the same A-coefficients (4.65), i.e.

$$\begin{aligned} g_{ljK_i}(r) &= \sum_{J_f} A^{J_iK_i}_{jJ_f} f_{J_flj}(r),\\ f_{J_flj}(r) &= \sum_{K_i>0} A^{J_iK_i}_{jJ_f} g_{ljK_i}(r).\end{aligned} \tag{4.70}$$

The intrinsic representation of the wave function can be rewritten as follows

$$\begin{aligned}\Psi_{J_iM_i}(\omega,\mathbf{r}',\mathbf{s}) &= \sum_{K_i>0}\frac{1}{\sqrt{2}}\Big[\mathscr{D}^{J_i*}_{M_i,K_i}(\omega)\Phi_{K_i}(\mathbf{r}',\mathbf{s}) + (-)^{J_i+K_i}\mathscr{D}^{J_i*}_{M_i,-K_i}(\omega)\Phi_{\tilde{K}_i}(\mathbf{r}',\mathbf{s})\Big]\\ &\equiv \sum_{K_i>0}\Psi^{J_i}_{M_iK_i}(\omega,\mathbf{r}',\mathbf{s}).\end{aligned} \tag{4.71}$$

The part describing the intrinsic proton motion is now given by the Nilsson single particle wave function for fermionic case or just the intrinsic boson wave function for cluster decays

$$\Phi_{K_i}(\mathbf{r}',\mathbf{s}) = \sum_{lj} \frac{g_{ljK_i}(r)}{r} \mathscr{Y}^{(ls)}_{jK_i}(\hat{r}',\mathbf{s}). \tag{4.72}$$

To derive the coupled system of equations for the intrinsic radial components g_{ljK_i} we use the expansion of the interaction (4.60) in the intrinsic system. The matrix element of the spherical harmonics is diagonal in the intrinsic projection K_i and acquires the value

$$\begin{aligned}
&\langle \mathscr{Y}^{(J_flj)}_{J_iM_i}(\omega,\hat{r},\mathbf{s})|T_{\lambda 0}(\hat{r}')|\mathscr{Y}^{(J'_fl'j')}_{J_iM_i}(\omega,\hat{r},\mathbf{s})\rangle \\
&= \sum_{K_i>0}\sum_{K'_i>0} A^{J_iK_i}_{jJ_f} \langle \mathscr{Z}^{(ljK_i)}_{J_iM_i}(\omega,\hat{r}',\mathbf{s})|T_{\lambda 0}(\hat{r}')|\mathscr{Z}^{(l'j'K'_i)}_{J_iM_i}(\omega,\hat{r}',\mathbf{s})\rangle A^{J_iK'_i}_{j'J'_f} \\
&= \sum_{K_i>0} A^{J_iK_i}_{jJ_f} \langle \mathscr{Y}^{(ls)}_{jK_i}(\hat{r}',\mathbf{s})|T_{\lambda 0}(\hat{r}')|\mathscr{Y}^{(l's)}_{j'K_i}(\hat{r}',\mathbf{s})\rangle A^{J_iK_i}_{j'J'_f}.
\end{aligned} \tag{4.73}$$

where (4.64) was used. In order to obtain the system of coupled equations for intrinsic components we multiply (2.63) by $A^{J_iK_i}_{jJ_f}$ and sum over J_f to get

$$\begin{aligned}
\frac{d^2 g_{ljK_i}(r)}{dr^2} &= \left\{ \frac{l(l+1)}{r^2} + \frac{2\mu}{\hbar^2}[V_0(r,\mathbf{s}) - E] \right\} g_{ljK_i}(r) + \frac{2\mu}{\hbar^2} \sum_{K'_i>0} W^{K_iK'_i}_{jJ_i} g_{ljK'_i}(r) \\
&+ \frac{2\mu}{\hbar^2} \sum_{\lambda>0} V_\lambda(r,\mathbf{s}) \sum_{l'j'} \langle \mathscr{Y}^{(ls)}_{jK_i}(\hat{r}',\mathbf{s})|T_{\lambda 0}(\hat{r}')|\mathscr{Y}^{(l's)}_{j'K_i}(\hat{r}',\mathbf{s})\rangle g_{l'j'K_i}(r),
\end{aligned} \tag{4.74}$$

where

$$W^{K_iK'_i}_{jJ_i} = \sum_{J_f} A^{J_iK_i}_{jJ_f} E_{J_f} A^{J_iK'_i}_{jJ_f}. \tag{4.75}$$

The above off-diagonal matrix elements give rise to the Coriolis coupling in the so-called non-adiabatic approach. The concrete form of these matrix elements is given in Ref. [23]. By neglecting all matrix elements (4.75) one obtains a system diagonal on the intrinsic projection. In this case K_i is conserved during the decay process and the wave function (3.19) has a given intrinsic projection. This approach is called adiabatic.

The decay width then becomes

$$\begin{aligned}
\Gamma = \sum_{J_flj} \hbar v_{J_f} |f_{J_flj}|^2 &= \sum_{J_fljK_i>0} \hbar v_{J_f} \left[\mathscr{F}_{K_i} A^{J_iK_i}_{jJ_f} \right]^2 \lim_{r\to\infty} |g_{ljK_i}(r)|^2 \\
&= \sum_{J_fljK_i>0} \hbar v_{J_f} \left[\mathscr{F}_{K_i} A^{J_iK_i}_{jJ_f} N'_{ljK_i} \right]^2,
\end{aligned} \tag{4.76}$$

where $\mathscr{F}_{K_i}$ is the preformation amplitude and N' is the matching constant (scattering amplitude), corresponding to the integration of the system (4.74) in the intrinsic system. We used that fact that at large distances, where one has only a spherical Coulomb interaction, the asymptotic form of the radial component is proportional to the Gamow state, i.e.

$$\lim_{r\to\infty} g_c(r) = N'_c H^{(+)}_{l_c}(\chi,\kappa_c r), \tag{4.77}$$

It is worthwhile to point out that both the Coulomb parameter χ_c and the wave number κ_c dependent upon the channel label $c = (ljK_i)$.

4.8 Adiabatic Approach

As we have already mentioned, in the adiabatic approach one neglects the Coriolis terms given by Eq. 4.75, in order to obtain a diagonal system with respect to the spin projection K_i. The description of the proton emission in a deformed potential within the adiabatic approach is a generalization of the Nilsson model [24] in the sense that the standard Nilsson model describes the discrete part of the spectrum, while in proton emission we are interested in positive energies. For boson emission one obtains a similar adiabatic description of the α-decay or heavy cluster emission.

We will consider, as usually, a common boson/fermion formalism. The only difference is the spin-orbit term, which disappears for boson emission. In the simplest approach one considers a spherical spin-orbit interaction $V_{so}(r)\mathbf{l}\cdot\boldsymbol{\sigma}$. The system of differential equations for the radial components (4.74) becomes

$$\frac{d^2 g_{lj}(r)}{d\rho^2} = \sum_{l'j'} \mathscr{A}_{lj;l'j'} g_{l'j'}(r), \tag{4.78}$$

where the matrix $\mathscr{A}_{lj;l'j'}$ is given by

$$\begin{aligned} \mathscr{A}_{lj;l'j'} = & \left[\frac{l(l+1)}{\rho^2} + \frac{V_0(r)}{E} + \frac{V_{so}(r)}{E}\langle \mathbf{l}.\boldsymbol{\sigma}\rangle - 1\right]\delta_{ll'}\delta_{jj'} \\ & + \sum_{\lambda>0} \frac{V_\lambda(r)\langle j', K_i; \lambda, 0 | j, K_i\rangle}{E \qquad \hat{j}} \langle lj||Y_{\lambda 0}||l'j'\rangle. \end{aligned} \tag{4.79}$$

For fermion and boson cases the reduced matrix elements of the central interaction are given in Appendix (14.4).

For simplicity here we will consider the case of an axially deformed mean field. The potential contains only even multipoles. Since parity is conserved the basis states for even and odd parities are respectively given by

$$|l\rangle = |0\rangle, |2\rangle, |4\rangle, |6\rangle, \ldots \quad \text{(even parity)},$$
$$|l\rangle = |1\rangle, |3\rangle, |5\rangle, |7\rangle, \ldots \quad \text{(odd parity)}, \tag{4.80}$$

for boson emission and

$$|lj\rangle = \left|0\frac{1}{2}\right\rangle, \left|2\frac{3}{2}\right\rangle, \left|2\frac{5}{2}\right\rangle, \left|4\frac{7}{2}\right\rangle, \left|4\frac{9}{2}\right\rangle, \ldots \quad \text{(even parity)},$$
$$|lj\rangle = \left|1\frac{1}{2}\right\rangle, \left|1\frac{3}{2}\right\rangle, \left|3\frac{5}{2}\right\rangle, \left|3\frac{7}{2}\right\rangle, \left|5\frac{9}{2}\right\rangle, \ldots \quad \text{(odd parity)} \tag{4.81}$$

for fermion emission.

In the case of the fermion emission, let us consider a deformed spin-orbit potential, which can be important for high angular momenta. The total interaction has the following form

$$V(\mathbf{r}', \mathbf{s}) = V_0(\mathbf{r}') + V_{so}(\mathbf{r}', \mathbf{s}). \tag{4.82}$$

where $V_0(\mathbf{r}')$ denotes the central (nuclear + Coulomb) part and $V_{so}(\mathbf{r}', \mathbf{s})$ the deformed spin-orbit interaction, given by (see Appendix (14.1))

$$V_{so}(\mathbf{r}', \mathbf{s}) = \nabla V_{so}(\mathbf{r}') \cdot [-i\nabla \times \boldsymbol{\sigma}], \; \boldsymbol{\sigma} = 2\mathbf{s}. \tag{4.83}$$

By using the multipole expansion of the central potential

$$V_0(\mathbf{r}') = V_0(r) + V_d(\mathbf{r}') = V_0(r) + \sum_{\lambda>0} V_{d\lambda}(r) Y_{\lambda 0}(\hat{r}'), \tag{4.84}$$

one obtains a closed expression of the matrix element

$$\langle ljK_i|V_0(\mathbf{r}')|l'j'K_i\rangle = \delta_{ll'}\delta_{jj'}V_0(r) + \sum_{\lambda>0} V_{d\lambda}(r)\frac{\langle j', K_i; \lambda, 0|j, K_i\rangle}{\hat{j}}\langle lj||Y_{\lambda 0}||l'j'\rangle. \tag{4.85}$$

The part containing the deformed spin-orbit interaction has a more complex ansatz. According to Eq. 14.86, in the spin-orbit matrix elements one also has to consider the radial part of the wave function

$$\langle ljK_i|V_{so}(\mathbf{r}', \mathbf{s})|l'j'K_i\rangle\frac{g_{l'j'}(r)}{r} = \delta_{ll'}\delta_{jj'}V_{so}(r)\langle \mathbf{l}\cdot\boldsymbol{\sigma}\rangle\frac{g_{lj}(r)}{r} + \sum_{\lambda>0}\frac{\langle j', K_i; \lambda, 0|j, K_i\rangle}{\hat{j}} \times\left[U^{(0)}_{\lambda 0}(l,j,l',j')\frac{g_{l'j'}(r)}{r} + U^{(1)}_{\lambda 0}(l,j,l',j')\frac{1}{r}\frac{dg_{l'j'}(r)}{dr}\right]. \tag{4.86}$$

The functions $U^{(k)}_{\lambda\mu}$ are given in Appendix (14.4).

By using the standard procedure to derive the coupled system of equations for the radial components one obtains

$$\frac{d^2 g_{lj}}{dr^2} = \left\{ \frac{l(l+1)}{r^2} + \frac{2\mu}{\hbar^2}[V_0(r) + V_{so}(r)\langle \mathbf{l} \cdot \boldsymbol{\sigma} \rangle - E] \right\} g_{lj}$$
$$+ \frac{2\mu}{\hbar^2} \sum_{l'j'} \sum_{\lambda > 0} \frac{\langle j', K_i; \lambda, 0 | j, K_i \rangle}{\hat{j}} \left[V_{d\lambda} \langle lj || Y_{\lambda 0} || l'j' \rangle + U_{\lambda 0}^{(0)}(l, j, l', j') \right] g_{l'j'}$$
$$+ \frac{2\mu}{\hbar^2} \sum_{l'j'} \sum_{\lambda > 0} \frac{\langle j', K_i; \lambda, 0 | j, K_i \rangle}{\hat{j}} U_{\lambda 0}^{(1)}(l, j, l', j') \frac{dg_{l'j'}}{dr}. \tag{4.87}$$

This system of second order equations has a form similar to (3.24)

$$\frac{d^2 \mathbf{g}}{dr^2} = (\mathscr{V}\mathscr{I} + \mathscr{C})\mathbf{g} + \mathscr{D}\frac{d\mathbf{g}}{dr}, \tag{4.88}$$

where $\mathscr{V}$ denotes the spherical part. Now the matrices $\mathscr{C}$ and $\mathscr{D}$ have the form given by (4.87). By using the notation

$$Z = \begin{bmatrix} \mathbf{g} \\ \mathbf{g}' \end{bmatrix}, \tag{4.89}$$

one obtains the first order system of equations given by

$$\frac{dZ}{dr} = \begin{bmatrix} 0 & \mathscr{I} \\ \mathscr{V}\mathscr{I} + \mathscr{C} & \mathscr{D} \end{bmatrix} Z. \tag{4.90}$$

which can be integrated by using the Runge–Kutta procedure described in Appendix (14.6).

4.9 Coupled Channels Calculations for Proton Emitters

The adiabatic approach is the most popular method to estimate proton half lives. Its validity was confirmed in Ref. [25], where it was shown that the paring interaction suppresses the influence of Coriolis terms.

In order to illustrate this approach let us compute the partial half life to the ground state, given by Eq. 4.76 with $J_f = 0$, i.e.

$$T_{lj} = \frac{\hbar \ln 2}{\Gamma_{lj}} = \frac{2j+1}{2} \frac{\ln 2}{v u_{K_i}^2 [N'_{ljK_i}]^2}, \tag{4.91}$$

where the scattering amplitude is

$$N'_{ljK_i} = \frac{1}{H_l^{(+)}(\kappa R)} \sum_{l'j'} \mathscr{K}_{lj;l'j'}(R) g_{l'j'K_i}(R). \tag{4.92}$$

In this calculation we have considered proton emission from ^{131}Eu ($\beta_2 = 0.331$). We have used the universal interaction, as given in Appendix (14.1), adjusting the depth of the potential to reproduce the experimental Q-value. In Table 4.1 we give the propagator matrix elements $\mathcal{K}_{lj;l'j'}$ corresponding to the Fermi level with $K_i = \frac{3}{2}^+$. The matching radius is R=10 fm. Thus, the lowest value of the spin in the parent nucleus is $J_i = \frac{3}{2}$. The channel components (l, j) are given in the first column. One sees that the propagator has small off-diagonal matrix elements. The values of the wave function components at the same radius and the logarithm of the partial half lives, respectively are given in the last two lines. The experimental value is $\log_{10}T_{exp} = -1.57$ (T in seconds). It corresponds to the decay channel $(l,j) = (2, \frac{3}{2})$, which is allowed by the condition $J_i = j$ imposed by conservation of the angular momentum, within a factor of three.

The radial components $g_{ljK_i}(r)$ are given in Fig. 4.1. It is worthwhile to point out that the component of the wave function corresponding to the emitted proton is very small (in absolute value) in this case.

A generalization of the Nilsson model in the continuum within the adiabatic approach was performed in Refs. [5, 26, 27], where the behaviour of Gamow resonances as a function of the quadrupole deformation was analyzed. In Ref. [5] the coupled system of equations (4.87) for a given projection m is solved in the adiabatic limit by using an equivalent first order system for the logarithmic derivative of the wavefunction, as described in Sect. 4.2. A Woods–Saxon potential plus the corresponding spin-orbit interaction with the universal set of parameters [28] was used. The proton spectrum of ^{154}Sm was investigated as a function of the quadrupole deformation parameter in the interval $-0.5 \leq \beta_2 \leq 0.5$. The basis included states up to $j = 27/2$ and gave results which are close to those provided by the diagonalization procedure SWBETA [28] for negative energies. For eigenstates with positive energy there were some differences, due to the restricted number of major shells considered in the diagonalization procedure. The behaviour of the width, i.e. Im(E), as a function of Re(E)

Table 4.1 The matrix elements of the propagator $\mathcal{K}(R)$ at the matching radius $R = 10$ fm for the $K_i = \frac{3}{2}^+$ state in ^{131}Eu

(l, j)	$(2, \frac{3}{2})$	$(2, \frac{5}{2})$	$(4, \frac{7}{2})$	$(4, \frac{9}{2})$	$(6, \frac{11}{2})$	$(6, \frac{13}{2})$	$(8, \frac{15}{2})$
$(2, \frac{3}{2})$	1.0234	0.0159	0.0167	0.0005	0.0006	0.0000	0.0000
$(2, \frac{5}{2})$	0.0159	1.0018	−0.0050	0.0192	−0.0003	0.0007	0.0000
$(4, \frac{7}{2})$	0.0360	−0.0113	0.9924	0.0042	0.0157	0.0002	0.0007
$(4, \frac{9}{2})$	0.0006	0.0418	0.0042	0.9930	−0.0015	0.0162	−0.0001
$(6, \frac{11}{2})$	0.0008	−0.0004	0.0481	−0.0050	0.9885	0.0018	0.0131
$(6, \frac{13}{2})$	0.0000	0.0010	0.0002	0.0497	0.0018	0.9922	−0.0007
$(8, \frac{15}{2})$	0.0000	0.0000	0.0008	−0.0001	0.0515	−0.0028	0.9879
$g_{ljK_i}(R)$	0.0079	−0.0603	0.0107	0.0035	−0.0042	0.0026	0.0009
$\log_{10}T_{lj}$	−1.12	−2.78	0.36	2.57	3.89	4.22	8.31

The partial half life T_{lj} is in seconds

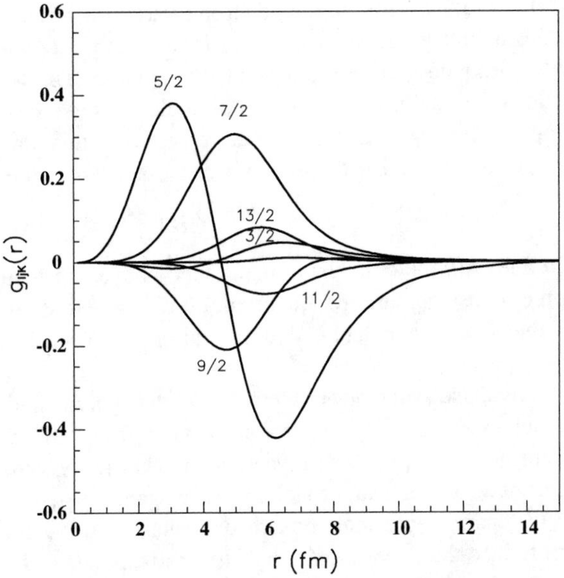

Fig. 4.1 Radial components of the proton wavefunction in ^{131}Eu. The spin values are indicated

for different deformations, was studied. It turned out that the energy corresponding to particles moving in high spin orbitals obey the rule

$$\mathrm{Im}(E) \sim \mathrm{Re}(E)^{l+1/2}. \tag{4.93}$$

Low spin orbitals have a more complex behaviour, due to the lower centrifugal barrier. For them, the "non-crossing trajectories rule", valid for bound states, sometimes can be violated. The emitter ^{113}Cs was also analyzed using the same approach [26]. The decay width was estimated according to Eq. 4.76 using for the spectroscopic amplitude the BCS estimate, i.e. $\mathscr{F}_K = u_K$. For transitions to the ground state, with $K_i = J_i = j$, one obtains $\sqrt{2/(2j+1)}$ for the transformation coefficient (4.65). As a check of the accuracy of the calculation as well as of the formalism itself it was verified that the total decay width coincides with twice the imaginary part of the resonant energy. Regarding the formalism this agreement is expected since the width is very small and, therefore, the resonance can be considered isolated. The partial width was defined for a large distance r as

$$\begin{aligned}\Gamma_{lj} &= -2E_n\mathscr{P}_{ij},\\ \mathscr{P}_{ij} &\equiv \frac{N_{lj}^2}{\sum_{l'j'} N_{l'j'}^2},\\ N_{lj}^2 &\equiv \frac{g_{lj}^2(r)}{|H_l^{(+)}(\chi,\kappa r)|^2}.\end{aligned} \tag{4.94}$$

The constant N_{lj} coincides with the matching coefficient previously defined by us in the intrinsic system, Eq. 4.91.

The calculation has shown a strong dependence of the Nilsson components upon deformations. This implies a strong variation of the half life as a function of the deformation parameters. Since variations in the angular momentum of the resonance also produce great variations in the half life (due to the centrifugal barrier) it is concluded that proton emission of nuclei close to the proton drip line is a powerful tool to investigate nuclear structure details in this rare region of the nuclear chart.

In Refs. [27, 29] the analysis was extended to the nuclei

$$^{109}\mathrm{I},\ ^{131}\mathrm{Eu},\ ^{141}\mathrm{Ho},\ ^{151}\mathrm{Lu}.$$

In line with the conclusion mentioned above, it became possible to make a rather precise assignment of quantum numbers and deformation, as summarized in Table 4.2. The calculated deformations are consistent with the predictions of Ref. [30].

All these calculations considered that parent and daughter nuclei are in the ground state. In Ref. [31] proton emission from ^{131}Eu to both ground and 2^+ excited state in ^{130}Sm [32] was investigated. The decay width was computed within the deformed adiabatic approach by using again Eq. 4.76. The branching ratio $\Gamma_{2^+}/\Gamma_{0^+}$ is in general dependent upon the deformation. It was found that the only state able to reproduce the experimental half life corresponds to $K_i = 3/2^+$, with a deformation larger than 0.27. This result is consistent with the empirical rule

$$E_{2^+} \approx \frac{1225}{A^{7/3}\beta_2^2}, \tag{4.95}$$

giving the value $\beta_2 \approx 0.34$ for the excitation energy $E_{2^+} = 121$ keV.

The authors also predicted for the decay of ^{141}Ho to the 2^+ state in ^{140}Dy that the branching ratio has a weak dependence upon the deformation. Thus, by using $\beta_2 = 0.29$ from the systematics of Ref. [30], the energy of the state ^{140}Dy (2_1^+) was found to be about 140 keV. By using the assignment in Table 4.2, i.e. $K_i = 7/2^-$, one obtains a branching ratio of about 5%, which is not reachable with present experimental facilities.

It is important to mention the extensive study performed in Ref. [33], concerning the dependence of the decay width on the single particle potential. Several potentials were considered, namely Becchetti and Greenlees [16], Cherpunov [34], Davids [23], Blomqvist and Wahlborn [35], Rost [36] and the universal parametrization [37]. The authors investigated ground state to ground state as well as ground state to

Table 4.2 Theoretical assignments [27, 29] in proton emitters as explained in the text

Nucleus	K_i	β_2
^{109}I	$1/2^+$	0.14
^{131}Eu	$5/2^+$	[0.3, 0.4]
^{141}Ho	$7/2^-$	[0.3, 0.4]
^{151}Lu	$5/2^-$	[−0.18, −0.14]
^{151*}Lu	$3/2^+$	[−0.18, −0.14]

The star indicates isomeric state

excited 2^+ state transitions in the deformed emitters ^{131}Eu and ^{141}Ho. The experimental data could be reproduced by using a deformed spin-orbit formfactor with a large radius parameter, $r_0 \in [1.24, 1.275]$ fm. It turns out that all the above potentials satisfy this requirement, except Becchetti–Greenlees where it is $r_0 = 1.17$ fm.

In Ref. [38] and then in [39, 40] the odd–even proton emitters

$$^{109}\text{I},\ ^{113}\text{Cs},\ ^{117}\text{La},\ ^{131}\text{Eu},\ ^{141,141*}\text{Ho}.$$

were studied within the non-adibatic approach. The Cherpunov parametrisation [34] including a spherical spin-orbit term was used for the interaction. The convergency of the calculations was achieved by using $\lambda_{\text{max}} = 8$ in the expansion of the potential and $J_f = 12$ for the wave function. The depth of the potential was adjusted to obtain the proton energy. A high accuracy was necessary in order to obtain the exact position of the energy in the complex plane. Thus, the decay width using (4.76) differs by less than 2% from twice the imaginary part of the resonant energy (in absolute value).

Half lives and branching ratios for the rotational bands in parent and daughter nuclei in the emitters ^{131}Eu and ^{141}Ho were calculated and the role of the deformation upon the decay width was analysed for all isotopes. Except for ^{109}I and ^{113}Cs, the deformation dependence is much weaker than the uncertainty due to the experimental proton Q-value. The intrinsic K-orbitals of the wave function were analyzed by using Eq. 4.70. From Table II of [39], where half lives and branching ratios for different intrinsic orbitals are given, one sees that a good agreement with available experimental data was obtained. However, an extensive comparison with the adiabatic method, also performed in this paper, showed that for some cases the results are close to each other, as for $^{141}\text{Ho}^*$ and ^{177}La, while in other cases, like ^{113}Cs, there are differences of up to a factor of 5.

In Ref. [11], the "direct" method based on the coupled channels formalism, was compared with the distorted wave (DW) approach by analyzing the spherical emitters

$$^{167,167*}\text{Ir},\ ^{147}\text{Tm}.$$

The comparison of the two methods gives practically the same results, with a precision of about 0.005%.

For the deformed nuclei

$$^{131}\text{Eu},\ ^{141}\text{Ho}$$

the precision achieved is only about 20% due to the truncation of the eigenfunction space in the initial function $f^{(\text{res})}$ entering Eq. 4.42. This feature is discussed in more detail in Ref. [23], where the authors apply the non-adiabatic coupled channels formalism, to describe transitions to ground and excited states in the same nuclei. Here the authors extrapolated the resonant function $f^{(\text{res})}$, obtained within the coupled channels procedure, by using the asymptotic form beyond the matching radius R. In this way the convergency with the DW method, which uses $r_{\text{int}} = 100$ fm in (4.42), was achieved. It was also pointed out the important role played by the deformed spin-orbit interaction when dealing with small components of the wave function.

They also performed an adiabatic calculation in the intrinsic system. For the asymptotic region the same extrapolation procedure was used. By rotating the wave function to the laboratory system, in order to use the DW formula (4.42), results which are close to the coupled channels approach were obtained. This is valid for low spin states, like the emission from ^{131}Eu ($3/2^+$), where the Coriolis correction is weak. For transitions like the one from ^{141}Ho ($7/2^-$), the adiabatic approach in the absence of pairing corrections gives results which are three times smaller than the ones provided by the coupled channels method.

The role of the pairing interaction on Coriolis terms was discussed in detail in Ref. [25]. Here, it was clarified why the results of adiabatic calculations describe very well most of the transitions to ground, as well as to excited states. First, it was performed an adiabatic calculation in the intrinsic system for the proton emission from ^{141}Ho. The resulting Nilsson wave functions were used as a basis to diagonalize the Coriolis interaction, i.e.

$$\Psi_{J_iM_i} = \sum_{K>0} a_K^{J_i} \Psi_{M_iK}^{J_i}, \tag{4.96}$$

where $\Psi_{M_iK}^{J_i}$ are defined by (4.71). The decay width was estimated by using Eq. 4.76 with the same deformations for parent and daughter nuclei and taking into account the mixing coefficients $a_K^{J_i}$. The results for half lives were consistent with those of Ref. [23], i.e. close to the experimental values, when the adiabatic approximation was used. However, when the Coriolis correction was included, the calculated decay widths were one order of magnitude larger than the corresponding experimental values.

In a second step the Coriolis matrix elements were corrected by the pairing term, i.e. $u_{K+1}v_K + v_{K+1}u_K$, and the half lives have become comparable with the values of the adiabatic calculation (and therefore close to the experimental values) for transitions to both the ground and excited states. This comparison cleary showed the importance of the residual pairing interaction upon proton emission.

The influence of the residual pairing interaction is still not fully clarified and further calculations, especially within the non-adiabatic approach where Coriolis terms are automatically included, are necessary.

4.10 Proton Emission from Rotational Odd–Odd Nuclei and from Rotational–Vibrational Odd–Even Nuclei

In the previous section, we described proton emission from rotational odd–even nuclei. On the other hand the wave function describing proton emission from rotational odd–odd nuclei has a similar structure with the proton emission from rotational–vibrational odd–even nuclei. The odd proton/neutron will be labelled with the letter "*p*/*n*".

We will consider the intrinsic system of coordinates first. The wave function for an odd–odd rotational nucleus is given by

$$\Psi_{J_iM_i}(\omega, \mathbf{r}'_p, \mathbf{r}'_n) = \sum_{K_i \geq 0} \frac{1}{\sqrt{2(1+\delta_{K_i,0})}} \times \left[\mathscr{D}^{J_i*}_{M_i,K_i}(\omega)\Phi_{K_i}(\mathbf{r}'_p, \mathbf{r}'_n) + (-)^{J_i+K_i}\mathscr{D}^{J_i*}_{M_i,-K_i}(\omega)\Phi_{\tilde{K}_i}(\mathbf{r}'_p, \mathbf{r}'_n)\right], \tag{4.97}$$

where we use the short-hand notation $(\mathbf{r}'_\tau, \mathbf{s}_\tau) \rightarrow \mathbf{r}'_\tau$. This relation is analogous to Eq. 4.71, but here Φ_{K_i} denotes the product of proton and neutron Nilsson functions, as defined by (4.72), i.e.

$$\Phi_{K_i}(\mathbf{r}'_p, \mathbf{r}'_n) = \Phi_{K_p}(\mathbf{r}'_p)\Phi_{K_n}(\mathbf{r}'_n), \tag{4.98}$$

where the initial intrinsic spin projection is $K_i = K_p \pm K_n$.

If one replaces neutron coordinate by a vibrational coordinate $\mathbf{r}'_n \rightarrow \alpha$, the wave function describes an odd–even rotational–vibrational nucleus, i.e.

$$\Phi_{K_i}(\mathbf{r}'_p, \alpha) = \Phi_{K_p}(\mathbf{r}'_p)\Phi_{K_n}(\alpha). \tag{4.99}$$

In order to compute the decay width of the odd proton, it is necessary to express its wave function in the laboratory system of coordinates with respect to $\mathbf{r}_p$, defined by the Euler angles ω. One obtains for the wave function (4.97) a superposition of terms corresponding to the rotational band built on the odd neutron state with a given spin projection $K_f = K_n$

$$\Psi_{J_iM_i}(\omega, \mathbf{r}_p, \mathbf{r}'_n) = \sum_{J_flj} \frac{f_{J_fK_flj}(r_p)}{r_p} \mathscr{Y}^{(J_fK_flj)}_{J_iM_i}(\omega, \hat{r}_p, \mathbf{r}'_n), \tag{4.100}$$

where we introduced the core-neutron-spin-orbit harmonics

$$\begin{aligned} \mathscr{Y}^{(J_fK_flj)}_{J_iM_i}(\omega, \hat{r}_p, \mathbf{r}'_n) &\equiv \left[\Psi^{J_f*}_{K_f}(\omega, \mathbf{r}'_n) \otimes \mathscr{Y}^{(ls)}_j(\hat{r}_p, \mathbf{s}_p)\right]_{J_iM_i}, \\ \Psi^{J_f}_{M_fK_f}(\omega, \mathbf{r}'_n) &\equiv \frac{1}{\sqrt{2}}\left[\mathscr{D}^{J_f*}_{M_fK_f}(\omega)\Phi_{K_f}(\mathbf{r}'_n) + (-)^{J_f+K_f}\mathscr{D}^{J_f*}_{M_f,-K_f}(\omega)\Phi_{\tilde{K}_f}(\mathbf{r}'_n)\right]. \end{aligned} \tag{4.101}$$

The laboratory radial wave function is connected with its intrinsic counterpart by

$$\begin{aligned} f_{J_fK_flj} &= \sum_{K_i \geq 0} A^{J_iK_i}_{jJ_fK_f} g_{lj,K_i-K_f}, \\ A^{J_iK_i}_{jJ_fK_f} &\equiv \frac{(-)^{J_i-J_f-K_i+K_f}}{\sqrt{1+\delta_{K_i,0}}}\langle J_i, K_i; j, K_f - K_i | J_f, K_f\rangle. \end{aligned} \tag{4.102}$$

The decay width of the proton is obtained by integrating on its angular coordinates, Euler angles and neutron coordinates. Due to the orthonormality of the angular functions in Eq. 4.100, for a given final spin of the residual core J_f one obtains the partial decay width as

$$\Gamma_{K_f} = \sum_{J_f lj} \hbar v_{J_f} \lim_{r\to\infty} |f_{J_f K_f lj}(r)|^2 = \sum_{J_f ljK_i} \hbar v_{J_f} \left[A^{J_i K_i}_{jJ_f K_f}\right]^2 \lim_{r\to\infty} |g_{l,j,K_i-K_f}(r)|^2$$
$$= \sum_{J_f ljK_i} \hbar v_{J_f} \left[A^{J_i K_i}_{jJ_f K_f} N'_{l,j,K_i-K_f}\right]^2, \tag{4.103}$$

where we introduced, as usually, the normalization constant in the intrinsic system. Thus, the formalism of the odd–even nuclei can be extended for odd–odd nuclei by considering the odd neutron as a spectator. In particular, if one neglects the Coriolis coupling, the system of coupled equations, written for intrinsic coordinates, becomes diagonal in K_i and the initial wave function (4.97) contains only one intrinsic projection. The difference for odd–odd emitters is that the initial spin J_i is an integer number, the final spin J_f is half-integer and the emitted proton carries the angular momentum j according to

$$|J_i - J_f| \le j \le J_i + J_f. \tag{4.104}$$

In Ref. [41] the emission from the odd–odd light proton emitter ^{58}Cu was analyzed. The adiabatic assumption, i.e. considering a given initial projection K_i in the intrinsic wave function (4.97), was used. The change of the deformation in the spectroscopic amplitude hindered the decay width by a factor of ≈ 30, however the experimental value was not reproduced. It was found that the Coriolis interaction mixing was not able to explain that discrepancy. It was then concluded that a possible explanation for the measured half life is a mixing of spherical (1%) and oblate (99%) shapes in the daughter nucleus.

In Ref. [42] the authors apply Eq. 4.103 with the spectroscopic factor $u^2_{K_p}$ and fixed initial intrinsic projection K_i to investigate proton emission from the odd–odd nuclei

$$^{112}\text{Cs},\ ^{140}\text{Ho},\ ^{150,150*}\text{Lu}.$$

They used the universal parametrization of the Woods–Saxon axially deformed potential. It was assumed that the initial spin has the lowest value, i.e. $J_i = |K_i|$, and that the daughter nucleus is left in the ground state with $J_f = |K_n|$. The initial projection was considered to be parallel or antiparallel to the proton spin, i.e. $K_i = K_p \pm K_n$. The decay width, as in the odd–even case, was found to be strongly dependent upon the quadrupole deformation as well as upon the single particle level occupied by the unpaired neutron. This leads to an assignment of quantum numbers and deformation according to Table 4.3.

Table 4.3 Theoretical assignments [42] in proton emitters as explained in the text

Nucleus	K_i	K_n	β_2
^{112}Cs	3^+	$3/2^+$	[0.1, 0.19]
^{140}Ho	8^+	$9/2^-$	[0.28,0.38]
^{150}Lu	2^+	$1/2^-$	[−0.17, −0.16]
^{150*}Lu	1^-	$1/2^-$	[−0.17, −0.16]

4.11 Emission from Rotational Triaxial Nuclei

The formalism developed for axially symmetric nuclei can be generalized in a straightforward way for triaxial emitters. In this section, we will consider a common angular function given by (4.55), in order to describe boson, as well as fermion emission. In the intrinsic system of coordinates the wave function of the initial nucleus is given by [43]

$$\Psi^{(\alpha)}_{J_iM_i}(\omega,\mathbf{r}',\mathbf{s}) = \frac{1}{\sqrt{2}}\sum_{K_i} c(\alpha,J_i,K_i) \times \left[\mathscr{D}^{J_i*}_{M_i,K_i}(\omega)\Phi^{(\alpha\rho=1)}(\mathbf{r}',\mathbf{s}) - (-)^{J_i+K_i}\mathscr{D}^{J_i*}_{M_i,-K_i}(\omega)\Phi^{(\alpha\rho=-1)}(\mathbf{r}',\mathbf{s})\right]. \tag{4.105}$$

For triaxial nuclei, the intrinsic angular momentum projection is no longer a good quantum number. Therefore, in the above equation the intrinsic wave function of the emitted light fragment is given by

$$\Phi^{(\alpha\rho)}(\mathbf{r}',\mathbf{s}) = \sum_{ljK} \frac{g^{(\alpha\rho)}_{ljK}(r)}{r}\mathscr{Y}^{(ls)}_{jK}(\hat{r}',\mathbf{s}). \tag{4.106}$$

We use the notation $\rho = \pm 1$ for the eigenvalues $\pm i$ of the symmetry operator $R_3(\pi)$. The coefficients $c(\alpha, J_i, K_i)$, labelled by the eigenstate number α of the triaxial core Hamiltonian, can be obtained by using the diagonalization procedure, i.e.

$$\sum_{k=1}^{3} \frac{\hat{J}_k^2}{2\mathscr{I}_k}\Psi^{(\alpha)}_{J_iM_i}(\omega,\mathbf{r},\mathbf{s}) = E^{(\alpha)}_{J_i}\Psi^{(\alpha)}_{J_iM_i}(\omega,\mathbf{r},\mathbf{s}). \tag{4.107}$$

In the laboratory system of coordinates one gets

$$\begin{aligned}\Psi^{(\alpha)}_{J_iM_i}(\omega,\mathbf{r},\mathbf{s}) = \frac{1}{\sqrt{2}}\sum_{J_f}\sum_{K_iKK_f}\sum_{lj}(-)^{J_i-J_f-K}c(\alpha,J_i,K_i)\langle J_i,K_i;j,-K|J_f,K_f\rangle \\ \times\left\{\left[\mathscr{D}^{J_f*}_{K_f}(\omega)\otimes\mathscr{Y}^{(ls)}_j(\hat{r},\mathbf{s})\right]_{J_iM_i}\frac{g^{(\alpha\rho=1)}_{ljK}(r)}{r}\right. \\ \left.+(-)^{J_f+K-K_i}\left[\mathscr{D}^{J_f*}_{-K_f}(\omega)\otimes\mathscr{Y}^{(ls)}_j(\hat{r},\mathbf{s})\right]_{J_iM_i}\frac{g^{(\alpha\rho=-1)}_{ljK}(r)}{r}\right\}.\end{aligned} \tag{4.108}$$

Here we used a formal summation over the three projections K_i, K, K_f, with the condition $K_f = K_i - K$. The above wave function can be rewritten as follows

$$\Psi^{(\alpha)}_{J_iM_i}(\omega,\mathbf{r},\mathbf{s}) = \sum_{J_flj}\frac{f^{(\alpha)}_{\alpha' J_flj}(r)}{r}\mathscr{Y}^{(\alpha' J_flj)}_{J_iM_i}(\omega,\hat{r},\mathbf{s}), \tag{4.109}$$

where we introduced the triaxial core-spin-orbit harmonics

$$\mathscr{Y}^{(\alpha' J_f lj)}_{J_i M_i}(\omega, \hat{r}, \mathbf{s}) \equiv \langle \omega, \hat{r}, \mathbf{s} | \alpha' J_f lj; J_i M_i \rangle = \left[\Phi^{(\alpha')}_{J_f}(\omega) \otimes \mathscr{Y}^{(ls)}_j(\hat{r}, \mathbf{s}) \right]_{J_i M_i}. \quad (4.110)$$

The core wave functions are eigenstates of the triaxial daughter Hamiltonian, i.e.

$$\mathbf{H}_d(\omega)\Phi^{(\alpha')}_{J_f M_f}(\omega) \equiv \sum_{k=1}^{3} \frac{\hat{J}_k^2}{2\mathscr{I}_k} \Phi^{(\alpha')}_{J_f M_f}(\omega) = E^{(\alpha')}_{J_f} \Phi^{(\alpha')}_{J_f M_f}(\omega), \quad (4.111)$$

and are given by [43]

$$\Phi^{(\alpha')}_{J_f M_f}(\omega) = \sum_{K_f \geq 0} c(\alpha', J_f, K_f) \frac{1}{\sqrt{2(1+\delta_{K_f,0})}} [\mathscr{D}^{J_f *}_{M_f K_f}(\omega) + (-)^{J_f + K_f} \mathscr{D}^{J_f *}_{M_f - K_f}(\omega)], \quad (4.112)$$

where α' labels the eigenvalue and K_f labels the intrinsic projections and the parity.

Let us consider a given quantum number ρ. The identification of Eqs 4.108 and 4.109, together with the orthonormality of the coefficients $c(\alpha', J_f, K_f)$, entering the wave function of the even–even core, allows one to obtain the expression for the radial components in the laboratory system

$$f^{(\alpha)}_{\alpha' J_f lj}(r) = \sum_K A^{\alpha J_i K}_{\alpha' j J_f} g^{(\alpha\rho)}_{ljK}(r). \quad (4.113)$$

where the transformation coefficients are given by

$$A^{\alpha J_i K}_{\alpha' j J_f} = \sum_{K_i K_f} \frac{(-)^{J_i - J_f - K}}{\sqrt{1+\delta_{K_f,0}}} c(\alpha, J_i, K_i) c(\alpha', J_f, K_f) \langle J_i, K_i; j, -K | J_f, K_f \rangle. \quad (4.114)$$

For transitions to the ground state, with $J_f = 0$, the coefficient $c(\alpha', J_f, K_f)$ is unity. Moreover, the angular momentum of the emitted light fragment is given by the initial spin $J_i = j$ and the summation in (4.108) is restricted to equal intrinsic projections, i.e. $K_i - K = K_f = 0$.

The decay width is a superposition of the partial widths corresponding to the different open channels, i.e.

$$\Gamma = \sum_{J_f lj} \hbar v_{J_f} \lim_{r \to \infty} |f^{(\alpha)}_{\alpha' J_f lj}(r)|^2. \quad (4.115)$$

By projecting out a given channel from the Schrödinger equation one obtains the following coupled channels system for the laboratory radial components

$$\frac{d^2 f^{(\alpha)}_{\alpha' J_f lj}(r)}{dr^2} = \left\{ \frac{l(l+1)}{r^2} + \frac{2\mu}{\hbar^2} \left[V_0(r, \mathbf{s}) + E^{(\alpha')}_{J_f} - E \right] \right\} f^{(\alpha)}_{\alpha' J_f lj}(r)$$
$$+ \frac{2\mu}{\hbar^2} \sum_{\lambda > 0, \nu} V_{\lambda\mu}(r, \mathbf{s}) \sum_{J'_f l' j'} \langle \mathscr{Y}^{(\alpha' J_f lj)}_{J_i M_i} | \sum_\nu D^{\lambda *}_{\nu\mu}(\omega) T_{\lambda\nu}(\hat{r}) | \mathscr{Y}^{(\alpha' J'_f l' j')}_{J_i M_i} \rangle f^{(\alpha)}_{\alpha' J'_f l' j'}(r). \quad (4.116)$$

Thus, by using the eigenstates of the triaxial core one avoids a very large non-axial basis, as it was done e.g. in Ref. [44].

In order to obtain the system of coupled equations for the intrinsic components, we multiply (4.116) by $A^{\alpha J_i K_i}_{\alpha' j J_j}$ and sum over J_f,

$$\frac{d^2 g^{(\alpha)}_{ljK}(r)}{dr^2} = \left\{ \frac{l(l+1)}{r^2} + \frac{2\mu}{\hbar^2}[V_0(r,\mathbf{s}) - E] \right\} g^{(\alpha)}_{ljK}(r)$$
$$+ \frac{2\mu}{\hbar^2} \sum_{K'>0} W^{\alpha K \alpha' K'}_{jJ_i} g^{(\alpha)}_{ljK'}(r)$$
$$+ \frac{2\mu}{\hbar^2} \sum_{\lambda>0,\mu} V_{\lambda\mu}(r,\mathbf{s}) \sum_{l'j'} \langle \mathscr{Y}^{(ls)}_{jK}(\hat{r}',\mathbf{s}) | T_{\lambda\mu}(\hat{r}') | \mathscr{Y}^{(l's)}_{j'K'}(\hat{r}',\mathbf{s}) \rangle g^{(\alpha)}_{l'j'K'}(r), \tag{4.117}$$

where the matrix elements are given in the Appendix (14.4) and a notation similar to the axial symmetric case was used, i.e.

$$W^{\alpha K \alpha' K'}_{jJ_i} = \sum_{J_f} A^{\alpha J_i K}_{\alpha' j J_f} E^{(\alpha')}_{J_f} A^{\alpha J_i K'}_{\alpha' j J_f}. \tag{4.118}$$

As we have already pointed out, in the so-called non-adiabatic approach, the off-diagonal matrix elements in W give rise to Coriolis couplings. In the non-axial case, even by neglecting the matrix W the system remains non-diagonal in the intrinsic projection K, as it can be seen from Eq. 4.117.

Proton emission from triaxial nuclei was investigated in Refs. [38, 45, 46]. In Ref. [45] the static triaxial deformation was introduced in the adiabatic coupled channels method in order to investigate proton emission from ^{141}Ho(7/2$^-$). The formalism uses a representation which is not the one given in (4.108), but taking Wigner functions coupled with the spin-orbit harmonics as a basis. A strong dependence of the total decay width was found, as well as of the branching ratio to the excited 2^+ state, upon the triaxiality, in an obvious correlation with the $K = 7/2^-$ component of the wave function. The experimental values support only a small triaxiality for this nucleus, namely $\gamma \leq 5^0$.

A similar conclusion was reached in Ref. [47] where triaxial oscillations of the nuclear surface were studied within the R-matrix approach, described later in Sect. 8.4. The calculations for ^{141}Ho (7/2$^-$) showed a strong dependence upon the triaxiality parameter of the decay widths to the ground and 2^+ excited states. However, this deformation does not improve the results in comparison with experimental data but rather goes in the opposite direction.

A different conclusion was reached in Ref. [46] where the role of the triaxiality was studied in the proton emission from ^{161}Re and ^{185}Bi. Here the dimension of the basis was significantly reduced by using the eigenfunctions of the triaxial rotator for both the parent and daughter nuclei. As in the calculation reviewed above, the decay width is very sensitive to the triaxiality. The actual triaxial parameters were obtained by minimizing the total energy, so that no free parameters were introduced. It was concluded that the experimental values are better reproduced when the computed triaxial deformation angle γ is considered. In this reference the anisotropy of emitted protons, defined by Eq. 3.33 was also investigated. It was concluded that for transitions, connecting ground states is an universal function, which does not depend upon nuclear structure details, in agreement with Ref. [48].

In this context we mention that the α-decay from triaxial nuclei was analyzed in Ref. [49], within the semiclassical Fröman approach [1] described below.

References

1. Fröman, P.P.: Alpha decay from deformed nuclei. Mat. Fys. Skr. Dan. Vid. Selsk. **1**(3) (1957)
2. Gyarmati, B., Vertse, T.: On the normalisation of Gamow functions. Nucl. Phys. A **160**, 523–528 (1971)
3. Vertse, T., Pál, K.F., Balogh, A.: GAMOW, a program for calculating the resonant state solution of the Radial Schrödinger Equation in an arbitrary optical potential. Comput. Phys. Commun. **27**, 309–322 (1982)
4. Ixaru, L., Rizea, M., Vertse, T.: Piecewiese perturbation methods for calculating Eigensolutions of complex optical potential. Comput. Phys. Commun. **85**, 217–230 (1995)
5. Ferreira, L.S., Maglione, E., Liotta, R.J.: Nucleon resonances in deformed nuclei. Phys. Rev. Lett. **78**, 1640–1643 (1997)
6. Kukulin, V.I., Krasnopol'sly, V.M., Horáček, J.: Theory of Resonances. Kluwer Academic Press, Dordrecht (1989)
7. Tanaka, N., Suzuki, Y., Varga, K.: Exploration of resonances by analytic continuation in the coupling constant. Phys. Rev. C **56**, 562–565 (1997)
8. Tanaka, N., Suzuki, Y., Varga, K., Lovas, R.G.: Unbound states by analytic continuation in the coupling constant. Phys. Rev. C **59**, 1391–1399 (1999)
9. Taylor, J.R.: Scattering Theory. Wiley, New York (1972)
10. Cattapan, G., Maglione, E.: From bound states to resonances: analytic continuation of the wave function. Phys. Rev. C **61**, 067301/1–4 (2000)
11. Davids, C.N., Esbensen, H.: Decay rates of spherical and deformed proton emitters. Phys. Rev. C **61**, 044302/1–5 (2000)
12. Satchler, G.R.: Direct Nuclear Reactions. Clarendon Press, Oxford (1983)
13. Bugrov, V.P., Kadmensky, S.G., Furman, V.I., Khlebostroev, V.G.: Multiparticle variant of proton and neutron radioactivity—the case of diagonal transitions. Yad. Fiz. **41**, 1123 (1985) [Sov. J. Nucl. Phys. **41**, 717–723 (1985)]
14. Bugrov, V.P., Kadmensky, S.G.: Proton decay of deformed nuclei. Yad. Fiz. **49**, 1562 (1989) [Sov. J. Nucl. Phys. **49**, 967–972 (1989)]
15. Kadmensky, S.G.: On absolute values of α-widths for heavy spherical nuclei. Z. Phys. A **312**, 113–120 (1983)
16. Becchetti, F.D. Jr., Greenlees, G.W.: Nucleon-nucleon optical-model parameters, $A > 40$, $E < 50$ MeV. Phys. Rev. **182**, 1190–1209 (1969)
17. Kadmensky, S.G., Bugrov, V.P.: Yad. Fiz. **59**, 424 (1996) [Phys. At. Nucl. **59**, 399 (1996)]
18. Gurvitz, S.A., Kalbermann, G.: Decay width and shift of a quasistationary state. Phys. Rev. Lett. **59**, 262–265 (1987)
19. Gurvitz, S.A.: New approach to tunneling problems. Phys. Rev. A **38**, 1747–1759 (1988)
20. Jackson, D.F., Rhoades-Brown, M.: Theories of alpha-decay. Ann. Phys. **105**, 151 (1977)
21. Berggren, T., Olanders, P.: Alpha decay from deformed nuclei: (I) formalism and application to ground-state cedays. Nucl. Phys. A **473**, 189–220 (1987)

22. Berggren, T.: Anisotropic alpha decay from oriented odd-mass isotopes of some light actinides. Phys. Rev. C **50**, 2494–2507 (1994)
23. Esbensen, H., Davids, C.N.: Coupled-channels treatment of deformed proton emitters. Phys. Rev. C **63**, 014315/1–13 (2000)
24. Nilsson, S.G.: Binding state of individual nucleons in strongly deformed nuclei. Kgl. Danske Videnskab. Selskab Mat. Fys. Medd. **29**(16) (1955)
25. Fiorin, G., Maglione, E., Ferreira, L.S.: Theoretical description of deformed proton emitters: nonadiabatic quasiparticle method. Phys. Rev. C **67**, 054302/1–4 (2003)
26. Maglione, E., Ferreira, L.S., Liotta, R.J.: Nucleon decay from deformed nuclei. Phys. Rev. Lett. **81**, 538–541 (1998)
27. Maglione, E., Ferreira, L.S., Liotta, R.J.: Proton emission from deformed nuclei. Phys. Rev. C **59**, R589–R592 (1999)
28. Cwiok, S., Dudek, J., Nazarewicz, W., Skalski, J., Werner, T.: Single-particle energies, wave functions, quadrupole moments and g-factors in an axially deformed Woods-Saxon potential with applications to the two-centre-type nuclear problem. Comput. Phys. Commun. **46**, 379–399 (1987)
29. Ferreira, L.S., Maglione, E.: ^{151}Lu: spherical or deformed? Phys. Rev. C **61**, 021304(R)/1–3 (2000)
30. Möller, P., Nix, R.J., Myers, W.D., Swiatecki, W.: Nuclear ground-state masses and deformations. At. Data Nucl. Data Tables **59**, 185–381 (1995)
31. Maglione, E., Ferreira, L.S.: Fine structure in proton emission from deformed ^{131}Eu. Phys. Rev. C **61**, 047307/1–3 (2000)
32. Sonzogni, A.A., Davids, C.N., Woods, P.J., et al.: Fine structure in the decay of the highly deformed proton emitter ^{131}Eu. Phys. Rev. Lett. **83**, 1116–1118 (1999)
33. Ferreira, L.S., Maglione, E., Fernandes, D.E.P.: Dependence of the decay widths for proton emission on the single particle potential. Phys. Rev. C **65**, 024323/1–9 (2002)
34. Cherpunov, V.A.: Yad. Fiz. **6**, 955 (1967)
35. Blomqvist, J., Wahlborn, S.: Shell model calculations in the Lead region with a diffuse nuclear potential. Ark. Fys. **16**, 545–566 (1960)
36. Rost, E.: Protron shell-model potentials for Lead and the stability of superheavy nuclei. Phys. Lett. **26** B, 184–187 (1968)
37. Dudek, J., Szymanski, Z., Werner, T., Faessler, A., Lima, C.: Description of high spin states in ^{146}Gd using the optimized Woods-Saxon potential. Phys. Rev. C **26**, 1712–1718 (1982)
38. Kruppa, A.T., Barmore, B., Nazarewicz, W., Vertse, T.: Fine structure in the decay of deformed proton emitters: nonadiabatic approach. Phys. Rev. Lett. **84**, 4549–4552 (2000)
39. Barmore, B., Kruppa, A.T., Nazarewicz, W., Vertse, T.: Theoretical description of deformed proton emitters: nonadiabatic coupled-channel method. Phys. Rev. C **62**, 054315/1–12 (2000)
40. Barmore, B., Kruppa, A.T., Nazarewicz, W., Vertse, T.: A new approach to deformed proton emitters: non-adiabatic coupled-channels. Nucl. Phys. A **682**, 256c–263c (2001)
41. Delion, D.S., Liotta, R.J., Wyss, R.: High-spin proton emitters inodd-odd nuclei and shape changes. Phys. Rev. C **68**, 054603(R)/1–5 (2003)
42. Ferreira, L.S., Maglione, E.: Odd-odd deformed proton emitters. Phys. Rev. Lett. **86**, 1721–1724 (2001)
43. Bohr, A., Mottelson, B.: Nuclear Structure. Benjamin, New York (1975)
44. Davids, C.N., Esbensen, H.: Decay rate of triaxially deformed proton emitters. Phys. Rev. C **69**, 043314/1–9 (2004)
45. Davids, C.N., Woods, P.J., Mahmud, H., et al.: Proton decay of the highly deformed odd-odd nucleus ^{130}Eu. Phys. Rev. C **69**, 011302(R)/1–3 (2004)
46. Delion, D.S., Wyss, R., Karlgren, D., Liotta, R.J.: Proton emission from triaxial nuclei. Phys. Rev. C **70**, 061301(R)/1–5 (2004)
47. Kruppa, A.T., Nazarewicz, W.: Gamow and R-Matrix approach to proton emitting nuclei. Phys. Rev. C **69**, 054311/1–11 (2004)
48. Kadmensky, S.G., Sonzogni, A.A.: Proton angular distributions from oriented proton-emitting nuclei. Phys. Rev. C **62**, 054601/1–5 (2000)
49. Rafiqullah, A.K.: Alpha decay of nonaxial nuclei. Phys. Rev. **127**, 905–913 (1962)

Chapter 5
Semiclassical Approach

5.1 Penetration Formula

5.1.1 Spherical Approach

The semiclassical, or Wentzell–Kramers–Brillouin (WKB), approach for spherical emitters can be derived by using the Gurwitz and Kälbermann procedure [1]. The decay width is given by the following expression

$$\Gamma_{WKB} = \lambda_0 F \frac{\hbar^2}{4\mu} P, \tag{5.1}$$

in terms of the penetration integral

$$P \equiv \exp\left[-2\int_{r_2}^{r_3} k(r)dr\right], \tag{5.2}$$

where λ_0 is the fragment preformation probability and F is the normalization factor given by the integration over the internal interval, i.e.

$$F^{-1} = \int_{r_1}^{r_2} \frac{dr}{k(r)} \cos^2\left[\int_{r_1}^{r} k(r')dr' - \frac{\pi}{4}\right]. \tag{5.3}$$

We denoted by $k(r)$ the wave number

$$k(r) = \sqrt{\frac{2\mu}{\hbar^2}|E - V(r)|}. \tag{5.4}$$

The radii $r_1 < r_2 < r_3$ are the classical turning points, given as solutions of the equation $k(r) = 0$ and E is the relative energy between emitted fragments (Q-value). For high lying strong oscillating states $\cos^2$ term can be replaced by $\frac{1}{2}$. Thus, one obtains the following relation

D. S. Delion, *Theory of Particle and Cluster Emission*, Lecture Notes in Physics, 819,
DOI: 10.1007/978-3-642-14406-6_5, © Springer-Verlag Berlin Heidelberg 2010

$$F^{-1} \approx \int_{r_1}^{r_2} \frac{dr}{2k(r)} = \frac{\hbar T}{4\mu}, \tag{5.5}$$

where T is the classical period of motion inside the barrier. In this way the preexponential factor in (5.1) becomes proportional to $\nu = \hbar/T$, which is called assault frequency. This expression is widely used in literature to estimate halflives within the semiclassical approach.

5.1.2 *Deformed Approach*

For deformed emitters one introduces an angle dependent penetration

$$\begin{aligned} P(\theta) &= \exp\left[-2\int_{r_2(\theta)}^{r_3(\theta)} k(r,\theta)dr\right], \\ k(r,\theta) &= \sqrt{\frac{2\mu}{\hbar^2}|E - V(r,\theta)|}, \end{aligned} \tag{5.6}$$

so that the total penetration factor is given by

$$P = \frac{1}{2}\int_0^{\pi} P(\theta)\sin\theta d\theta. \tag{5.7}$$

A very useful approximation of the penetration matrix through a deformed barrier in Eq. 4.18 is given by the so-called Fröman approach [2]. If the interaction has only quadrupole deformation, one gets for the α-decay process from a ground state of an even–even nucleus to some member of the rotational band, i.e. $P(J_i = 0) \to D(J_f = l) + \alpha(l)$, the following expression

$$\begin{aligned} P_{ll'}^{1/2}(R) &= \sqrt{\kappa R}\left[\mathcal{H}_{ll'}^{(+)}(R)\right]^{-1} \approx \frac{\sqrt{\kappa R}}{G_l(\chi,\rho)}\mathbf{K}_{ll'}(R) \\ &\approx \frac{\sqrt{\kappa R}}{G_0(\chi,\rho)}\exp\left[-\frac{l(l+1)}{\chi}\sqrt{\frac{\chi}{\kappa R} - 1}\right]\mathbf{K}_{ll'}(R), \end{aligned} \tag{5.8}$$

where the Fröman matrix $\mathbf{K}_{ll'}(R)$, depending upon the matching radius R, is nothing else than the WKB approximation of the propagator matrix $\mathcal{K}_{ll'}(R)$ given by Eq. 4.14, i.e.

$$\mathbf{K}_{ll'}(R) = \int_0^\pi \Theta_{l0}(\theta) e^{BP_2(\cos\theta)} \Theta_{l'0}(\theta) \sin\theta d\theta$$
$$B \equiv \frac{2}{5}\chi\beta_2\left(2 - \frac{\kappa R}{\chi}\right)\sqrt{\frac{5}{4\pi}\frac{\kappa R}{\chi}\left(1 - \frac{\kappa R}{\chi}\right)}. \quad (5.9)$$

Here $P_{\lambda=2}(\cos\theta)$ is the Legendre polynomial.

An useful approximation can be derived by considering that the most important component of the internal wave function is the monopole one $f_0^{(int)}(r)$. Thus, the decay width becomes factorized as follows

$$\Gamma = \Gamma_0(R)D(R)$$
$$\Gamma_0(R) = 2P_0(R)\frac{\hbar^2}{2\mu R}\left[f_0^{(int)}(R)\right]^2 \quad (5.10)$$
$$D(R) = \sum_l \left\{\exp\left[-\frac{l(l+1)}{\chi}\sqrt{\frac{\chi}{\kappa R} - 1}\right]\mathbf{K}_{l0}\right\}^2.$$

The deformation effects are included in the function *D*(*R*), which slightly depends upon the radius, as can be seen in Fig. 5.1a, b. We notice that the effect of the deformation becomes very important for large deformations, giving a correction of about 4–5 times for $\beta_2 = 0.3$ with respect to the spherical decay width Γ_0.

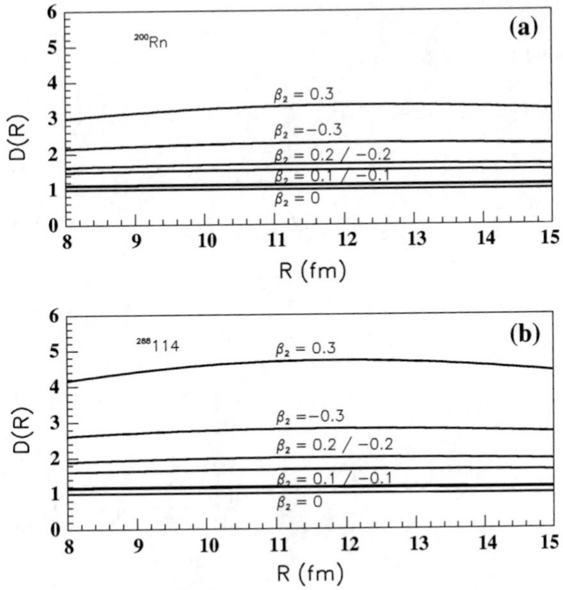

Fig. 5.1 **a** Dependence of the deformation factor *D*(*R*), defined by (5.10), upon the radius for different quadrupole deformations. The decay process is $^{200}Rn \rightarrow {}^{196}Po + \alpha$. **b** The same as in **a** but for the decay process $^{288}114 \rightarrow {}^{284}112 + \alpha$

5.2 Cluster Model (CM)

As we already mentioned, all phenomenological descriptions suppose that the fragment dynamics is described by an inter-fragment potential, defined for all distances and the wave function has the factorized cluster-like form (Eq. 2.59) in terms of core-angular harmonics (Eq. 2.60). From this point of view all phenomenological approaches can be considered as "cluster models".

The simplest α-cluster model was applied in Ref. [3] and concerned the α-core dynamics described by the potential in Fig. 5.2

$$\begin{aligned} V(r) &= -V_N + \frac{C}{R}, \quad r < R_0 \\ &= \frac{C}{r}, \quad r \geq R, \end{aligned} \tag{5.11}$$

with $C = 2(Z-2)e^2$.

According to (5.1), the semiclassical decay width, becomes

$$\Gamma = \lambda_0 \frac{\hbar K}{2\mu R} \exp\left[-2\int_R^{C/Q} k(r)dr\right], \tag{5.12}$$

where K and k are the wave numbers in the internal and barrier regions, respectively

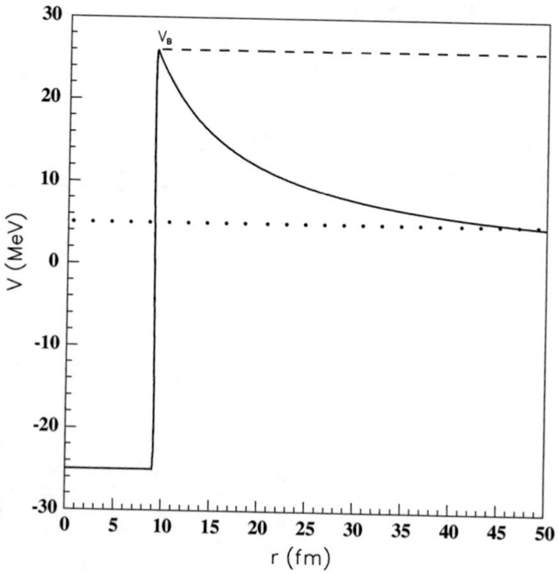

Fig. 5.2 The α-core potential (5.11) with $V_N = 25$ MeV (*solid line*) and its barrier value (*dashed line*). Q-value is denoted by a *dotted line*

$$K = \left[\frac{2\mu}{\hbar}\left(Q + V_N - \frac{C}{R}\right)\right]^{1/2}$$
$$k(r) = \left[\frac{2\mu}{\hbar}\left(\frac{C}{r} - Q\right)\right]^{1/2}. \tag{5.13}$$

It turns out that it is more profitable to fix V_N and to change the radius R for each individual decay in order to produce a resonant state at the Q-value, which satisfies the Bohr–Sommerfeld condition with $L = 0$, i.e.

$$\int_0^R dr\sqrt{\frac{2\mu}{\hbar^2}\left(Q + V_N - \frac{C}{R}\right)} = R\sqrt{\frac{2\mu}{\hbar^2}\left(Q + V_N - \frac{C}{R}\right)} = (G+1)\frac{\pi}{2}. \tag{5.14}$$

This relation determines the radius R. The penetration integral can be performed analytically, i.e.

$$2\int_R^{C/Q} k(r)dr = 2\sqrt{\frac{2\mu}{\hbar^2}}\frac{C}{\sqrt{Q}}\left[\frac{\pi}{2} - \arcsin x - x\sqrt{1-x^2}\right], \tag{5.15}$$

where $x = \sqrt{RQ/C}$. By using the following set of parameters

$$\lambda_0 = 1, \quad V_N = 135.6\ \text{MeV}, \quad G = 22\ (N \le 126), \quad G = 24\ (N > 126),$$

it was possible to obtain an agreement within a factor of two for most even–even α-emitters.

In this context, we mention that this problem was for the first time solved in Ref. [4, 5]. Later on, a simple method to solve Schrödinger equation with complex energies in a similar square well was developed in Ref. [6].

In Ref. [7], the cluster model was applied to a potential fitting the double folded interaction (Eq. 2.115), namely

$$V_N(r) = -V_0\frac{1 + \cosh(\mathrm{R/a})}{\cosh(r/a) + \cosh(R/a)} \tag{5.16}$$

in order to describe 409 α-decays from even–even and odd–mass α-emitters. The radius of the interaction R was fixed to reproduce the Q-value. The Bohr–Sommerfeld condition reads

$$\int_{r_1}^{r_2} k(r)dr = (G - l + 1)\frac{\pi}{2},$$
$$G \equiv 2n + l, \tag{5.17}$$

where r_1, r_2 were determined as the first two solutions of the equation $k(r) = 0$ (turning points). By using the following set of parameters

$$\lambda_0 = 1 \text{ (even-even)}, \quad \lambda_0 = 0.6 \text{ (odd−mass)}, \quad V_N = 162.3 \text{ MeV}$$
$$a = 0.40 \text{ fm}, \quad G = 20\ (N \leq 126), \quad G = 22\ (N > 126),$$

it was possible to obtain an agreement with experimental data within a factor of three.

In Ref. [8] the potential (5.16) was used to compute decay widths for the emission of ^{14}C, ^{24}Ne and ^{28}Mg. The used parameters were

$$R = 6.7\,\text{fm}, \quad a = 0.75\,\text{fm}$$
$$G = 68\ (^{14}\text{C}), \quad G = 106\ (^{14}\text{Ne}), \quad G = 120\ (^{14}\text{Mg}).$$

It was obtained an agreement with experimental data within one order of magnitude for the first two decays and two orders for the last one.

In Ref. [9], the same WKB formalism, previously used in α and cluster decays, was applied to analyze proton emission. The proton emission widths was estimated according to (5.1), by including the spectroscopic factor $S_p = \lambda_0 = u_{lj}^2$, i.e.

$$\Gamma_p = S_p \Gamma_{\text{WKB}}. \tag{5.18}$$

The spherical potential (Eq. 2.133) with Becchetti-Greenlees optical model parametrisation [10] was used. This interaction describes proton scattering data. Then, the radii of nuclear and Coulomb parts were changed in order to satisfy the Bohr–Sommerfeld condition (5.17). The quantum numbers of the quasibound proton state n, l, j (radial quantum number, angular momentum and spin) were estimated by using the simplest spherical harmonic oscillator shell model scheme and a pure single-particle state, i.e. the spectroscopic factor was taken $S_p = 1$. The following nuclei were analyzed:

$$^{109}\text{I}, {}^{113}\text{Cs}, {}^{147}\text{Tm}, {}^{150,151}\text{Lu}, {}^{156}\text{Ta}, {}^{160}\text{Re}.$$

A satisfactory agreement (within a factor of three) was obtained, except for the first nucleus. It was found that the decay widths are very sensitive to the number of nodes n, which was modified by changing the radial parameters. Thus, it was concluded that proton emission is an important tool to determine the quantum numbers of the mother nucleus. By using this potential for all analyzed nuclei, it was obtained an agreement between theory and experiment within a factor of two, according to the Table III of Ref. [9]. This discrepancy can be corrected by assuming a spectroscopic factor $S_p \approx 0.5$, which corresponds to the BCS estimate.

In Ref. [11] the deformed double folding α-core interaction, introduced in Sect. 2.8, was corrected by a density dependent term within the so-called density dependent cluster model (DDCM). The penetration through the deformed Coulomb barrier was estimated according to (5.6) and (5.7). An extensive analysis of 485 even–even and odd–mass α-emitters in the region $Z = 52-110$ was performed. Most of computed decay widths were reproduced within a factor of five.

Double folding interaction in the spherical approach was used in Ref. [12] in order to make a systematic investigation of the α-decay width in superheavy nuclei within a similar cluster model. Based on the discontinuity of the volume integral the Author predicted a double shell closure at $Z = 132$, $N = 194$.

In order to describe all decay processes, from proton to cluster emission, in Ref. [13] an universal potential was proposed. It has the following form

$$V(r) = V_N(r) + V_C(r), \tag{5.19}$$

where $V_C(r)$ is the Coulomb cluster-core potential and

$$V_N(r) = \lambda[Z_c v_p(r) + N_c v_n(r)]. \tag{5.20}$$

Here Z_c/N_c are proton/neutron numbers of the cluster and $v_\tau(r)$ are single particle potentials generated by the core, obtained within the mean field model. The folding factor λ was obtained by considering the Bohr–Sommerfeld condition (5.17) with $l = 0$. By using the Wildermuth rule for the Bohr–Sommerfeld parameter, i.e. $G = \sum_{i=1}^{A_c} g_i$, where g_i is the oscillator quantum number of the cluster nucleon orbiting in the core, it was possible to reproduce α-decay half lives within a factor of two. Diproton and various cluster decay half lives were reproduced within one order of magnitude.

5.3 Super Asymmetric Fission Model (SAFM)

In Ref. [14] the α-decay was described as a superasymmetric fission process. Then in Ref. [15] a Numerical Super Asymmetric Fission Model (NSAFM) was proposed. This method supposes that the fissioning system is described by two intersected spheres of radii R_1 (daughter), R_2 (α-particle). In cylindrical coordinates (ρ, z) the surface equation of the fissioning system is given by the following equation

$$\begin{aligned} \rho^2 &= R_1^2 - z^2, \quad z \leq z_c \\ &= R_2^2 - (z - R)^2, \quad z > z_c, \end{aligned} \tag{5.21}$$

where z_c gives the position of the separation plane between the fragments and R is the distance between the centers. By supposing that the size of the smaller cluster R_2 is fixed, the values of R_1 and z_c before separation are given by the condition

$$R_2^2 - (z_c - R)^2 = R_1^2 - z_c^2, \tag{5.22}$$

and the volume conservation

$$\left[2R_1^3 + z_c(3R_1^2 - z_c^2)\right] + \left[z_c^3 - 3Rz_c^2 - 3(R^2 - R_2^2)z_c + R_2^2(2R_2 + 3R) - R^3\right] = 4. \tag{5.23}$$

The energy of the system is given by

$$E(R) = E_{Y+E}(R) + E_{\text{corr}}(R), \tag{5.24}$$

where

$$E_{Y+E} = E_Y + E_C + E_V - E_\infty, \quad R<R_t. \tag{5.25}$$

Here E_Y is folded Yukawa plus exponential term, E_C Coulomb energy, E_V volume energy and E_∞ the energy of two spheres at infinity. The term E_{corr} takes into account phenomenologically the shell correction [15]. The decay width within the WKB approach was estimated according to a relation similar to (5.1), namely

$$\Gamma = \frac{n}{\ln 2}\int_{R_1}^{R_2} \exp\left[-\frac{2}{\hbar}\sqrt{2\mu(E(R)-E')}dR\right], \tag{5.26}$$

where $n = \omega/2\pi$ is the assault frequency and $E' = Q_\alpha + E_{\text{vib}}$ is the initial excitation energy, with $E_{\text{vib}} = \frac{1}{2}\hbar\omega$ being the zero-point vibration energy. It turns out that the potential energy $E(R)$, until the touching configuration R_t, can be very well approximated by a parabola, like in Fig. 2.7. Based on this observation soon an analytical version (ASAFM) was introduced.

According to Ref. [16], the half life of a parent nucleus (A, Z) against the splitting into a cluster (A_e, Z_e) and a daughter (A_d, Z_d) is calculated by using the semiclassical approximation (5.1), i.e.

$$T = \frac{h\ln 2}{2E_v}\exp(K_{ov} + K_s). \tag{5.27}$$

The action integral is expressed as follows

$$K = K_{ov} + K_s = \frac{2}{\hbar}\int_{R_1}^{R_t} \sqrt{2B(R)E(R)}dR + \frac{2}{\hbar}\int_{R_t}^{R_2} \sqrt{2B(R)E(R)}dR, \tag{5.28}$$

with $B = \mu$ and $E(R)$ replaced by $[E(R) - E_{\text{corr}}] - Q$. E_{corr} is a correction energy similar to the Strutinsky shell correction, taking also into account the fact that Myers-Swiatecki's liquid drop model (LDM) overestimates fission barrier heights, and the effective inertia in the overlapping region is different from the reduced mass. The turning points of the WKB integral are:

$$R_1 = R_i + (R_t - R_i)\left[(E_v + E^*)/E_b^0\right]^{1/2} \tag{5.29}$$

$$R_2 = R_t E_c\left\{1/2 + \left[1/4 + (Q + E_v + E^*)E_l/E_c^2\right]^{1/2}\right\}/(Q + E_v + E^*) \tag{5.30}$$

where E^* is the excitation energy concentrated in the separation degree of freedom, $R_i = R_0 - R_e$ is the initial separation distance, $R_t = R_e + R_d$ is the touching point separation distance, $R_j = r_0 A_j^{1/3}$ ($j = 0, e, d$; $r_0 = 1.2249$ fm) are the radii of parent, emitted and daughter nuclei, and $E_b^0 = E_i - Q$ is the barrier height before correction. The interaction energy at the top of the barrier, in the presence of a non-negligible angular momentum, $l\hbar$, is given by:

$$E_i = E_c + E_l = e^2 Z_e Z_d / R_t + \hbar^2 l(l+1)/(2\mu R_t^2) \tag{5.31}$$

The two terms of the action integral K, corresponding to the overlapping (K_{ov}) and separated (K_s) fragments, are calculated by analytical formulas (approximated for K_{ov} and exact for K_s in case of separated spherical shapes within LDM):

$$K_{ov} = 0.2196(E_b^0 A_e A_d/A)^{1/2}(R_t - R_i)\left[\sqrt{1-b^2} - b^2 \ln\frac{1+\sqrt{1-b^2}}{b}\right] \tag{5.32}$$

$$K_s = 0.4392[(Q + E_v + E^*)A_e A_d/A]^{1/2} R_b J_{rc}; b^2 = (E_v + E^*)/E_b^0 \tag{5.33}$$

$$\begin{aligned} J_{rc} &= (c)\arccos\sqrt{(1-c+r)/(2-c)} - [(1-r)(1-c+r)]^{1/2} \\ &\quad + \sqrt{1-c}\ln\left[\frac{2\sqrt{(1-c)(1-r)(1-c+r)} + 2 - 2c + cr}{r(2-c)}\right] \end{aligned} \tag{5.34}$$

where $r = R_t/R_b$ and $c = rE_c/(Q + E_v + E^*)$. In the absence of the centrifugal contribution ($l = 0$), one has $c = 1$. The quantity m used in some previous publications is expressed simply as $m = 1 - c$.

One considers $E_v = E_{corr}$ in order to get smaller number of parameters. It is obvious that, due to the exponential dependence, any small variation of E_{corr} induces a large change of T, and thus it plays a more important role compared to the preexponential factor variation due to E_v. Shell and pairing effects are included in $E_{corr} = a_i(A_e)Q$ ($i = 1, 2, 3, 4$ for even–even, odd–even, even–odd, and odd–odd parent nuclei). For a given cluster radioactivity, one has four values of the coefficients a_i, the largest for even–even parent and the smallest for the odd-odd one. The shell effects for every cluster radioactivity is implicitly contained in the correction energy due to its proportionality with the Q-value, which is maximum when the daughter nucleus has a magic number of neutrons and protons.

In Ref. [17] ASAFM was applied to analyze all possible decay modes, including α-decay and heavy cluster emission. It was possible to have a reasonable agreement with existing experimental data and to predict all energetically allowed cluster decays. This simple model was very successful and some of the predicted decays were confirmed experimentally.

In Refs. [18, 19] the SAFM was generalized, namely the internal part of the barrier was estimated by a double folding potential in order to analyze α and heavy cluster decay widths.

5.4 Effective Liquid Drop Model (ELDM)

The Effective Liquid Drop Model (ELDM) was developed in Ref. [20], in order to estimate the potential between emitted fragments. It was applied to proton radioactivity, but also to analyze cold fission, α and cluster decay [21]. The geometry of the emission process is given by two intersected spheres, similar to SAFM of the previous section. The model uses the WKB penetration integral (5.1), where the factor λ_0 was understood as the assault frequency parameter. Two main additional ingredients are important:

1. The constant mass μ in (5.4) is replaced by a radial function $m(r)$. Two alternative ways are used to define the mass parameter, namely

   a. the Werner-Wheeler's inertia coefficient, defined by

$$\frac{1}{2}m_{\rm WW}\left(\frac{dr}{dt}\right)^2=\frac{1}{2}\int\rho\mathbf{v}^2dr, \tag{5.35}$$

where ρ is the mass density of the system, and

   b. an effective mass parameter defined as

$$m_{\rm eff}=\mu\alpha^2. \tag{5.36}$$

The density ρ and the parameter α are evaluated by considering a variable mass asymmetry shape (VMAS) with a constant radius of the lighter fragment "C"

$$R_C=\bar{R}_C\equiv\frac{Z_C}{Z_P}R_P, \tag{5.37}$$

or a constant mass asymmetry shape (CMAS), where the masses of the fragments are conserved.

2. The potential barrier has a surface term $V^{(S)}(r)$, in addition to the Coulomb energy $V^{(C)}(r)$ and centrifugal barrier $V_l(r)$. The surface energy is defined as

$$V^{(S)}=\sigma_{\rm eff}(S_D+S_C), \tag{5.38}$$

where $S_{D/C}$ are the surface of the overlapping spherical segments, while the effective surface tension $\sigma_{\rm eff}$ is given by the Q-value of the process, i.e.

$$\begin{aligned}Q&=\frac{3}{5}e^2\left(\frac{Z_P^2}{R_P}-\frac{Z_D^2}{\bar{R}_D}-\frac{Z_C^2}{\bar{R}_C}\right)+4\pi\sigma_{\rm eff}(R_P^2-\bar{R}_D^2-\bar{R}_C^2),\\ \bar{R}_D&=\frac{Z_D}{Z_P}R_P,\quad \bar{R}_C=\frac{Z_C}{Z_P}R_P.\end{aligned} \tag{5.39}$$

The mass parameter $\alpha=m(r)/\mu\neq 1$ and the surface energy $V^{(S)}\neq 0$ were evaluated in the so-called molecular phase, where the fragments overlap. This approach depends only upon the angular momentum l and not on the total spin j

because the spin-orbit term, and therefore the shell effects, are taken into account in an effective way, through the Q-value and the surface term in (5.39).

A similar approach to estimate barriers if α-decay, together with corrections induced by the proximity potential is given in Ref. [22].

5.5 Fragmentation Theory (FT)

Fragmentation theory (FT) [23, 24] is able to compute the probability of a binary splitting from a given parent nucleus in terms of mass and charge asymmetry coordinates, respectively defined as

$$\eta = \frac{A_1 - A_2}{A}, \quad \eta_Z = \frac{Z_1 - Z_2}{Z}. \tag{5.40}$$

Potential energy surface (PES), as a function of these variables and relative distance between centers $V(\eta, \eta_Z, R)$, can be estimated by using various methods. One of the most popular approaches is to consider the liquid drop model plus shell model corrections, given by the Two Center Schell Model (TCSM) [25–28]. This potential can be inserted in a Schrödinger-like equation, by using a kinetic term with respect to the mass coordinate η. The solutions of this equation describe the probability to create different combinations of fragments.

One of the most successful applications of the FT is the Preformed Cluster Model (PCM) [29, 30]. Here one computes the preformation amplitude to create the mass splitting (A_1, A_2) with a given charge asymmetry η_Z and at a fixed radius R, by solving the following Schrödinger equation

$$\left[-\frac{\hbar^2}{2\sqrt{B_{\eta\eta}}}\frac{\partial}{\partial\eta}\frac{1}{\sqrt{B_{\eta\eta}}}\frac{\partial}{\partial\eta} + V(\eta,\eta_Z,R)\right]\psi^{(n)}_{R\eta_Z}(\eta) = E^{(n)}_{R\eta_Z}\psi^{(n)}_{R\eta_Z}(\eta). \tag{5.41}$$

The probability to find a pair in a mass asymmetry space is given by

$$P_0(A_1) = |\psi^{(0)}_{R\eta_Z}(A_1)|^2\sqrt{B_{\eta\eta}}\frac{2}{A}. \tag{5.42}$$

By solving (5.41) one makes a simplification concerning the radius R, by considering the touching configuration of emitted fragments. The potential $V(\eta, \eta_Z, R)$ has a typical parabolic-like minimum with respect to the charge asymmetry coordinate η_Z and the calculations are performed in this "charge equilibrium" point. The remaining potential $V(\eta, R)$ has several minima corresponding to different splittings. For instance, in Ref. [29] the potential for ^{222}Ra at the touching configuration has minima corresponding to

$$^4\text{He}, ^{10}\text{Be}, ^{14}\text{C}, ^{48,50}\text{Ca}, ^{70}\text{Ni}, ^{88}\text{Kr}.$$

The quantum numbers $n = 0, 1, 2, \ldots$ correspond to vibrational states in this potential. By considering the ground state with $n = 0$ it is assumed that the process has a non-dissipative character. The mass parameter $B_{\eta\eta}$ was estimated within the classical model of Kröger and Scheid [31]. The radial potential is given by the nuclear V_N plus Coulomb V_C components, by considering the normalization at infinity, i.e. by subtracting the binding energies of fragments

$$V(\eta, r) = V_N(R) + V_C(R) + \frac{Z_1 Z_2 e^2}{R} - B(A_1, Z_1) - B(A_2, Z_2). \tag{5.43}$$

As a nuclear interaction V_N the proximity potential [32] was used. The total decay width is given by the following product

$$\Gamma = P_0(A_1)\hbar \nu P, \tag{5.44}$$

where the assault frequency

$$\nu = \frac{v}{R_0} = \sqrt{\frac{2E_2}{\mu R_0^2}}, \tag{5.45}$$

is defined in terms of the parent radius R_0 and the kinetic energy of the emitted cluster E_2. The penetrability P is considered as the WKB integral

$$P = \int_{R_1}^{R_2} \left[-\frac{2}{\hbar}\sqrt{2\mu(V(R) - Q)} \right]. \tag{5.46}$$

FT/PCM was successfully applied to describe α-decay, cluster radioactivity and cold fission [33, 34].

References

1. Gurvitz, S.A., Kalbermann, G.: Decay width and shift of a quasistationary state. Phys. Rev. Lett. **59**, 262–265 (1987)
2. Fröman, P.P.: Alpha decay from deformed nuclei. Mat. Fys. Skr. Dan. Vid. Selsk. **1** (3) (1957)
3. Buck, B., Merchand, A.C., Perez S.M.: Ground state to ground state alpha decays of heavy even-even nuclei. J. Phys. G **17**, 1223–1235 (1991)
4. Preston, M.A.: The theory of alpha-radioactivity. Phys. Rev. **71**, 865–877 (1947)
5. Preston, M.A.: The electrostatic interaction and low energy particles in alpha-radioactivity. Phys. Rev. **75**, 90–99 (1949)
6. Pierronne M., Marquez, L.: On the complex energy Eigenvalue theory of alpha decay. Z. Phys. A **286**, 19–25 (1978)
7. Buck, B., Merchand, A.C., Perez, S.M.: α Decay calculations with a realistic potential. Phys. Rev. C **45**, 2247–2253 (1992)
8. Buck, B., Merchand, A.C.: A consistent cluster model treatment of exotic decays and alpha decays from heavy nuclei. J. Phys. G **15**, 615–635 (1989)

9. Buck, B., Merchand, A.C., Perez, S.M.: Ground state proton emission from heavy nuclei. Phys. Rev. C **45**, 1688–1692 (1992)
10. Becchetti, F.D., Jr., Greenlees, G.W.: Nucleon-nucleon optical-model parameters, A > 40, E < 50 MeV. Phys. Rev. **182**, 1190–1209 (1969)
11. Xu, C., Ren, Z.: Global calculation of α-decay half-lives with a deformed density-dependent cluster model. Phys. Rev. C **74**, 014304/1–10 (2006)
12. Mohr, P.: α-Nucleus potentials, α-decayhalf-lives, and shell closures for superheavy nuclei. Phys. Rev. C **73**, 031301(R)/1–5 (2006)
13. Xu, F.R., Pei, J.C.: Mean-field cluster potentials for various clusterdecays. Phys. Lett. B **642**, 322–325 (2006)
14. Poenaru, D.N., Ivaşcu, M., Săndulescu, A.: Alpha decay as a fision like process. J. Phys. G **5**, L169–L173 (1979)
15. Poenaru, D.N., Mazilu, D., Ivaşcu, M.: Deformation energies for nuclei with different charge-to-mass ratio. J. Phys. G **5**, 1093–1106 (1979)
16. Poenaru, D.N., Ivaşcu, M., Săndulescu, A., Greiner, W.: Atomic nuclei decay modes by spontaneous emission of heavy ions. Phys. Rev. C **32**, 572–581 (1985)
17. Poenaru, D.N., Greiner, W., Depta, K., Ivaşcu, M., Mazilu, D., Săndulescu, A.: Calculated half-lives and kinetic energies for spontaneousfission of heavy ions from nuclei. At. Data Nucl. Data Tables **34**, 423–538 (1986)
18. Basu, D.N.: Folding model analysis of alpha radioactivity. J. Phys. G **29**, 2079–2085 (2003)
19. Basu, D.N.: Spontaneous heavy cluster emission rates using microscopic potentials. Phys. Rev. C **66**, 027601/1–4 (2002)
20. Guzmán, F., Gonçalves, M., Tavares, O.A.P., Duarte, S.B., García, F., Rodríguez, O.: Proton radioactivity from proton-rich nuclei. Phys. Rev. C **59**, R2339–R2342 (1999)
21. Duarte, S.B., Tavares, O.A.P., Guzmán, G., Dimarco, A., García, F., Rodríguez, O., Gonçanves, M.: Half-lives for proton emission, alpha decay, cluster radioactivity, and cold fission processes calculated in a unified theoretical framework. At. Data. Nucl. Data Tables **80**, 235–299 (2002)
22. Royer, G., Zbiri, K., Bonilla, C.: Entrance channels and alpha decay half-lives of the heaviest elements. Nucl. Phys. A **730**, 355–367 (2007)
23. Maruhn, J.A., Greiner, W.: Theory of fission-mass distribution demonstrated for ^{226}Ra, ^{236}U, ^{258}Fm. Phys. Rev. Lett. **32**, 548–551 (1974)
24. Gupta, R.K., Scheid, W., Greiner, W.: Theory of charge dispersion in nuclear fission. Phys. Rev. Lett. **35**, 353–356 (1975)
25. Maruhn, J.A., Greiner, W.: The asymmetric two-center shell model. Z. Phys. **251**, 431–457 (1972)
26. Greiner, W., Park, J.Y., Scheid, W.: Nuclear Molecules. World Scientific, Singapore (1994)
27. Mirea, M.: Superasymmetric two-center shell model for spontaneous heavy-ion emission. Phys. Rev. C **54**, 302–314 (1996)
28. Gherghescu, R.A.: Deformed two-center shell model. Phys. Rev. C **67**, 014309/1–20 (2003)
29. Malik, S.S., Gupta, R.K.: Theory of cluster radioactivite decay and of cluster formation in nuclei. Phys. Rev. C **39**, 1992–2000 (1989)
30. Gupta, R.K., Singh, S., Puri, R.K., Săndulescu, A., Greiner, W., Scheid, W.: Influence of the nuclear surface diffuseness on exotic cluster decay half-live times. J. Phys. G **18**, 1533–1542 (1992)
31. Kröger, H., Scheid, W.: Classical models for the mass transfer in heavy-ion collisions. J. Phys. G **6**, L85–L88 (1980)
32. Blocki, J., Randrup, J., Swiatecki, W.J., Tsang, C.F.: Proximity forces. Ann. Phys. (NY) **105**, 427–462 (1977)
33. Gupta, R.K., Bir, D., Balasubramaniam, M., Scheid, W.: Cold fission versus exotic cluster decay in ^{234,236,238}U nuclei. J. Phys. G **26**, 1373–1388 (2000)
34. Gupta, R.K., Dhaulta, S., Kumar, R., Balasubramaniam, M.: Closed-shell effects from the stability and instability of nuclei against cluster decays in the mass regions 130–158 and 180–198. Phys. Rev. C **68**, 034321/1–10 (2003)

Chapter 6
Fine Structure of Emission Processes

6.1 α-Decay Fine Structure

By definition the fine structure refers to all partial decay widths of excited states in emitted fragments. We speak about a *single fine structure* if one fragment is in a ground state (e.g. in α-decay) and *double fine structure* if both fragments are excited (e.g. in fission processes). The most sensitive test of nuclear structure details in emission processes is the investigation of the fine-structure. At present there are a lot of high precision data on α-decay to excited states in even–even nuclei, see e.g. [1]. We define α-decay fine structure by the logarithm of the ratio between decay widths (or intensities) to ground and J^+ states, i.e.

$$I_J \equiv \log_{10} \frac{\Gamma_0}{\Gamma_J}. \tag{6.1}$$

In order to remove the effect of the Coulomb barrier and to compare the corresponding formation probabilities one defines the hindrance factor (HF) as follows

$$\mathrm{HF}_J \equiv \frac{\Gamma_0}{\Gamma_J}\frac{P_J}{P_0}, \tag{6.2}$$

where P_J is the standard penetrability defined by (2.88) in Chap. 2, i.e.

$$P_J \equiv \frac{\kappa_J r}{G_J^2(\kappa_J r)}, \tag{6.3}$$

in terms of the irregular Coulomb function $G_J(\kappa_J r)$ at a certain radius r inside the barrier. We point out that the above relation (6.2) is actually the diagonal approximation of Eq. 4.22 in Chap. 4 and gives a very good estimate of the relative wave function on the nuclear surface. It is important to stress the fact that HF practically does not depend upon the radius r, because in (6.2) the ratio of penetrabilities has a weak dependence upon this variable. In our calculations we considered the geometrical touching radius, defined by $r = R_0 = 1.2(A_D^{1/3} + A_\alpha^{1/3})$.

D. S. Delion, *Theory of Particle and Cluster Emission*, Lecture Notes in Physics, 819,
DOI: 10.1007/978-3-642-14406-6_6,

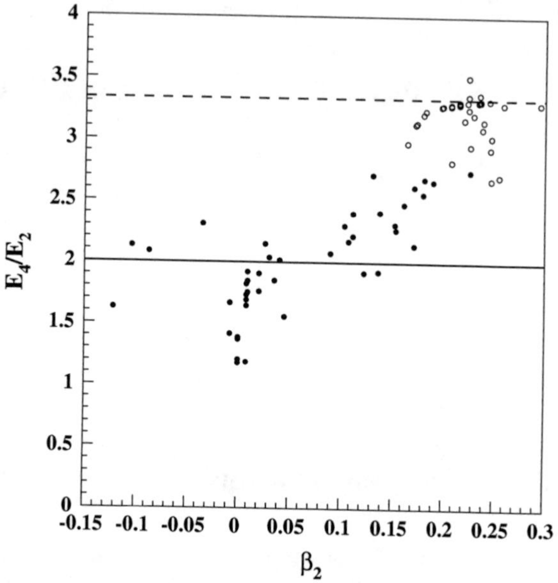

Fig. 6.1 **a** The ratio E_4/E_2 versus the quadrupole deformation for nuclei α-emitters. The *solid line* denotes vibrational limit, while the *dashed one* rotational limit. *Solid circles* correspond to vibrational and transitional nuclei while *open circles* to rotational nuclei

As it was shown in Ref. [2] the relation (6.2) gives close results to the standard Rasmussen definition [3].

The behaviour of the energy spectrum for all α-emitters with known energy values of $J = 2$ and $J = 4$ levels is given in Fig. 6.1.

Here we plotted the ratio E_4/E_2 versus the quadrupole deformation β_2. One clearly sees that the vibrational region is given by $|\beta_2| \leq 0.1$ (solid circles), while the rotational region by $|\beta_2| > 0.2$ (open circles). The solid line denotes the vibrational limit, while the dashed one the rotational limit. Thus, the transitional region is defined by the region of deformations $0.1 < |\beta_2| < 0.2$.

A clear difference between vibrational–transitional and rotational regions is given in Fig. 6.2a, where we plotted the logarithm of the intensity ratio $I \equiv I_2$, (6.1) versus the energy E_2. All values of the rotational region with $E_2 \leq 0.1$ MeV are concentrated around $I \approx 0.5$, while for the other nuclei one has an increasing trend, given by the following linear ansatz

$$I_{\text{exp}} = 5.121\, E_2 + 0.357, \quad \sigma = 0.523, \tag{6.4}$$

shown by the solid line. Here σ denotes the mean error. In Fig. 6.2b we plotted the same quantity, but versus the neutron number. One clearly sees the smooths behaviour of I in the rotational region with $N \geq 132$. The dependence of experimental values $\log_{10}\text{HF}_{\text{exp}}$ (6.2) versus E_2 is given in Fig. 6.3a and versus neutron number in Fig. 6.3b. Once again one clearly sees the difference between the vibrational–transitional region and the rotational one, characterized by small HF's.

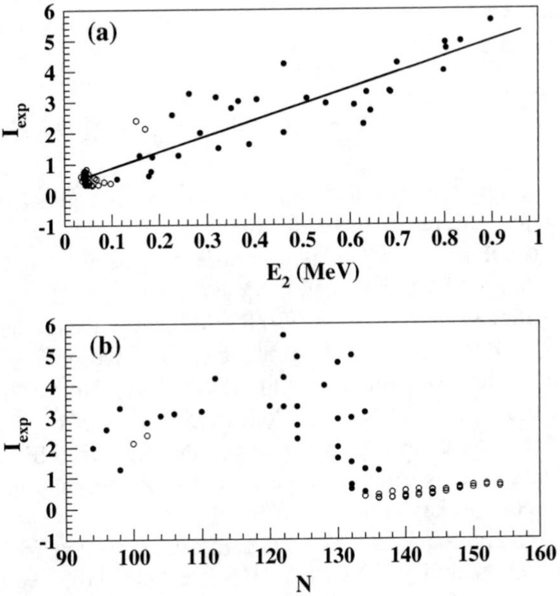

Fig. 6.2 **a** The experimental values of the logarithm of the intensity ratio (6.1) versus the energy E_2. *Solid circles* correspond to vibrational and transitional nuclei while *open circles* to rotational nuclei. The *straight line* fits the values of vibrational–transitional emitters in Table 6.1. **b** The experimental values of the logarithm of the intensity ratio (6.1) versus the neutron number

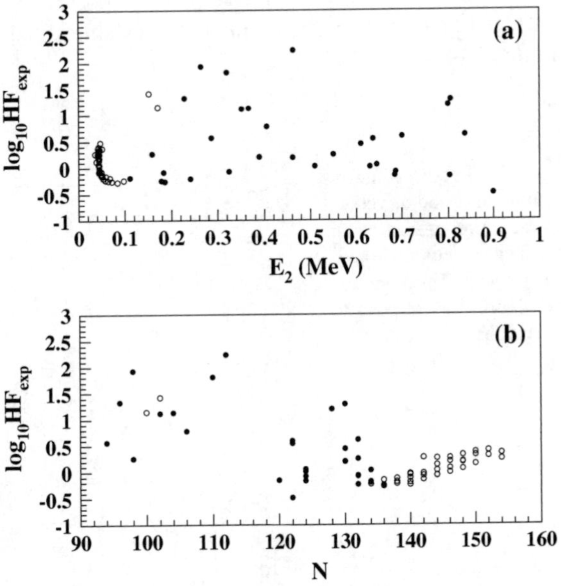

Fig. 6.3 **a** The experimental values of the logarithm of the hindrance factor (6.2) versus the energy E_2. *Solid circles* correspond to vibrational and transitional nuclei while *open circles* to rotational nuclei. **b** The experimental values of the logarithm of the hindrance factor (6.2) versus the neutron number

HF can also be defined as the ratio of reduced widths squared connecting the ground states and ground to excited states with the angular momentum J, i.e.

$$\mathrm{HF}_J = \frac{\gamma_0^2}{\gamma_J^2}. \tag{6.5}$$

Thus, by using Eq. 2.108 in Chap. 2, the logarithm of the HF for transitions to the lowest $J^{\pi} = 2^{+}$ state becomes proportional with the excitation energy of the daughter nucleus

$$\log_{10} \mathrm{HF} = \frac{\log_{10} e^2}{\hbar\omega} E_2 + \log_{10} \frac{A_0^2}{A_2^2}. \tag{6.6}$$

It is worth mentioning that this relation is equivalent with the Boltzman distribution for the reduced width to the excited state γ_2. In Refs. [4, 5] such a dependence was postulated in order to describe HF's.

In Fig. 6.4 it is plotted the logarithm of the HF versus the excitation energy for rotational nuclei with $E_2 < 0.1$ MeV, by using the same notations given by Eq. 2.109 in Chap. 2. As a rule the HF's have small values and therefore the wave functions have similar amplitudes $A_0 \approx A_2$. Notice that the region 4 in Fig. 6.4a, with $Z > 82$, $126 < N < 126$, contains most of rotational emitters (33 out of 41). Moreover, our analysis has shown that here the HF has an almost constant value along various isotopic chains, due to the fact that the energy range is very short (about 100 keV).

In Fig. 6.5 it is given the same quantity, but for transitional and vibrational nuclei, with $E_2 > 0.1$ MeV. Here the situation looks to be different and more complex, with respect to rotational nuclei. The best example is given by the same region 4 in Fig. 6.5c, where the slope has a positive value, as predicted by Eq. 6.6 in Chap. 6. Notice that this region contains almost half of the analyzed vibrational emitters (14 out of 31).

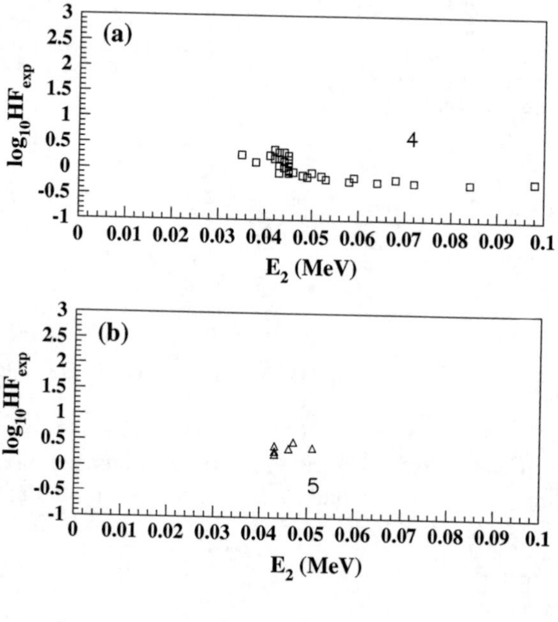

Fig. 6.4 The logarithm of the hindrance factor versus the excitation energy of the daughter nucleus for rotational nuclei. The *symbols and numbers* correspond to the regions given by Eq. 2.109 in Chap. 2

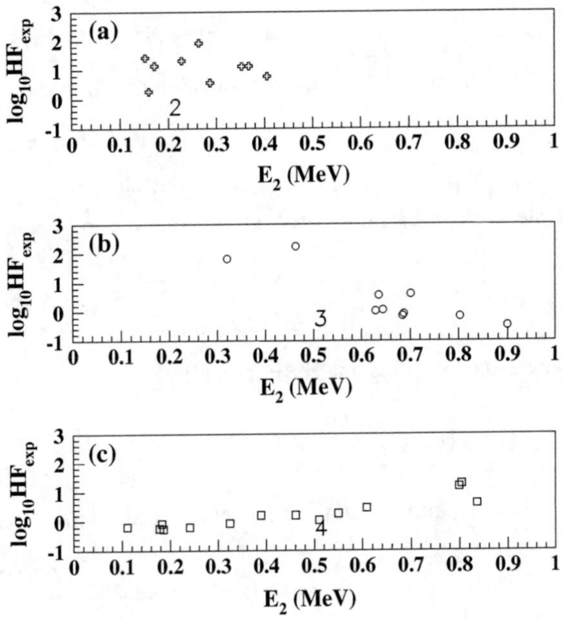

Fig. 6.5 The same as in Fig. 6.4, but for transitional and vibrational nuclei

6.2 Coupled Channels Description of the α-Decay Fine Structure

The general framework to describe fine structure is the coupled channels approach in the laboratory system. Let us consider an α-decay process

$$P \to D(J) + \alpha, \tag{6.7}$$

where J denotes the spin of the state of the even–even daughter nucleus, i.e $J = 0, 2, 4, 6, \ldots$. The Hamiltonian describing the α-decay in the laboratory system of coordinates is written as follows

$$H = -\frac{\hbar^2}{2\mu}\nabla_r^2 + H_D(\mathbf{x}) + V_0(\mathbf{r}) + V_d(\mathbf{x}, \mathbf{r}), \tag{6.8}$$

where μ is the reduced mass of the dinuclear system and $H_D(\mathbf{x})$ describes the intrinsic dynamics of the daughter nucleus, depending upon the internal coordinate $\mathbf{x}$. As usually, we separated the spherical $V_0(\mathbf{r})$ from the deformed part $V_d(\mathbf{x}, \mathbf{r})$ of the inter-fragment interaction. The wave function is given by the following superposition

$$\Psi(\mathbf{x}, \mathbf{r}) = \frac{1}{r}\sum_J f_J(r)\mathcal{Y}_J(\mathbf{x}, \hat{r}), \tag{6.9}$$

where we introduced the core-angular harmonics by

$$\mathscr{Y}_J(\mathbf{x},\hat{r}) = [\Phi_J(\mathbf{x}) \otimes Y_J(\hat{r})]_0. \qquad (6.10)$$

The internal wave function satisfies the following eigenvalue equation

$$H_D \Phi_{JM}(\mathbf{x}) = E_J \Phi_{JM}(\mathbf{x}). \qquad (6.11)$$

By using the orthonormality of core-angular harmonics one obtains the coupled system of differential equations for radial components in a standard way

$$\frac{d^2 f_J(r)}{d\rho_J^2} = \sum_{J'} \mathscr{A}_{JJ'}(r) f_{J'}(r), \qquad (6.12)$$

where the coupling matrix is given by

$$\mathscr{A}_{JJ'}(r) = \left[\frac{J(J+1)}{\rho_J^2} + \frac{V_0(r)}{E-E_J} - 1\right]\delta_{JJ'} + \frac{1}{E-E_J}\langle \mathscr{Y}_J | V_d | \mathscr{Y}_{J'} \rangle. \qquad (6.13)$$

Here we introduced the channel reduced radius and momentum, respectively by

$$\rho_J = \kappa_J r, \quad \kappa_J = \sqrt{\frac{2\mu(E-E_J)}{\hbar^2}}. \qquad (6.14)$$

6.3 α-Decay Fine Structure in Rotational Nuclei

The first computations of the α-decay widths in rotational nuclei by using the coupled channels method were performed in Ref. [6]. Later on in Ref. [7] an approximate analytical method was developed in order to estimate the α-decay intensities to rotational levels. By using the collective model in the strong coupling scheme in Ref. [8] relative α-decay intensities of favoured transitions in odd-mass nuclei (i.e. without the change of the odd nucleon state) were estimated. A coupled channel estimate of the α-decay rates from ^{212}Po to various excited states in ^{208}Pb was performed in Ref. [9]. Later on the generalization to odd-mass nuclei was given in Ref. [10].

In Ref. [11] HF's were estimated in rotational nuclei by using the Fröman approach [12] for the barrier penetration and a simple phenomenological ansatz for the preformation factor. The fine structure and shape coexistence in the neutron deficient Hg–Po region was investigated in Ref. [13], by using an α-core model. The α-core potential was estimated by using double folding procedure in Refs. [14, 15] and more recently in [16]. In all these works it was concluded that the strength of the nucleon–nucleon force should be quenched (i.e. the Coulomb barrier should increase) in order to describe the right relation between the half life and Q-value, as we already pointed out in Sect. 2.9 in Chap. 2. This kind of potential was used to estimate ground state to ground state half lives within the spherical approach in

Ref. [17]. A rather complete systematic analysis was performed in Ref. [18], by using the deformed folding potential, in order to describe transitions to ground and rotational 2^+ states. Finally, let us mention that in Ref. [19] the densities in the double folded α-core potential were computed within the relativistic mean field theory.

This section is based on the results of Ref. [2], where a systematic analysis of rotational α-emitters was performed. We suppose that the investigated nuclei have axial symmetry. The internal coordinate is the Euler angle $\mathbf{x} \equiv \omega$, defining the symmetry axis of the daughter nucleus. The wave functions of the ground rotational band are spherical harmonics $\Phi_{JM}(\omega) = Y_{JM}(\omega)$ and the core-angular function (6.10) is given by

$$\mathcal{Y}_\lambda(\omega, \hat{r}) = [Y_\lambda(\omega) \otimes Y_\lambda(\hat{r})]_0. \tag{6.15}$$

Therefore the motion of the core is compensated by the rotation of the α-particle.

The deformed part of the interaction is expanded as follows

$$V_d(\omega, \mathbf{r}) = \sum_{\lambda > 0} V_\lambda(r) \sqrt{\frac{4\pi}{2\lambda + 1}} \mathcal{Y}_\lambda(\omega, \hat{r}) \tag{6.16}$$

where the multipole formfactors are given by the double folding procedure. As usually in our computations we use the M3Y nucleon–nucleon (2.116) in Chap. 2 plus Coulomb force [20]. Notice that in Eq. (6.16) we used the expansion in terms of usual harmonics instead of Wigner functions. For small distances the fragments are strongly overlapped and their identity is lost into the parent nucleus. Thus, the radial wave function should vanish in this region. Therefore to the double folded potential we add a repulsive core, depending upon one independent parameter. It simulates the Pauli principle and adjusts the energy of the system to the experimental Q-value. In Ref. [2] we showed that the total half life and the partial decay widths does not depend upon the shape of this repulsive potential. Moreover, in Sect. 12.2 in Chap. 12 we will show in Fig. 12.3 that this kind of pocket-like potential is predicted within the Two Center Shell Model.

The matrix element $\langle \mathcal{Y}_J | V_d | \mathcal{Y}_{J'} \rangle$, entering Eq. 6.13, is given in terms of the Clebsch–Gordan coefficient as follows

$$\langle \mathcal{Y}_J | V_d(r) | \mathcal{Y}_{J'} \rangle = \sum_{\lambda > 0} V_\lambda(r) \sqrt{\frac{2J + 1}{4\pi(2J' + 1)}} [\langle J, 0; \lambda, 0 | J', 0 \rangle]^2. \tag{6.17}$$

We describe the density of the daughter nucleus by an axially deformed Woods–Saxon shape (2.118) in Chap. 2, by using quadrupole and hexadecapole deformation parameters of the nuclear surface (2.128) in Chap. 2

$$\rho_{D,\tau}(\mathbf{r}_D) = \frac{\rho^{(0)}_{D,\tau}}{1 + e^{[r_D - R_{D,\tau}]/a}}, \tag{6.18}$$

where the central densities are normalized by the total number of protons ($\tau = \pi$) and neutrons ($\tau = \nu$). We consider a standard value of the diffusivity

$a = 0.5$ fm and the nuclear radius $R_{D,\tau} = r_\tau^{(0)} A_D^{1/3}$. The deformations β_2, β_4 are given by the systematics of Ref. [20]. The α-particle density is given by a Gaussian distribution, i.e.

$$\rho_\alpha(\mathbf{r}_\alpha) = \frac{4}{b^3 \pi^{3/2}} e^{-(r_\alpha/b)^2}, \tag{6.19}$$

where $b = 1.19$ fm is the α-particle size parameter [16].

In Ref. [2] we multiplied the double folding integral (6.16) by a strength parameter v_a, in order to achieve the well known relation between the half life and Q-value. This relation is mainly given by the ratio between the Q-value and the height of the repulsive barrier. This factor multiplying the double folded α-daughter potential simulates the spectroscopic factor multiplying the wave function, given by the initial not-corrected potential [14, 15]. It is connected with the fact that the parameters of the used interaction were fitted from scattering experiments involving already existing particles [20, 22]. In decay processes this factor simulates the probability that α-particle is born on the nuclear surface. Anyway, the relative decay widths weakly depend upon this factor [2] Thus, in this analysis we will consider the multipole components of the potential

$$\overline{V}_\lambda(r) = v_a V_\lambda(r). \tag{6.20}$$

They are plotted in Fig. 6.6. By dots we indicated the monopole component $\lambda = 0$, by dashes $\lambda = 2$ and by dot-dashes $\lambda = 4$ components, respectively. The horizontal solid line indicates the Q-value of the process, which in our case is ^{232}Pu $\rightarrow$ ^{228}U + α. In order to obtain the experimental value of the half life the quenching factor has the value $v_a = 0.62$.

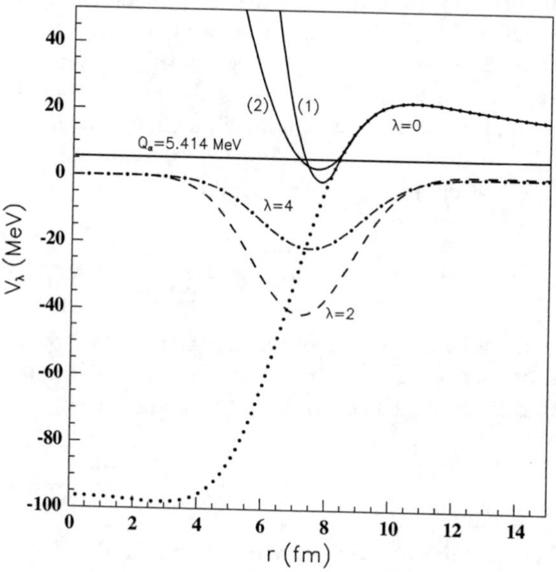

Fig. 6.6 The radial components of the renormalized α-nucleus potential (6.20) for $\lambda = 0$ (*dots*), $\lambda = 2$ (*dashes*) and $\lambda = 4$ (*dot-dashes*). The solid pocket-like curves (*1*) and (*2*) are the monopole parts of the interaction (6.21), giving the same Q-value. Their parameters are (*1*) $c = 90.117$ (MeV fm^{-2}), $Q_\alpha + v_0 = 10.272$ (MeV) and (*2*) $c = 30.296$ (MeV fm^{-2}), $Q_\alpha + v_0 = -3.816$ (MeV). The *horizontal line* denotes the Q-value. The decay process is ^{232}Pu $\rightarrow$ ^{228}U + α

This interaction is able to describe the α-daughter system for large distances $r > r_m$. In order to describe the internal two-body dynamics, we use a repulsive core, taking into account the fact that an α-particle exists only on the nuclear surface. Indeed, microscopic computations, see e.g. Figure 2.a of the Ref. [23], suggest that the α-particle wave function is peaked in the region of the nuclear surface and it has a Gaussian-like shape.

The main reason for this behaviour is connected with the fact that many single particle orbitals satisfying Talmi–Moshinsky rule contribute to the preformation amplitude. They have a destructive effect inside the nucleus, but a coherent effect on the surface. *Thus, the Wildermuth rule to built an α-particle from two proton and two neutron orbitals is a very crude approximation.*

Moreover, the monopole component represents more than 90% of the total decay width. Such a wave function corresponds to a shifted harmonic oscillator potential, which we have to consider in the internal region $r \leq r_m$, i.e.

$$\begin{aligned} \overline{V}_0(r) &= v_a V_0(r), \quad r > r_m \\ &= c(r - r_0)^2 - v_0, \quad r \leq r_m. \end{aligned} \tag{6.21}$$

In Fig. 6.6 this interaction is given by two pocket-like curves. The curves labeled by (1) and (2) give the same Q-value. Their parameters are:

(1) $c = 90.117$ (MeV fm^{-2}), $Q_\alpha + v_0 = 10.272$ (MeV) and
(2) $c = 30.296$ (MeV fm^{-2}), $Q_\alpha + v_0 = -3.816$ (MeV).

We considered the quantity $Q_\alpha + v_0$ because it is the excitation energy inside the pocket-like interaction. From this figure, it is clear that if one considers a deformed part for repulsive multipoles with $\lambda \neq 0$ the effect can be renormalized by the monopole repulsion in the internal region $r < r_0 = 1.2A_D^{1/3}$.We stress on the fact that only three parameters, namely v_a, v_0, c, are independent, because the radii r_0, r_m are determined by using the matching conditions

$$\begin{aligned} v_a V_0(r_m) &= c(r_m - r_0)^2 - v_0 \\ v_a \frac{dV_0(r_m)}{dr} &= 2c(r_m - r_0). \end{aligned} \tag{6.22}$$

These conditions allow us to write down a single equation determining the matching radius

$$v_a V_0(r_m) = \frac{v_a^2}{4c}\left[\frac{dV_0(r_m)}{dr}\right]^2 - v_0, \tag{6.23}$$

for some given combination of v_a, v_0, c. The parameters of the repulsive cores (1) and (2) in Fig. 6.6 were chosen to give the best fit simultaneously for Q_α, Γ_J, $J = 0, 2, 4$.

In Fig. 6.7a, we plotted the radial components of the wave function for $J = 0$ (solid line), $J = 2$ (dashes) and $J = 4$ (dot-dashes), corresponding to the pocket-like

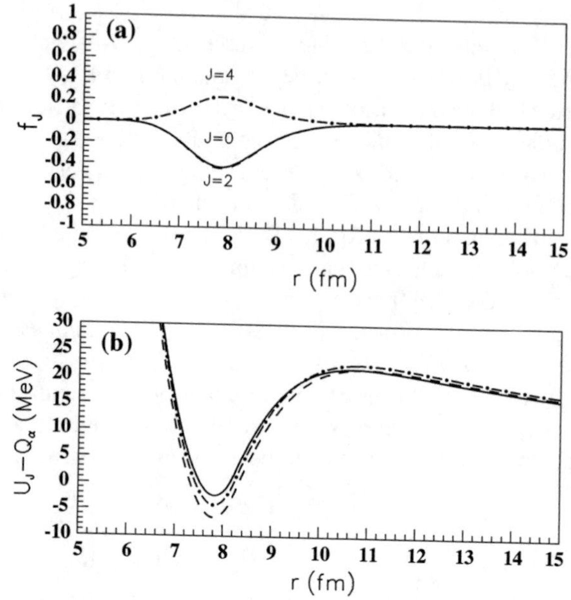

Fig. 6.7 **a** The radial components of the α-nucleus wave function inside the pocket-like potential for $J = 0$ (*solid line*), $J = 2$ (*dashes*) and $J = 4$ (*dot-dashes*). **b** The radial dependence of the diagonal α-nucleus matrix elements plus the centrifugal barrier (6.24) for $J = 0$ (*solid line*), $J = 2$ (*dashes*) and $J = 4$ (*dot-dashes*). The repulsive core is labeled by (*1*) in Fig. 6.6. The decay process is $^{232}\mathrm{Pu} \rightarrow {}^{228}\mathrm{U} + \alpha$

repulsion (1), in Fig. 6.6. In Fig. 6.7b, we give the radial dependencies of the diagonal terms [with the same symbols as in (a)], corresponding to the α-daughter potential plus the centrifugal barrier, i.e.

$$\begin{aligned} U_J(r) &= \langle \mathscr{Y}_J | \overline{V}(r) | \mathscr{Y}_J \rangle + \frac{\hbar^2 J(J+1)}{2\mu r^2} \\ &= \frac{1}{\sqrt{4\pi}} \sum_\lambda \overline{V}_\lambda(r) [\langle J,0;\lambda,0|J,0\rangle]^2 + \frac{\hbar^2 J(J+1)}{2\mu r^2}. \end{aligned} \tag{6.24}$$

Thus, in the spherical case, where the components with $\lambda > 0$ vanish, the decay widths to excited states are entirely determined by the corresponding centrifugal barriers.

A very important observation is related to the shape of the pocket-like potentials (1) and (2) in Fig. 6.6. The repulsive strength c and the quantity $Q_\alpha + v_0$ are strongly related and therefore the repulsive core is characterized by one independent parameter. Indeed, by increasing c one should simultaneously increase the excitation energy $Q_\alpha + v_0$, in order to obtain the same Q-value and therefore the total half life. Moreover, our computations has shown that the total half-life and the fine structure, defined by (6.1), is weakly affected by simultaneously changing the parameters of the repulsive potential for this decay process.

The importance of the quenching strength v_a is shown in Fig. 6.8. We plotted here by a solid line the Q-value, by dots $\log_{10} T$, by a dashed line I_2 and by a dot-dashed line I_4, as a function of v_a, by considering fixed the parameters of the repulsive potential, i.e. $Q_\alpha + v_0 = 10.272$, $c = 90.117$. One sees a strong dependence of the first two quantities and a weaker variation for I_4, while I_2 is practically

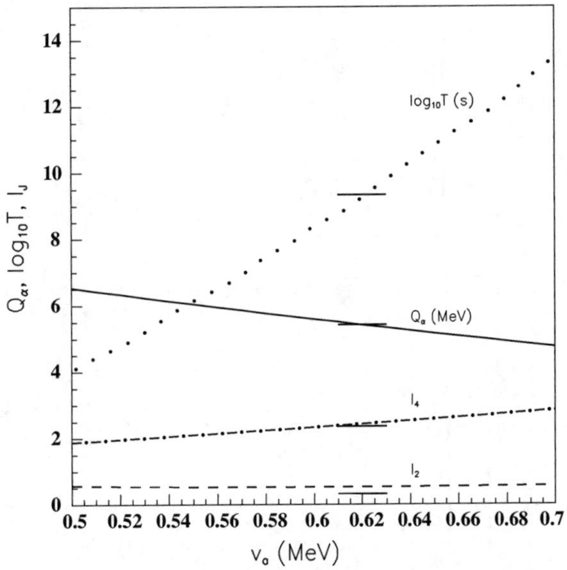

Fig. 6.8 The Q-value (*solid line*), the logarithm of the total half-life (*dots*), I_2 (*dashes*) and I_4 (*dot-dashes*) versus the attraction strength v_a. The other potential parameters are $c = 90.117$ (MeV fm^{-2}), $Q_\alpha + v_0 = 10.272$ (MeV). The corresponding experimental values are shown by short *horizontal lines*. The decay process is $^{232}Pu \rightarrow {}^{228}U + \alpha$

a constant. It is important to stress that at the value $v_a \approx 0.62$ one obtains simultaneously the best fit with the experimental data (shown by short horizontal lines) for all considered quantities.

Thus, we will consider in our calculations concerning the fine structure $v_a = 1$, because the relative intensities weakly depend upon this parameter. Once this parameter is fixed, we can adjust the Q-values for different decays by using one parameter, namely the repulsive depth v_0, because for a given quenching strength v_a the repulsive strength c has a definite value. As mentioned above, we will consider as an independent parameter the sum $Q_\alpha + v_0$, because it gives the energy of the first excitation in the pocket-like potential.

We analyzed all α-decays for rotational nuclei with known ratios I_2 and I_4, and one where I_4 was only given as a limit. The experimental data, namely the excitation energies, total half lives and Q-values, are taken from the compilation [1]. We also compared fine structure intensities for $J = 2^+$, $J = 4^+$ states and total half lives, with respective uncertainties, with the ENSDF database. The deformation parameters were taken from the systematics in Ref. [21]. These data for daughter nuclei are given in Table 6.1.

We have also plotted various quantities versus the number "n", labeling isotopic chains with fixed charge number Z of the daughter nucleus, given in the second column of this Table. The α-decay fine structure is given by the intensities I_J, defined in Eq. 6.1.

In Fig. 6.9a we give the experimental quadrupole intensities I_2 (dark circles) versus "n" and the results of our calculations (open circles), including predictions for some light emitters.

In Fig. 6.9b are plotted the corresponding quadrupole deformation parameters β_2 versus "n". One sees a clear correlation between calculated intensities and

Table 6.1 Experimental and computed data for α-decays from rotational nuclei: quadrupole deformation, Q-value, excitation energies (in keV), logarithm of the relative intensities, logarithm of the total half-life (in seconds), phenomenological and microscopic spectroscopic factors

n	Z	A	β_2	Q_α	E_2	E_4	I_2^{exp}	I_2^{th}	I_4^{exp}	I_4^{th}	lgT^{exp}	lgT^{th}	S	S_{gs}
1	70	170	0.295	2584	84.255	277.430	–	1.422	–	5.859	22.799	21.595	0.063	0.018
2	74	168	0.208	5254	199.300	562.300	–	1.510	–	4.638	3.283	2.289	0.101	0.019
3	74	170	0.226	4900	156.720	462.330	–	1.267	–	4.250	5.342	4.093	0.056	0.019
4	74	182	0.259	2846	100.106	329.427	–	1.479	–	6.528	22.800	21.017	0.016	0.031
5	76	176	0.246	5285	135.100	395.500	–	0.982	–	3.632	4.271	2.884	0.041	0.024
6	76	178	0.247	4977	131.600	397.700	–	0.990	–	3.921	5.764	4.482	0.052	0.025
7	76	180	0.238	4618	132.110	408.620	–	1.063	–	4.362	7.786	6.577	0.062	0.028
8	76	182	0.239	4352	127.000	400.400	–	1.075	–	4.654	9.899	8.290	0.025	0.031
9	76	184	0.229	4033	119.800	383.770	–	1.124	–	4.903	12.483	10.647	0.015	0.035
10	76	186	0.220	3272	137.159	434.087	–	1.687	–	7.090	19.312	17.578	0.018	0.039
11	78	178	0.254	5997	171.000	569.999[a]	2.138	1.053	–	3.736	1.869	0.463	0.039	0.022
12	78	180	0.265	5662	152.000	506.666[a]	2.394	0.998	–	3.717	3.447	1.914	0.029	0.025
13	78	182	0.255	5236	154.900	419.080	–	1.070	–	4.259	5.663	4.034	0.023	0.028
14	78	184	0.247	4740	162.970	435.960	–	1.245	–	5.080	8.688	6.881	0.016	0.031
15	88	224	0.164	5520	84.373	250.783	0.424	0.811	2.503	2.570	7.781	7.097	0.207	0.003
16	88	226	0.172	4770	67.670	211.540	0.513	0.545	2.803	3.492	12.376	11.582	0.161	0.003
17	88	228	0.180	4083	63.823	204.680	0.557	0.632	3.054	3.900	17.647	16.940	0.196	0.003
18	90	224	0.164	6804	98.000	326.666[a]	0.383	0.750	–	2.289	2.903	1.871	0.093	0.003
19	90	226	0.173	5993	72.200	226.430	0.324	0.426	2.249	3.058	6.255	5.197	0.087	0.003
20	90	228	0.182	5414	57.762	186.828	0.334	0.392	2.356	3.081	9.337	8.442	0.127	0.003
21	90	230	0.198	4859	53.200	174.100	0.400	0.408	2.553	3.326	12.889	11.893	0.101	0.003
22	90	232	0.207	4573	49.460	162.250	0.455	0.415	2.692	3.351	14.869	13.854	0.097	0.003
23	90	234	0.215	4270	49.550	163.000	0.577	0.460	3.006	3.505	17.149	16.328	0.151	0.002

(continued)

Table 6.1 (continued)

n	Z	A	β_2	Q_α	E_2	E_4	I_2^{exp}	I_2^{th}	I_4^{exp}	I_4^{th}	lgT^{exp}	lgT^{th}	S	S_{gs}
24	92	228	0.191	6716	59.000	196.666	0.308	0.289	–	2.810	3.949	2.724	0.060	0.003
25	92	230	0.199	6310	51.720	169.500	0.327	0.272	2.230	2.867	5.723	4.472	0.056	0.003
26	92	232	0.207	5867	47.580	156.540	0.351	0.272	2.478	2.990	7.955	6.650	0.050	0.003
27	92	234	0.215	5593	43.498	143.352	0.389	0.269	2.830	2.974	9.442	8.109	0.046	0.002
28	92	236	0.215	5256	45.244	149.478	0.429	0.313	2.938	3.032	11.316	10.134	0.066	0.002
29	92	238	0.215	4984	44.915	148.390	0.513	0.343	3.396	3.039	13.071	11.936	0.073	0.002
30	92	240	0.224	4666	45.000	151.000	0.619	0.385	–	3.086	15.403	14.212	0.064	0.002
31	94	234	0.216	6620	46.000	153.333	0.358	0.221	–	2.796	5.352	3.841	0.031	0.002
32	94	236	0.215	6398	44.630	147.450	0.391	0.225	3.136	2.841	6.369	4.846	0.030	0.002
33	94	238	0.215	6216	44.080	146.000	0.456	0.238	3.326	2.798	7.149	5.698	0.035	0.002
34	94	240	0.223	5902	42.824	141.690	0.510	0.258	3.541	2.750	8.757	7.293	0.034	0.002
35	94	242	0.224	5475	44.540	147.300	0.664	0.313	–	2.757	11.177	9.738	0.036	0.002
36	94	244	0.224	5162	44.200	155.000	0.657	0.337	3.032	2.856	13.079	11.719	0.044	0.002
37	94	246	0.235	5208	46.000	155.000	–	0.370	–	2.674	12.163	11.367	0.160	0.002
38	96	236	0.215	7719	45.000	150.000	0.327	0.175	–	2.558	1.821	0.230	0.026	0.002
39	96	238	0.215	7516	35.000	116.666	0.602	0.149	–	2.401	2.321	0.938	0.041	0.002
40	96	240	0.224	7329	38.000	126.666	0.477	0.167	–	2.457	3.221	1.615	0.025	0.002
41	96	242	0.224	6862	42.130	137.000	0.585	0.210	2.723	2.509	5.109	3.525	0.026	0.002
42	96	244	0.234	6361	42.965	142.348	0.611	0.238	2.301	2.614	7.460	5.838	0.024	0.002
43	96	246	0.234	6128	42.852	142.010	0.752	0.266	2.451	2.527	8.616	7.005	0.024	0.002
44	96	248	0.234	6217	43.400	143.600	0.729	0.281	2.545	2.384	7.935	6.544	0.041	0.002
45	96	250	0.225	5926	43.000	143.333	0.689	0.315	–	2.327	9.227	8.061	0.068	0.002
46	98	242	0.224	8374	45.000	150.000	0.602	0.166	–	2.344	0.078	−1.335	0.039	0.002
47	98	244	0.234	8002	41.000	136.666	0.602	0.161	–	2.353	1.588	−0.155	0.018	0.002

(continued)

Table 6.1 (continued)

n	Z	A	β_2	Q_α	E_2	E_4	I_2^{exp}	I_2^{th}	I_4^{exp}	I_4^{th}	lgT^{exp}	lgT^{th}	S	S_{gs}
48	98	246	0.234	7557	44.000	146.666	0.689	0.204	–	2.307	3.342	1.491	0.014	0.002
49	98	248	0.235	7153	41.530	137.810	0.748	0.227	1.938	2.207	4.961	3.121	0.014	0.002
50	98	250	0.245	7307	42.721	141.875	0.777	0.234	2.016	2.136	4.067	2.435	0.023	0.002
51	98	252	0.236	7027	45.720	151.740	0.753	0.281	–	2.098	5.067	3.612	0.035	0.002
52	100	248	0.235	8549	44.000	146.666	0.477	0.178	–	2.138	0.593	−1.241	0.015	0.002
53	100	250	0.235	8226	44.000	145.000	–	0.206	–	2.050	1.753	−0.172	0.012	0.002
54	100	252	0.245	8581	46.600	155.333	0.826	0.214	–	1.962	0.466	−1.403	0.014	0.002
55	102	252	0.236	8995	46.400	153.800	–	0.202	–	1.934	0.301	−1.903	0.006	0.002
56	104	256	0.247	9923	51.000	170.000	0.689	0.205	–	1.810	−2.143	−3.921	0.017	0.002

[a] These energies are not measured, but a simple perfect rotor approximation we used in our calculations

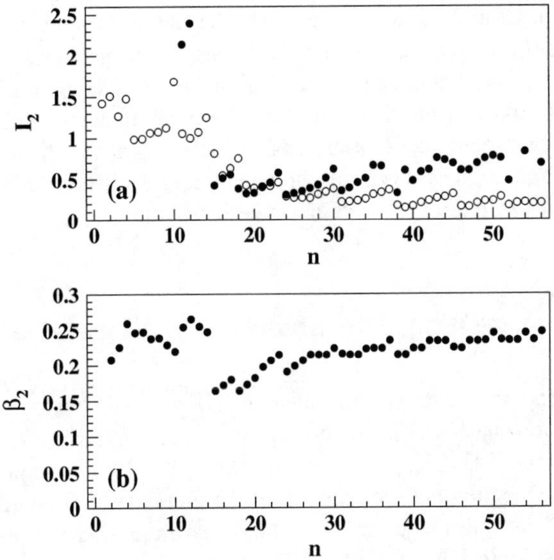

Fig. 6.9 **a** The experimental quadrupole intensities I_2, given by Eq. 6.1, (*dark circles*) versus "n" and the calculated values (*open circles*). **b** Quadrupole deformation parameters β_2 versus "n"

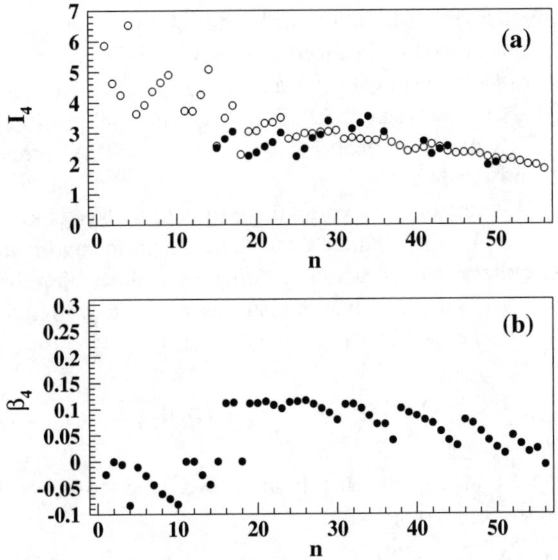

Fig. 6.10 The same as in Fig. 6.9, but for the hexadecapole quantities

deformations, already mentioned in Ref. [24]. At the same time we notice an abrupt change for both quantities, corresponding to the magic numbers $Z_0 = 82$, $N_0 = 126$, corresponding to $n = 15$.

Similar plots are given in Fig. 6.10a, b for the hexadecapole intensities I_4 and deformation parameters β_4, respectively, versus "n". A similar connection between intensities and deformations is observed by crossing the above magic numbers, but in the right region the correlation is less pronounced than in the quadrupole case. In any case, the abrupt change for both the intensity and the deformation parameter is also present for hexadecapole quantities.

Concluding, the rotational model is able to qualitatively explain the gross features of the fine structure. The theoretical intensities I_J are proportional to the corresponding deformations β_J, including sharp changes around $Z = 82$. We made predictions concerning the fine structure of those α-emitters where the intensities to excited levels were not measured yet.

6.4 α-Decay Fine Structure in Vibrational Nuclei

The first attempts to calculate HF's for 2^+ states in vibrational nuclei within the quasiparticle random-phase approximation (QRPA) were performed in Refs. [25–27]. Later on, in Ref. [28, 29] an explanation was given for the connection between the HF of the first excited 0^+ state and the neutron number in Pb isotopes in terms of proton–neutron pairing vibrations. Recent calculations to estimate HF's to excited 0^+ states were performed in [30] within the Hartree–Fock–Bogoliubov approach. In Ref. [4], the preformation factor of excited 0^+ states in neutron deficient Pb and Po isotopes was supposed to be proportional with a Boltzman distribution with respect to the excitation energy. This ansatz is consistent with the reduced width given by Eq. 2.108 in Chap. 2. In the last decade, the α-decay spectroscopy was used to investigate 0^+ and 2^+ excited states in the Pb [31–38] and U region [39]. Some experimental results concerning the fine structure of 2^+ states were analyzed by using the QRPA formalism in Refs. [40, 41] and recently in [42].

This section is based on the results of the systematic coupled channels analysis of Ref. [43]. In vibrational nuclei the intrinsic coordinate in the daughter nucleus is the collective quadrupole coordinate $\alpha_{2\mu}$, describing the oscillations of the nuclear surface. Thus, the Hamiltonian describing α-particle emission to a low-lying 2^+ state in the daughter nucleus is given by

$$\mathbf{H} = -\frac{\hbar^2}{2\mu}\Delta_r + \mathbf{H}_D(\alpha_2) + V_0(r) + V_d(\alpha_2, \mathbf{r}), \tag{6.25}$$

where $\mathbf{H}_D(\alpha_2)$ is the Hamiltonian describing vibrations. The α-daughter interaction has the following form

$$V_d(\alpha_2, \mathbf{r}) = V_2(r) \sum_\mu \alpha_{2\mu} Y^*_{2\mu}(\hat{r}). \tag{6.26}$$

The monopole part of the interaction is given by the same ansatz as in Eq. 6.21, estimated within the double folding approach with M3Y particle–particle interaction [21, 22, 44]. The second line is the repulsive core, mocking the Pauli effect and fixing the energy of the first resonant state to the experimental Q-value.

Concerning the $\lambda = 2$ formfactor is the linear term of the expansion, i.e.

$$V_2(r) = -v_2(r - r_0)\frac{dV_0(r)}{dr}. \tag{6.27}$$

The wave function is given by a superposition similar to (6.9), i.e.

$$\Psi(\alpha_2, \mathbf{r}) = \frac{1}{r}\sum_{J=0,2} f_J(r)\mathscr{Y}_J(\alpha_2, \hat{r}), \tag{6.28}$$

where the core-angular harmonics is given by

$$\mathscr{Y}_J(\alpha_2, \hat{r}) \equiv [\Phi_J(\alpha_2) \otimes Y_J(\hat{r})]_0. \tag{6.29}$$

$\Phi_J(\alpha_2)$ is the Jth eigenstate of the vibrational Hamiltonian $H_D(\alpha_2)$. In the above wave function, we considered only one phonon excitations, because low-lying 2^+ states have large electromagnetic E2 rates to the ground state. In principle, they can also couple with two-phonon components through the quadrupole operator α_2. An extensive analysis of this coupling was recently performed in several references within a microscopic approach (see [45] and the references therein). There, an analysis of not only energy spectra, but also of electromagnetic and β decays has shown that, in general, this coupling is relatively weak. On the other hand, we are interested to estimate only the hindrance factor, i.e the ratio between the ground-state and $J = 2$ components and not the absolute decay widths. Our results in Table 6.2 cleary show that the experimental values can be reproduced by using only one free parameter, namely the coupling constant C_v, defined below by Eq. 6.30. Thus, the inclusion of two-phonon components may affect the absolute values of the ground state and one-phonon components, but should not affect their ratio. Moreover, the α-decay intensities to two-phonon states in vibrational and transitional nuclei are so small, that they cannot be detected at the moment. This is supported by the available systematics [1]. Therefore, this is another argument that the coupling of these components should be relatively small. The above conclusion is also supported by the coupled channels analysis in rotational nuclei, where $J = 4$ intensities are very small but still measurable. They correspond to a relatively small wave function component, in spite of the fact that the $J = 0$ and 2 components are comparable.

By using the orthonormality of angular functions entering the superposition (6.28), one obtains, in a standard way, the coupled system of differential equations for radial components (6.12). The matrix element (6.13) has only off diagonal non-vanishing values, given by

$$\begin{aligned} \langle \mathscr{Y}_2 | V_d | \mathscr{Y}_0 \rangle &= V_2(r)\frac{1}{\sqrt{4\pi}}\langle \Phi_2 || \alpha_2 || \Phi_0 \rangle \\ &\equiv -C_v(r - r_0)\frac{dV_0(r)}{dr}, \end{aligned} \tag{6.30}$$

depending upon a new constant $C_v = \frac{v_2}{\sqrt{4\pi}}\langle \Phi_2 || \alpha_2 || \Phi_0 \rangle$.

Table 6.2 Charge and mass numbers of the daughter vibrational/transitional nuclei, quadrupole deformations [21], Q-values and E_2 energies (keV), experimental [1] and computed values of I (6.1), logarithm of experimental [1] and computed half lives (s), phenomenological spectroscopic factor S (2.142) in Chap. 2 and microscopic spectroscopic factor S_{gs} (9.60) in Chap. 9, potential parameters C_v (MeV^{-1}) and c (MeV fm^{-2})

n	Z	A	β_2	Q_α	E_2	E_4	I_2^{exp}	I_2^{th}	lgT^{exp}	lgT^{th}	S	S_{gs}	C_v	c
1	72	156	0.035	6064	858.000	1587.000	–	4.908	−0.979	−2.490	0.031	0.015	0.050	2000
2	72	158	0.107	5674	476.360	1033.330	–	2.960	0.479	−0.865	0.045	0.015	0.075	2000
3	72	160	0.152	5279	389.400	898.260	–	2.621	2.199	1.018	0.066	0.016	0.150	2000
4	72	162	0.180	4856	285.000	729.500	–	2.202	4.740	3.098	0.023	0.016	0.150	2000
5	74	160	0.089	6477	609.900	1264.600	–	3.780	−1.670	−3.285	0.024	0.017	0.024	2000
6	74	164	0.161	5818	332.700	823.700	–	2.242	0.679	−0.644	0.048	0.018	0.055	2000
7	74	166	0.181	5539	252.000	675.700	–	1.847	1.938	0.634	0.050	0.018	0.069	2000
8	76	170	0.171	6184	286.700	749.900	1.996	1.995	0.068	−1.336	0.039	0.021	0.043	2000
9	76	172	0.190	5887	227.770	606.170	2.584	2.590	1.199	−0.259	0.035	0.021	0.011	2000
10	76	174	0.226	5573	158.600	435.000	1.270	1.273	2.625	1.310	0.048	0.022	0.100	2000
11	78	174	0.153	6578	394.200	891.800	–	2.588	−0.562	−2.053	0.032	0.021	0.030	2000
12	78	176	0.171	6258	264.000	564.100	3.267	3.272	0.731	−0.935	0.022	0.022	0.005	2000
13	80	180	−0.122	6775	434.300	706.700	–	2.788	0.328	−2.088	0.004	–	0.027	2000
14	80	182	−0.122	6471	351.800	–	2.794	2.792	0.951	−0.954	0.012	–	0.017	2000
15	80	184	−0.130	6111	366.510	–	3.018	3.046	2.431	0.428	0.010	–	0.016	2000
16	80	186	−0.130	5698	405.330	–	3.075	3.065	4.250	2.306	0.011	–	0.030	2000
17	82	204	−0.008	5216	899.171	1274.000	5.620	6.556	7.961	5.562	0.004	–	0.150	2000
18	82	206	−0.008	5408	803.049	–	4.914	5.507	7.079	4.502	0.003	–	0.160	2000
19	82	210	0.000	7833	799.700	1097.700	3.983	3.986	−3.785	−5.428	0.023	–	0.025	2000
20	82	212	0.000	6906	804.900	1117.000	4.721	4.728	−0.839	−2.427	0.026	–	0.021	2000

(continued)

Table 6.2 (continued)

n	Z	A	β_2	Q_α	E_2	E_4	I_2^{exp}	I_2^{th}	lgT^{exp}	lgT^{th}	S	S_{gs}	C_v	c
21	82	214	0.009	6115	836.000	–	4.959	4.955	2.270	0.824	0.036	–	0.068	2000
22	84	194	0.026	7349	319.800	686.500	3.154	3.162	−1.184	−2.693	0.031	0.008	0.006	2000
23	84	196	0.136	7043	463.120	890.990	4.222	4.236	−0.009	−1.654	0.023	0.008	0.005	2000
24	84	198	0.122	6774	604.940	1158.390	–	3.600	1.045	−0.649	0.020	0.009	0.032	2000
25	84	200	0.009	6546	665.900	1276.900	–	3.831	2.009	0.229	0.017	0.010	0.050	2000
26	84	202	0.009	6384	677.300	1248.900	–	3.852	2.732	0.939	0.016	0.010	0.080	2000
27	84	204	0.009	6261	684.342	1200.660	3.328	3.889	3.373	1.451	0.012	0.010	0.160	2000
28	84	206	−0.018	6159	700.660	–	4.252	4.256	3.954	1.761	0.006	0.009	0.060	2000
29	84	208	−0.018	6385	686.526	–	3.301	3.787	3.155	0.817	0.005	1.000	0.150	2000
30	84	210	0.000	9208	1181.400	1426.700	–	6.571	−6.569	−8.395	0.015	1.000	0.003	2000
31	84	212	0.045	8200	727.330	1132.530	–	4.464	−4.347	−5.773	0.037	0.002	0.008	2000
32	84	214	−0.008	7263	609.316	1015.050	2.896	2.898	−1.456	−2.723	0.054	0.002	0.090	2000
33	84	216	0.020	6405	549.760	968.940	2.943	3.031	1.745	0.595	0.071	0.002	0.150	2000
34	84	218	0.039	5590	509.700	–	3.107	3.390	5.519	4.427	0.081	0.003	0.150	2000
35	86	202	−0.104	7416	504.000	1073.100	–	3.172	−0.620	−2.270	0.022	0.010	0.019	2000
36	86	204	−0.087	7273	542.900	1131.400	–	3.345	0.137	−1.778	0.012	0.010	0.021	2000
37	86	206	−0.044	7156	575.300	–	–	3.523	0.585	−1.394	0.010	0.009	0.022	2000
38	86	208	−0.026	7032	635.800	–	3.301	3.304	1.185	−0.916	0.008	0.008	0.065	2000
39	86	210	−0.026	7273	643.800	–	2.698	3.048	0.391	−1.768	0.007	1.000	0.150	2000
40	86	212	0.000	9526	1273.800	1501.500	–	6.968	−6.740	−8.471	0.019	1.000	0.003	2000
41	86	214	0.008	8546	694.700	1141.200	–	4.346	−4.592	−6.001	0.039	0.002	0.007	2000
42	86	216	0.008	7592	461.400	840.500	1.996	2.124	−1.745	−2.956	0.062	0.002	0.150	2000

(continued)

Table 6.2 (continued)

n	Z	A	β_2	Q_α	E_2	E_4	I_2^{exp}	I_2^{th}	lgT^{exp}	lgT^{th}	S	S_{gs}	C_v	c
43	86	218	0.040	6679	324.320	653.180	1.502	1.524	1.559	0.572	0.103	0.003	0.420	50
44	86	220	0.111	5789	240.986	533.690	1.273	1.265	5.500	4.731	0.170	0.003	0.200	50
45	86	222	0.137	4871	186.211	448.370	1.231	1.221	10.703	10.031	0.213	0.003	0.140	50
46	88	208	−0.104	7952	520.200	–	–	3.302	−1.521	−3.375	0.014	0.010	0.014	2000
47	88	210	−0.053	7826	603.000	–	–	3.724	−1.066	−2.945	0.013	0.009	0.014	2000
48	88	212	−0.035	8071	629.300	1454.300	2.265	2.158	−0.585	−3.405	0.002	1.000	0.420	50
49	88	214	0.008	9849	1382.400	1639.300	–	7.380	−6.963	−8.554	0.026	1.000	0.002	2000
50	88	216	0.008	8953	688.200	1164.100	–	4.206	−5.013	−6.437	0.038	0.002	0.007	2000
51	88	218	0.020	8127	388.900	741.100	1.628	1.301	−2.650	−3.673	0.095	0.002	0.400	50
52	88	220	0.103	7304	178.470	410.070	0.619	0.624	0.021	−0.929	0.112	0.003	0.150	50
53	88	222	0.130	6452	111.120	301.390	0.520	0.515	3.262	2.490	0.169	0.003	0.170	50
54	90	218	0.008	9500	689.600	1194.200	–	4.209	−6.000	−7.091	0.081	0.002	0.006	2000
55	90	220	0.030	8620	373.300	759.800	–	2.513	−3.155	−4.832	0.021	0.002	0.014	2000
56	90	222	0.111	7715	183.300	439.800	0.753	0.749	−0.456	−1.608	0.070	0.003	0.400	50

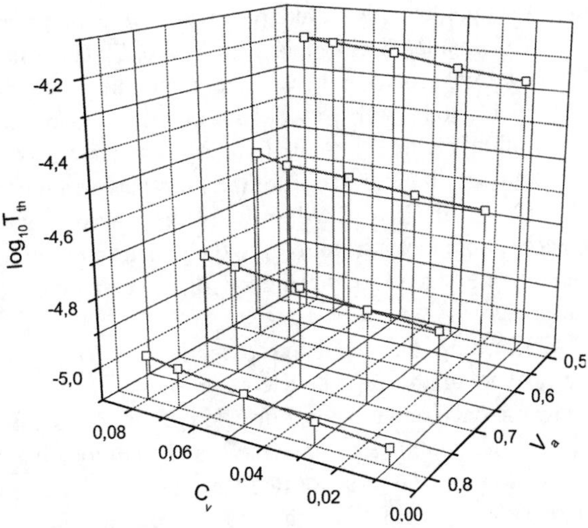

Fig. 6.11 Logarithm of the half life (in seconds) versus the attractive parameter v_a and vibrational coupling strength C_v

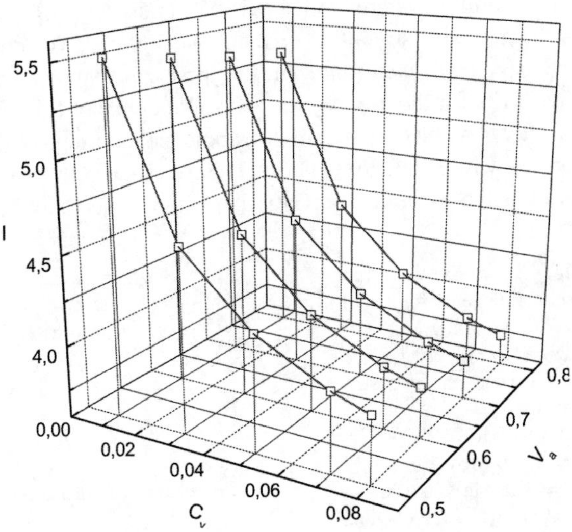

Fig. 6.12 Logarithm of the intensity ratio (6.1) versus the attractive parameters v_a and vibrational coupling strength C_v

We analyze the dependence of the fine structure (6.1) and hindrance factors (6.2) versus various parameters. Our model mainly depends upon four parameters, namely v_a, C_v given by the attractive part of the potential and c, v_0 defining the repulsive core. In Fig. 6.11 we plotted the dependence between the logarithm of the half life $T = \hbar \ln 2/\Gamma$ and the two parameters of the attractive part of the interaction v_a and C_v. One can see that it strongly depends upon v_a, while the dependence upon C_v is very weak. On the other hand, one sees an opposite behavior in Fig. 6.12, where we plotted the dependence between the logarithm of the intensity ratio I (6.1) with respect to the same parameters. This quantity practically does not depend upon v_a and has a strong dependence on C_v.

Our goal is to investigate the fine structure given by (6.1), which mainly depends upon the vibrational parameter C_v, defining the coupling strength between the two considered channels with $J = 0, 2$. Thus, in order to simplify the calculation we will consider $v_a = 1$, corresponding to a "pure" α-cluster model. Thus, our calculations will provide the components of the wave function giving the penetration factor P_J. Actually the α-cluster exists on the nuclear surface with some probability and therefore the ratio between theoretical and experimental half lives (2.142) in Chap. 2 will provide this α-particle preformation probability, or the spectroscopic factor, multiplying the penetrability given by our potential [46]. This ratio, being directly connected with v_a, will give us the information about this parameter.

Concerning the repulsion potential, in Ref. [24] we have shown that the parameters of the repulsive core c and v_0, used to determine the Q-value, are related to each other. Thus, our calculation will depend only upon two parameters, namely the coupling parameter C_v and the repulsive strength c.

We analyze the α-decay fine structure for those emitters with the known value of (6.1). Our calculations showed that this quantity can be reproduced for vibrational nuclei only by considering a strong repulsive core, at variance with rotational emitters where we used a soft repulsion [24]. We consider a constant value of the repulsive strength $c = 2{,}000$ MeV fm^{-2}. The components $J = 0$ (solid line) and $J = 2$ (dashed line) of the first zero nodes wave function are given in Fig. 6.13a for the α-decay to ^{216}Po. We consider this eigenstate as a decaying state based on results of the microscopic calculations. Indeed, the calculations performed within QRPA [41] have shown that both ground and 2^+ formation probabilities have basically Gaussian shapes, as can be seen in Fig. 6.2 of this

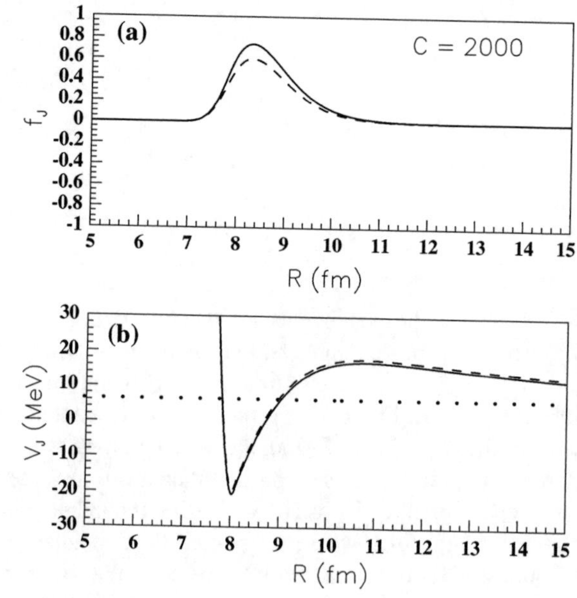

Fig. 6.13 **a** The wave function components for $J = 0$ (*solid line*) and $J = 2$ (*dashed line*) for the α-decay to ^{216}Po with the potential parameters $C_v = 0.15$ MeV^{-1}, $c = 2{,}000$ MeV fm^{-2}, $v_0 = 65.6$ MeV. **b** The components of the diagonal matrix elements of the potential (6.31) for $J = 0$ (*solid line*) and $J = 2$ (*dashed line*)

reference. Here, the hindrance factor is estimated as a ratio between the integral of the corresponding preformation amplitudes squared, computed in the region beyond the last maximum. Small oscillations, present at larger distances [47], do not affect the value of this ratio.

The corresponding values of the diagonal matrix elements

$$U_J(r) = V_0(r) + \frac{\hbar^2 J(J+1)}{2\mu r^2}, \tag{6.31}$$

are also plotted in Fig. 6.13b, by using the same symbols.

By using only one parameter C_v and with $c = 2{,}000$ MeV fm^{-2} it was possible to reproduce the experimental values $I_{\rm exp}$ of most measured vibrational nuclei. The results of our calculations are given in Table 6.2.

Concerning some transitional nuclei in Table 6.2 it turn out that their decay properties are closer to rotational nuclei, investigated in the previous section, were we used a soft repulsive core. It turns out that only by using the same repulsive strength as for rotational emitters, i.e. $c = 50$ MeV fm^{-2}, it is possible to reproduce their fine structure.

It is interesting to point out the fact that the inverse of the vibrational coupling strength $1/C_v$ is proportional with the logarithm of the HF squared for all analyzed transitions, as it is shown in Fig. 6.14. Here the solid line represents the corresponding fit, i.e.

$$\frac{1}{C_v} = 44.379(\log_{10}\mathrm{HF})^2, \quad \sigma = 8.657. \tag{6.32}$$

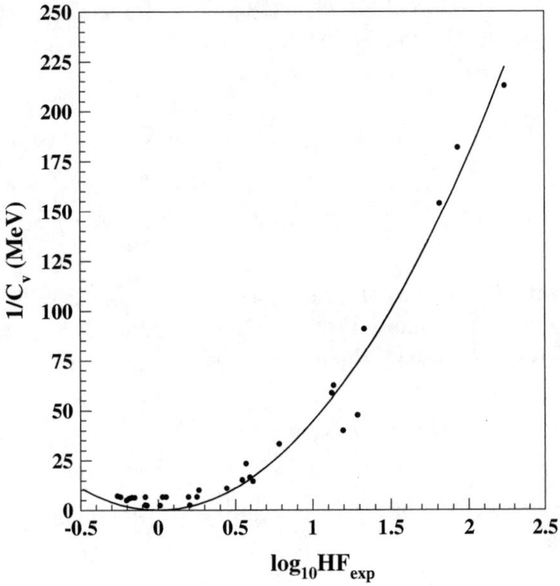

Fig. 6.14 The inverse of the vibrational coupling parameter versus the logarithm of the experimental HF. The parabola is the fit given by (6.32)

The relation (6.4) is able to describe the logarithm of the intensity ratio (6.1) within a factor of five. At the same time from Fig. 6.14 one sees that Eq. 6.32, connecting the vibrational strength and HF, is fulfilled with a much better accuracy by all emitters. By using the definitions (6.1) and (6.2) one derives the following relation

$$I = \frac{1}{\sqrt{44.379 C_v}} - \log_{10} \frac{P_2}{P_0}. \tag{6.33}$$

The intensity I provided by our calculation depends upon the vibrational strength C_v. In this way, it becomes possible to simultaneously determine I and C_v, i.e. to obtain selfconsistency between the above relation and the value given as a result of the coupled channels calculation. Based on this fact, we predicted the fine structure I for several α-decays from vibrational nuclei. The results are also given in Table 6.2.

6.5 Proton Emission in Vibrational Nuclei

The fine-structure for proton emitters is more complex, due to the fact that the initial spin of the parent nucleus is different from zero. The Hamiltonian describing proton emission to a low-lying 2^+ state in the daughter nucleus can be written as follows

$$\mathbf{H} = -\frac{\hbar^2}{2\mu}\Delta_r + \mathbf{H}_D(\alpha_2) + V_0(r, \mathbf{s}) + V_d(\alpha_2, \mathbf{r}), \tag{6.34}$$

in terms of the collective quadrupole coordinates $\alpha_{2\mu}$. Here, $V_0(r, \mathbf{s})$ is the spherical single proton interaction (2.133) in Chap. 2, while the last terms denotes the particle–vibration coupling interaction, given by

$$\begin{aligned} V_d(\alpha_2, \mathbf{r}) &= V_2(r) \sum_\mu \alpha_{2\mu} Y^*_{2\mu}(\hat{r}) \\ V_2(r) &= -R_N \frac{dV_N(r)}{dr} - R_C \frac{dV_C(r)}{dr}. \end{aligned} \tag{6.35}$$

where R_N and R_C are the nuclear and Coulomb radii associated with the corresponding potentials. The wave function describing the proton coupled with surface vibrations can be written as follows

$$\Psi_{J_i M_i}(\alpha_2, \mathbf{r}, \mathbf{s}) = \sum_{\lambda l j} \frac{f_{\lambda l j}(r)}{r} \mathcal{Y}^{(\lambda l j)}_{J_i M_i}(\alpha_2, \hat{r}, \mathbf{s}), \tag{6.36}$$

where $\mathcal{Y}$ is the quadrupole-spin-orbital harmonics given by

$$\mathcal{Y}^{(\lambda lj)}_{J_iM_i}(\alpha_2, \hat{r}, \mathbf{s}) = \left[\Phi_\lambda(\alpha_2) \otimes \mathcal{Y}^{(ls)}_j(\hat{r}, \mathbf{s})\right]_{J_iM_i}. \tag{6.37}$$

The coupled system of equations for radial components is derived, as usually, by projecting out a given channel, i.e.

$$\begin{aligned}\frac{d^2 f_{\lambda lj}(r)}{dr^2} &= \left\{\frac{l(l+1)}{r^2} + \frac{2\mu}{\hbar^2}\left[V_0(r,\mathbf{s}) + E_\lambda - E\right]\right\} f_{\lambda lj}(r) \\ &+ \frac{2\mu}{\hbar^2}\sum_{\lambda' l' j'}\left\langle \mathcal{Y}^{(\lambda lj)}_{J_iM_i}\right| \sum_\mu \alpha_{2\mu} Y^*_{2\mu}(\hat{r}) \left|\mathcal{Y}^{(\lambda' l' j')}_{J_iM_i}\right\rangle f_{\lambda' l' j'}(r).\end{aligned} \tag{6.38}$$

The matrix element of the interaction is

$$\begin{aligned}\left\langle \mathcal{Y}^{(2lj)}_{J_iM_i}\right| \sum_\mu \alpha_{2\mu} Y^*_{2\mu}(\hat{r}) \left|\mathcal{Y}^{(0l'J_i)}_{J_iM_i}\right\rangle &= \sqrt{\frac{5}{4\pi}}\alpha_2^{(0)}\left\langle J_i, \frac{1}{2}; 2, 0 \middle| j, \frac{1}{2}\right\rangle, \\ \alpha_2^{(0)} &\equiv \langle 2\mu|\alpha_{2\mu}|00\rangle.\end{aligned} \tag{6.39}$$

As usually, at large distances the radial component $f_{\lambda lj}(r)$ is the matching constant $N_{\lambda lj}$ times the Gamow state. The partial decay width is proportional to this matching constant squared, as shown in Eq. 2.83 in Chap. 2.

The interaction of the emitted proton with the oscillations of the nuclear surface was investigated in Refs. [48–50]. In Ref. [48], the particle–vibration modes were included within the above described coupled channels formalism. The resonant states $1h_{11/2}$, $3s_{1/2}$ in ^{161}Re and $2d_{3/2}$ in ^{160}Re weighted by BCS spectroscopic factors were considered. The experimental half lives for these states can be reproduced by considering the coupling with the quadrupole phonon using a phonon frequency $4\hbar\omega_2 \geq 0.6\,\text{MeV}$ and deformation $\beta_2 \approx 0.18$. This approach removed the discrepancy between the experimental data and optical potential calculations for proton emission from the $2d_{3/2}$ state. The octupole phonon was also considered in a similar way but the results were not satisfactory.

In Ref. [49], a similar particle–phonon coupled channels formalism was applied to describe the fine structure in proton emission from the odd–even emitters

$$^{145,145*,147,147*}\text{Tm},\ ^{131}\text{Eu},\ ^{141}\text{Ho},\ ^{151,151*}\text{Lu},$$

$$^{161,161*}\text{Re},\ ^{165,167,167*}\text{Ir},\ ^{171,171*}\text{Au},\ ^{171,171*}\text{Tl},$$

and odd–odd emitters

$$^{146,146*}\text{Tm},\ ^{150,150*}\text{Lu},\ ^{160}\text{Re},\ ^{164,166,166*}\text{Ir},\ ^{170}\text{Au}.$$

In these nuclei the odd neutron was considered a spectator. The used interaction was the same as in Ref. [51] and the decay widths were computed using the DW formalism. The agreement for most of odd–even (Table III) and odd–odd (Table IV) emitters is remarkable. Still for ^{164,166,167*}Ir, ^{177,177*}Tl, and ^{170}Au the deviation is up to a factor of 5.

In Ref. [50], the influence of the particle–phonon interaction on the half life of the isomeric state $1d_{3/2}$ in ^{151*}Lu was described. At variance with Ref. [52], were the same emitter was considered as a pure oblate rotor, the model was able to explain the reduction of the spectroscopic factor necessary to reproduce the experimental value.

6.6 Double Fine Structure in Cold Fission

Cold (neutron-less) fission is the heaviest member of the cold emission family, because both fragments have comparable sizes. An intense experimental activity was performed in order to investigate cold binary and ternary fission process of ^{252}Cf [53–59]. It involved modern facilities, as the Gammasphere and Eurogam, which were able to identify this rare process using the triple γ-rays coincidence technique.

The measurements confirmed the idea that this process is a natural extension of the cluster radioactivity [60, 61]. Among the most important papers dedicated to this subject, we mention here [62–68]. Recently several theoretical approaches were devoted to this subject [69–72]. They are based on the classical treatment of the three-body problem and semiclassical barrier penetration theory. A very convincing theoretical evidence that cold fission has a sub-barrier character was the WKB penetration calculation, using a double folding potential with M3Y plus Coulomb nucleon–nucleon forces. This simple estimate was able to reproduce the gross features of the binary cold fragmentation isotopic yields of ^{252}Cf [73].

The yields of rotational states were extracted from the intensities of γ-rays emitted in coincidence during the deexcitation of fragments for ^{104}Mo–^{148}Ba and ^{106}Mo–^{146}Ba [54]. It was shown that the cold fission population is centered around the low-lying 2^+ and 4^+ states and the states higher than 6^+ are practically not populated. This proves the assumption concerning the cold rearrangement of nucleons during the cold fission.

We describe the fission process of an even–even nucleus with $J_i = 0$ using the stationary Schrödinger equation [73]. The Hamiltonian can be written as follows

$$\mathbf{H} = -\frac{\hbar^2}{2\mu}\nabla_R^2 + \mathbf{H}_1(\Omega_1) + \mathbf{H}_2(\Omega_2) + V(\Omega_1, \Omega_2, \mathbf{R}), \tag{6.40}$$

where μ is the reduced mass of the dinuclear system and $\mathbf{H}_k$ are the Hamiltonians describing the rotation of the fragments. We denote by $\mathbf{R} \equiv (R, \Omega)$ the distance between the centers of the fragments, because we keep the notation $\mathbf{r}$ for the radius of the light particle in Section dedicated to the ternary fission. We also use Ω_k to denote Euler angles of the symmetry axis of fragments in the laboratory system of coordinates.

As usually, we estimate the interaction between nuclei in terms of the double folding between the nuclear densities [44], by using the M3Y nucleon–nucleon

[20] plus Coulomb force. By expanding the nuclear densities in multipoles, one obtains the deformed part of the interaction as in Eq. 2.123 in Chap. 2, i.e.

$$V_d(\Omega_1,\Omega_2,\mathbf{R}) = \sum_{\lambda_1\lambda_2\lambda_3} \frac{4\pi}{\hat{\lambda}_1\hat{\lambda}_2} V_{\lambda_1\lambda_2\lambda_3}(R)\mathscr{Y}_{\lambda_1\lambda_2\lambda_3}(\Omega_1,\Omega_2,\hat{R}), \tag{6.41}$$

where the term $(\lambda_1\lambda_2\lambda_3) = (000)$ is excluded from summation. Here, the angular part of the wave function has the following ansatz

$$\mathscr{Y}_{\lambda_1\lambda_2\lambda_3}(\omega_1,\omega_2,\hat{r}) = \left\{[Y_{\lambda_1}(\Omega_1)\otimes Y_{\lambda_2}(\Omega_2)]_{\lambda_3}\otimes Y_{\lambda_3}(\hat{R})\right\}_0. \tag{6.42}$$

For the internal part, we use a repulsive potential simulating Pauli principle and adjusting the Q-value. The most favourable fissioning configuration is the pole-to-pole (p–p) one, with $\Omega_1 = \Omega_2 = \hat{R}$, where the Coulomb barrier has the lowest possible value. It is given in Fig. 6.15 by a solid line for the fission process ^{252}Cf $\rightarrow$ ^{104}Mo + ^{148}Ba. From this figure, it becomes clear that this configuration gives a much less barrier than for instance the monopole component, given by a dot-dashed line. For comparison in the same figure the pure Coulomb potential is drawn by a dashed line.

If the rotational states of fragments belong to the ground band, then $\Phi_{JM} = Y_{JM}$ and the wave function is given by a similar superposition, i.e.

$$\Psi(\Omega_1,\Omega_2,\mathbf{R}) = \frac{1}{R}\sum_{J_1J_2l} f_{J_1J_2l}(R)\mathscr{Y}_{J_1J_2l}(\Omega_1,\Omega_2,\hat{R}), \tag{6.43}$$

It is interesting to derive the expression of the angular function in the so-called "molecular" intrinsic system of coordinates, defined by the Euler angle $\hat{r}$ as follows

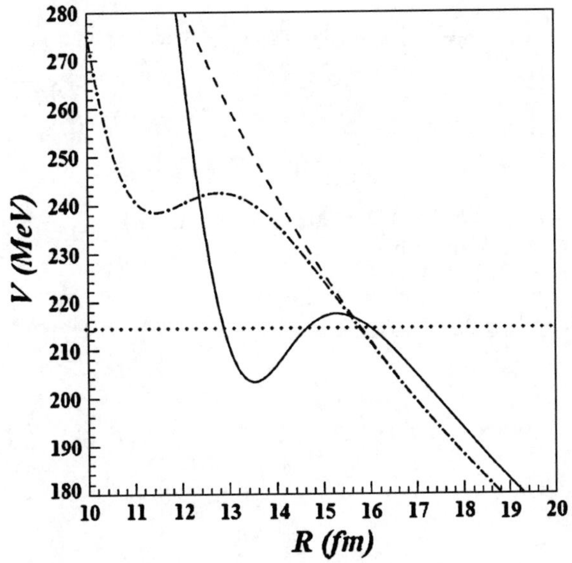

Fig. 6.15 The potential for the pole–pole configuration (*solid line*) in the cold fission process ^{252}Cf $\rightarrow$ ^{104}Mo + ^{148}Ba. The monopole component of the folding potential (*dot-dashed line*), the Coulomb potential (*dashed line*) and the Q-value of the process (*dotted line*)

$$\Omega'_k = \hat{R}^{-1}\Omega_k, \quad k = 1,2. \tag{6.44}$$

One obtains

$$\mathcal{Y}_{J_1J_2l}(\Omega'_1,\Omega'_2,\hat{R}) = \frac{(-)^l}{\sqrt{4\pi}}\left[Y_{J_1}(\Omega'_1)\otimes Y_{J_2}(\Omega'_2)\right]_{l0}. \tag{6.45}$$

This expression can also be obtained by replacing $\hat{R}=0$ in Eq. 6.42. Thus, the interaction (6.41), as well as the wave function (6.43) does not depend on the whole orientation of the system in space, given by Ω.

The wave functions describing the intrinsic rotations, satisfy the following eigenvalue equations

$$\mathbf{H}_k Y_{J_kM_k}(\Omega_k) = E_{J_k} Y_{J_kM_k}(\Omega_k), \quad k = 1,2. \tag{6.46}$$

By using the orthonormality of angular functions entering superposition (6.43), one obtains in a standard way the coupled system of differential equations for radial components (3.15) in Chap. 3, i.e.

$$\frac{d^2 f_{J_1J_2l}(R)}{d\rho_{12}^2} = \sum_{J'_1J'_2l'} \mathcal{A}_{J_1J_2l;J'_1J'_2l'}(R) f_{J'_1J'_2l'}(R), \tag{6.47}$$

where the coupling matrix is given by (3.16) in Chap. 3, i.e.

$$\begin{aligned}\mathcal{A}_{J_1J_2l;J'_1J'_2l'}(R) = &\left[\frac{l(l+1)}{\rho_{12}^2} + \frac{V_0(R)}{E_{12}} - 1\right]\delta_{ll'}\delta_{J_1J'_1}\delta_{J_2J'_2}\\ &+\frac{1}{E_{12}}\langle\mathcal{Y}_{J_1J_2l}|V_d(R)|\mathcal{Y}_{J'_1J'_2l'}\rangle.\end{aligned} \tag{6.48}$$

Here we introduced the usual short-hand notations

$$\begin{aligned}\rho_{12} &= \kappa_{12}r, \quad \kappa_{12} = \sqrt{\frac{2\mu E_{12}}{\hbar^2}},\\ E_{12} &= E - E_{J_1} - E_{J_2},\end{aligned} \tag{6.49}$$

where E_{J_1}, E_{J_2} are the internal energies of the emitted fragments. The matrix element is given by

$$\begin{aligned}\langle\mathcal{Y}_{J_1J_2l}|V_d(R)|\mathcal{Y}_{J'_1J'_2l'}\rangle = &\sum_{\lambda_1\lambda_2\lambda_3} V_{\lambda_1\lambda_2\lambda_3}(R)(-)^{\lambda_1+\lambda_2-\lambda_3+l}\frac{\hat{J}'_1\hat{J}'_2\hat{l}\hat{l}'\hat{\lambda}_1\hat{\lambda}_2}{(4\pi)^{3/2}}\\ &\times\langle l,0;l',0|\lambda_3,0\rangle\langle J'_1,0;\lambda_1,0|J_1,0\rangle\langle J'_2,0;\lambda_2,0|J_2,0\rangle\begin{Bmatrix} J_1 & J_2 & l\\ J'_1 & J'_2 & l'\\ \lambda_1 & \lambda_2 & \lambda_3\end{Bmatrix},\end{aligned} \tag{6.50}$$

where the bracket denotes Clebsch–Gordan recoupling coefficient, curly bracket 9-j symbol and $\hat{J} \equiv \sqrt{2J+1}$.

The double fine structure is given by the partial decay widths $\Gamma_{J_1J_2l}$, entering the total decay width

$$\Gamma = \sum_{J_1J_2l} \Gamma_{J_1J_2l} = \sum_{l} \hbar v_{J_1J_2l} \lim_{R\to\infty} |f_{J_1J_2l}(R)|^2 = \sum_{J_1J_2l} \hbar v_{J_1J_2l} |N_{J_1J_2l}|^2, \qquad (6.51)$$

where $v_{J_1J_2l}$ is the center of mass velocity at infinity in the channel $c = (J_1, J_2l)$, i.e.

$$v_{J_1J_2l} = \frac{\hbar\kappa_{12}}{\mu}. \qquad (6.52)$$

We investigate the following cold fission process

$$^{252}\text{Cf} \to {}^{104}\text{Mo} + {}^{148}\text{Ba}. \qquad (6.53)$$

The emitted fragments are neutron rich unstable nuclei. Their deformations can be determined from electromagnetic transitions, but these values are not available at this moment. Therefore we choose ground state deformations from the systematics given in Ref. [21]. The corresponding values are $\beta_2^{(1)} = 0.349, \beta_4^{(1)} = 0.030$ for ^{104}Mo and $\beta_2^{(2)} = 0.236, \beta_4^{(2)} = 0.131$ for ^{148}Ba. We mention that in Ref. [73] it was clearly pointed out the important role played by hexadecapole deformations on the penetration process.

The most favourable fissioning configuration, where the Coulomb barrier has the lowest possible value, is the pole-to-pole one, with $\Omega_1 = \Omega_2 = \Omega$, i.e.

$$V_{p-p}(R) = V(\Omega, \Omega, R, \Omega). \qquad (6.54)$$

As we pointed out, the above expression does not depend upon Ω. Before the scission point, this potential rapidly increases around this configuration in the direction of angular variables [75]. By approaching the scission point, the potential becomes gradually flatter, and for large distances where the fragments are separated the interaction is given only by the Coulomb term. In this region, the two nuclei are left in excited rotational states. Thus, in our calculation we consider the pole-to-pole interaction $V_{p-p}(R)$ as a spherical component, until its intersection with the true spherical part $V_0(R)$, as in Fig. 6.15. Beyond this point, which is close to the touching configuration, $V_0(R)$ becomes energetically more favourable and the two fragments start to rotate separately. The energy of the resonant state is given by the Q-value, i.e. $E = 214.67$ MeV. We adjust it, by using the strength of a repulsive core, like in Ref. [72]. This is a standard procedure used to describe α or heavy cluster decays. It is important to point out that $\Gamma/Q < 10^{-4}$, and therefore the roots of the system, giving the resonant energies are practically real numbers. It is import to stress on the fact that this pocket-like potential corresponds to the second minimum of the double humped potential in standard nuclear fission [77]. This shape of the potential is confirmed by Two Center Shell Model calculations [78].

We investigate the role of the nuclear shape on the double fine structure of fragments. The nuclear densities of fragments are parametrized in the intrinsic system of coordinates by deformed Fermi distributions as follows

$$
\begin{aligned}
\rho^{(k)}(\mathbf{r}_k) &= \frac{\rho_0^{(k)}}{1+\exp[(r_k - c^{(k)})/a^{(k)}]}, \quad k = 1,2, \\
c^{(k)} &= c_0^{(k)} \left[1 + \sum_{\lambda \geq 2} \beta_\lambda^{(k)} Y_{\lambda 0}(\Omega_k') \right],
\end{aligned}
\tag{6.55}
$$

where the nuclear radius is given by

$$
\begin{aligned}
c_0^{(k)} &= (r_0 + w^{(k)}) A^{1/3}, \\
r_0 &= (d - 1/d) A^{-1/3}, \quad d = 1.28 A^{1/3} + 0.8 A^{-1/3} - 0.76.
\end{aligned}
\tag{6.56}
$$

Here, we dropped the isospin index. For $w^{(k)} = 0$, one obtains the standard liquid drop expression. A special role is played by the neutron density shape, due to the large excess of neutrons in both fragments. For proton densities, we consider a standard diffusivity $a_p^{(k)} = 0.5$ fm and skin parameter $w_p^{(k)} = 0$. The only free parameters, according to (6.55) remain neutron diffusivity $a_n^{(k)}$ and skin parameter $w_n^{(k)}$.

Firstly we studied the influence of the neutron diffusivity on the decay widths by considering $w_n^{(k)} = 0$. Our calculation have shown that partial decay widths remain practically unchanged for a constant value of the sum $a_n^{(1)} + a_n^{(2)}$. We will present our results for a common value $a_n \equiv a_n^{(1)} = a_n^{(2)}$.

It turns out that a small increase of this parameter from $a_n = 0.50$ fm up to $a_n = 0.53$ fm changes the behaviour of radial components beyond the external turning point. The radial probabilities in this region are proportional with the partial decay widths. As a criterium selecting the resonant state we use the angular distribution of fragments around the pole-to-pole configuration. The first resonance inside the pocket-like potential has a much narrower distribution than higher states and therefore it is the best candidate describing the fission process. The 15 radial components of the wave function in Table 6.3 $f_l(R)$, $l \to (l, I_1, I_2)$ are plotted in Fig. 6.16a for $a_n = 0.50$ fm and in Fig. 6.16b for $a_n = 0.53$ fm. They are mainly concentrated in the internal part of the potential and have zero nodes in this region. One can clearly see the increase of the oscillatory tail in the case (b) with respect to (a).

We expect an important dependence of partial rotational widths upon the neutron diffusivity. To this purpose we investigate partial yields for each fragment, defined as follows

Table 6.3 The rotational basis (I_1, I_2, l)

No.	I_1	I_2	l
1	0	0	0
2	2	2	0
3	4	4	0
4	2	0	2
5	0	2	2
6	2	2	2
7	4	4	2
8	4	2	2
9	2	4	2
10	4	0	4
11	0	4	4
12	2	2	4
13	4	4	4
14	4	2	4
15	2	4	4

$$
\begin{aligned}
y_0^{(k)} &= 100\left[\Gamma_0^{(k)} + \Gamma_2^{(k)} + \Gamma_4^{(k)}\right]/\Gamma, \quad k = 1,2, \\
y_2^{(k)} &= 100\left[\Gamma_2^{(k)} + \Gamma_4^{(k)}\right]/\Gamma, \\
y_4^{(k)} &= 100\Gamma_4^{(k)}/\Gamma,
\end{aligned}
\tag{6.57}
$$

where the decay widths for each fragment are respectively given by

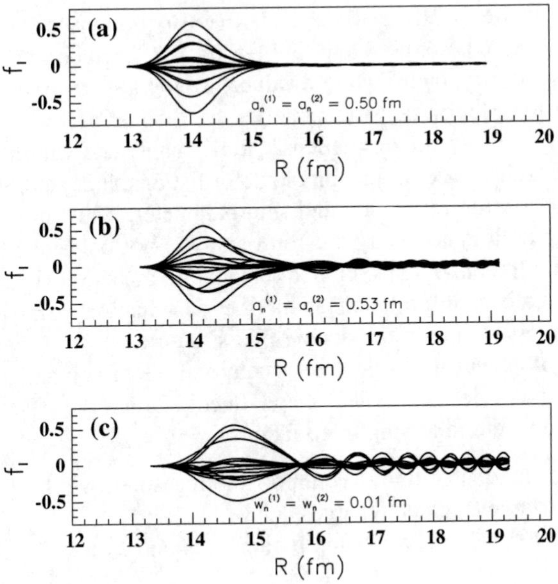

Fig. 6.16 The radial wave function components $f_l(R)$ for neutron density parameters in Eq. 6.55 $a_n = 0.50$ fm, $w_n = 0$ (**a**), $a_n = 0.53$ fm, $w_n = 0$ (**b**) and $w_n = 0.01$ fm, $a_n = 0.5$ fm (**c**). The basis states are given in Table 6.3. The fission process is $^{252}\text{Cf} \rightarrow {}^{104}\text{Mo} + {}^{148}\text{Ba}$

$$\Gamma_I^{(1)} = \sum_{II'} \Gamma_{II'}, \quad \Gamma_I^{(2)} = \sum_{II'} \Gamma_{II'I}. \tag{6.58}$$

Here the upper index (1) denotes ^{104}Mo, while (2) ^{148}Ba. The summed yields (6.57) can be measured by using the triple γ-rays coincidence, except $y_0^{(k)}$ measured by charge detectors. In Fig. 6.17a the summed yields/fragment (6.57) as a function of a_n are given. By solid lines we give the values of ^{104}Mo and by dashes of ^{148}Ba for different angular momenta.

At this moment only the following relative yields can be measured

$$\gamma_4^{(k)} = 100\frac{y_4^{(k)}}{y_2^{(k)}}, \quad k = 1, 2. \tag{6.59}$$

The dependence of relative hexadecapole yields versus a_n is given in Fig. 6.17b. From comparable values their ratio increases by three times over the investigated interval. It is important to point out that by changing neutron diffusivity the relative hexadecapole yields approach the experimental values, namely $\gamma_4^{(1)} = 80 \pm 20, \gamma_4^{(2)} \le 15$ [54].

Secondly, let us investigate the influence of the neutron skin parameter $w_n = w_n^{(1)} = w_n^{(2)}$ by keeping a constant diffusivity $a_n = 0.5$ fm. The results remain again qualitatively unchanged for a constant sum of fragment skin parameters. We change this parameter from 0 up to 0.01 fm. This corresponds to an increase of the neutron radius by $0.05\,\text{fm} \approx 0.01\,A^{1/3}$. In Fig. 6.16c, we give the radial components for $w_n = 0.01$ fm. The oscillatory tails are even larger in this case. In Fig. 6.17c, the yields (6.57) are given as functions of w_n. From Fig. 6.17d, one can see a strong dependence of the relative yields (6.59) upon the skin parameter.

In Ref. [79], relativistic Hartree–Bogoliubov calculations for Sn isotopes are performed, i.e. between the investigated fragments. For a similar proton–neutron asymmetry, by using these values, we found a larger diffusivity and skin parameter of the neutron density, namely $a_n \approx 0.55\,\text{fm}$ and $w_n \approx 0.04\,\text{fm}$, respectively. For the considered deformations, the resonant state vanishes at these values, but by a small decrease of quadrupole deformations, it is possible to obtain these values of the neutron diffusivity and skin parameter. Thus, in order to measure the density profile it is necessary a careful analysis not only the absolute yields (6.57) of the cold fission, but also of electromagnetic transitions in each rotational band, giving a realistic information about the deformation parameters, which also play an important role in the double fine structure.

It is known that the fine structure of the proton and α-emission is very sensitive to the nuclear mean field shape. In cold fission the effect is even more pronounced, due to the following three factors:

(1) the double fine structure of the measured widths,
(2) large fragment charges,
(3) a relative low Coulomb barrier, about 3 MeV.

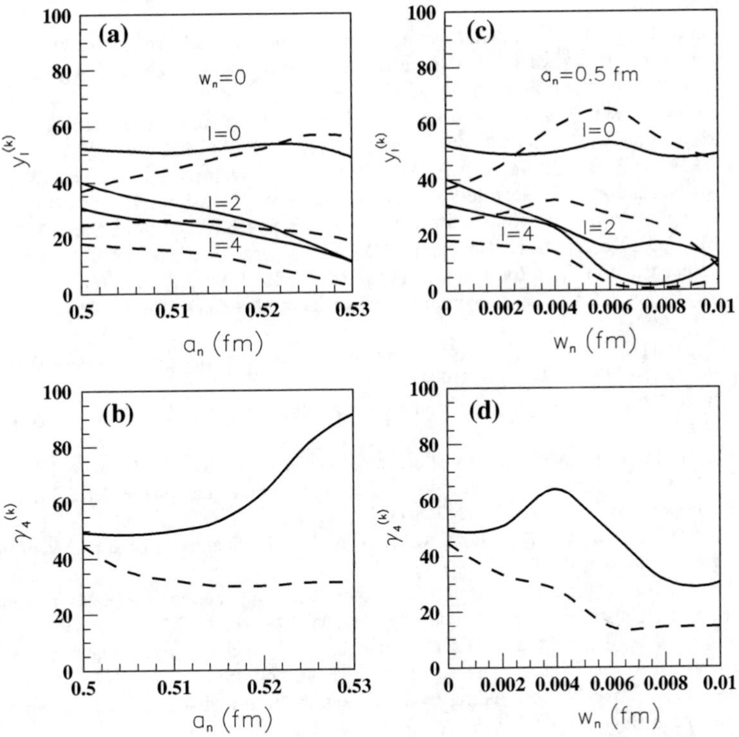

Fig. 6.17 **a** The yields/fragment $y_l^{(k)}$ defined by Eq. (6.57) versus the neutron diffusivity parameter $a_n \equiv a_n^{(k)}$. The skin parameters are $w_p^{(k)} = w_n^{(k)} = 0$. By *solid lines* the yields of the ^{106}Mo ($k = 1$) are given and by *dashes* those of ^{146}Ba ($k = 2$). **b** The relative hexadecapole yields $\gamma_4^{(k)}$ defined by Eq. 6.59 versus the diffusivity parameter a_n. The skin parameters are $w_p^{(k)} = w_n^{(k)} = 0$. The *solid line* corresponds to ^{106}Mo ($k = 1$) and *dashed line* to ^{146}Ba ($k = 2$). **c** The same as in **a** versus the skin parameter $w_n \equiv w_n^{(k)}$ for $a_n = 0.5$ fm. **d** The same as in **b** versus the skin parameter w_n for $a_n = 0.5$ fm

References

1. Akovali, Y.A.: Review of alpha-decay data from double-even nuclei. Nucl. Data Sheets **84**, 1–114 (1998)
2. Delion, D.S., Peltonen, S., Suhonen, J.: Systematics of the α-decay to rotational states. Phys. Rev. C **73**, 014315/1–10 (2006)
3. Rasmussen, J.O.: Alpha-decay barrier penetrabilities with an exponential nuclear potential: even–even nuclei. Phys. Rev. **113**, 1593–1598 (1959)
4. Xu, C., Ren, Z.: α Transitions to coexisting 0^+ states in Pb and Po isotopes. Phys. Rev. C **75**, 044301/1–5 (2007)
5. Wang, Y.Z., Zhang, H.F., Dong, J.M., Royer, G.: Branching ratios of α decay to excited states in even-even nuclei. Phys. Rev. C **79**,014316/1–5 (2009)
6. Rasmussen, J.O., Segal, B.: Alpha decay of spheroidal nuclei. Phys. Rev. **103**, 1298–1308 (1956)

7. Chasman, R.R., Rasmussen, J.O.: Alpha decay of deformed even–even nuclei. Phys. Rev. **112**, 512–518 (1958)
8. Săndulescu, A., Iosifescu, M.: On favoured alpha transitions. Nucl. Phys. **26**, 209–216 (1961)
9. Rauscher, E.A., Rasmussen, J.O., Harada, K.: Coupled-channel, alpha decay rate theory applied to ^{212m}Po. Nucl. Phys. A **94**, 33–51 (1967)
10. Soinski, A.J., Rasmussen, J.O., Rausher, E.A., Raich, D.G.: Coupled-channel α-decay theory for odd-mass nuclei: ^{283}Es and ^{255}Fm. Nucl. Phys. A **291**, 386–400 (1977)
11. Radi, H.M.A., Shihab-Eldin, A.A., Rasmussen, J.O., Oliveira L.F.: Relation of α-decay rotational signatures to nuclear deformation changes. Phys. Rev. Lett. **41**, 1444–1446 (1978)
12. Fröman, P.P.: Alpha decay from deformed nuclei. Mat. Fys. Skr. Dan. Vid. Selsk. **1**, 3 (1957)
13. Richards, J.D., Berggren, T., Bingham, C.R., Nazarewicz, W., Wauters, J.: α Decay and shape coexistence in the α-rotor model. Phys. Rev. C **56**, 1389–1397 (1997)
14. Neu, R., Hoyler, F.: Isoscalar transition rates from inelastic alpha scattering. Phys. Rev. C **46**, 208–219 (1992)
15. Abele, H., Staudt, G.: α-^{16}O and α-^{15}N Optical potentials in the range between 0 and 150 MeV. Phys. Rev. C **47**, 742–756 (1993)
16. Dao Khoa, T.: α-Nucleus potential in the double-folding model. Phys. Rev. C **63**, 034007/1–15 (2001)
17. Basu, D.N.: Folding model analysis of alpha radioactivity. J. Phys. G **29**, 2079–2085 (2003)
18. Ni, D., Ren, Z.: Systematic calculation of α decay within a generalized density-dependent cluster model. Phys. Rev. C **81**, 024315/1–10 (2010)
19. Gambhir, Y.K., Bhagwat, A., Gupta, M., Jain, A.K.: α Radioactivity of superheavy nuclei. Phys. Rev. C **68**, 044316/1–6 (2003)
20. Bertsch, G., Borysowicz, J., McManus, H., Love, W.G.: Interactions for inelastic scattering derived from realistic potentials. Nucl. Phys. A **284**, 399–419 (1977)
21. Möller, P., Nix, R.J., Myers, W.D., Swiatecki, W.: Nuclear ground-state masses and deformations. At. Data Nucl. Data Tables **59**, 185–381 (1995)
22. Satchler, G.R., Love, W.G.: Folding model potentials from realistic interactions for heavy-ion scattering. Phys. Rep. **55**, 183–254 (1979)
23. Delion, D.S., Săndulescu, A., Greiner, W.: Evidence for α clustering in heavy and superheavy nuclei. Phys. Rev. C **69**, 044318/1–19 (2004)
24. Delion, D.S., Liotta, R.J., Wyss, R.: Theories of proton emission. Phys. Rep. **424**, 113–174 (2006)
25. Săndulescu, A., Dumitrescu, O.: Alpha decay to vibrational states. Phys. Lett. **19**, 404–407 (1965)
26. Săndulescu, A., Dumitrescu, O.: Alpha decay and the structure of β-vibrational states. Phys. Lett. B **24**, 212–216 (1967)
27. Cristu, M.I., Dumitrescu, O., Pyatov, N.I., Săndulescu, A.: Alpha decay and the structure of the $K^{\pi} = 0^{+}$ states in the Th–U region. Nucl. Phys. A **130**, 31–40 (1969)
28. Delion, D.S., Florescu, A., Huise, M., Wauters, J., Van Duppen, P., ISOLDE Collaboration, Insolia, A., Liotta, R.J.: Microscopic description of alpha decay to intruder 0^{+} states in Pb, Po, Hg and Pt isotopes. Phys. Rev. Lett. **74**, 3939–3942 (1995)
29. Delion, D.S., Florescu, A., Huise, M., Wauters, J., Van Duppen, P., ISOLDE Collaboration, Insolia, A., Liotta, R.J.: Alpha decay as a probe for phase transitions in nuclei. Phys. Rev. C **54**, 1169–1176 (1996)
30. Karlgren, D., Liotta, R.J., Wyss, R., Huyse, M., Van de Vel, K., Van Duppen, P.: α-Decay hindrance factors: a probe of mean field wave functions. Phys. Rev. C **73**, 064304/1–10 (2006)
31. Wauters, J., et al.: The alpha-branching ratios of the ^{188,190,192}Pb isotopes. Z. Phys. A **342**, 277–282 (1992)
32. Wauters, J., et al.: Fine structure in the α decay of ^{202}Rn. Observation of a low-lying state in ^{198}Po. Z. Phys. A **344**, 29–33 (1992)
33. Wauters, J., et al.: Alpha-decay study of ^{188}Pb and $^{180-182}$Hg. Z. Phys. A **345**, 21–27 (1993)

34. Wauters, J., et al.: α Decay properties of neutron-defficient Polonium and Radon nuclei. Phys. Rev. C **47**, 1447–1454 (1993)
35. Wauters, J., et al.: Alpha decay of ^{186}Pb and ^{184}Hg: the influence of mising of 0^+ states on α-decay transition probabilities. Phys. Rev. C **50**, 2768–2773 (1994)
36. Wauters, J., et al.: Fine structure in the alpha decay of even–even nuclei as an experimental proof for the stability of the $Z = 82$ magic shell at the very-deficient neutron side. Phys. Rev. Lett. **72**, 1329–1332 (1994)
37. Bijnens, N., et al.: Intruder states and the onset of deformation in the neutron-deficient even–even polonium isotopes. Phys. Rev. Lett. **75**, 4571–4574 (1995)
38. Allatt, R.G., et al.: Fine Structure in ^{192}Po α decay and shape coexistence in ^{188}Pb. Phys. Lett. B **437**, 29–34 (1998)
39. Liang, C.F., et al.: Alpha decay of ^{231}U to levels in ^{227}Th. Phys. Rev. C **49**, 2230–2232 (1994)
40. Delion, D.S., Insolia, A., Liotta, R.J.: Pairing correlations and quadrupole deformation effects on the ^{14}C decay. Phys. Rev. Lett. **78** 4549–4552 (1997)
41. Delion, D.S., Săndulescu, A., Mişicu, S., Cârstoiu, F., Greiner, W.: Quasimolecular resonances in the binary cold fission of ^{252}Cf. Phys. Rev. C **64**, 041303(R)/1–5 (2001)
42. Peltonen, S., Delion, D.S., Suhonen, J.: Systematics of the alpha decay to vibrational states. Phys. Rev. C **71**, 044315/1–9 (2005)
43. Peltonen, S., Delion, D.S., Suhonen, J.: Folding description of the fine structure of alpha decay to 2^+ vibrational and transitional states. Phys. Rev. C **75**, 054301/1–9 (2007)
44. Carstoiu, F., Lombard, R.J.: A new method of evaluating folding type integrals. Ann. Phys. (N.Y.) **217**, 279–303 (1992)
45. Delion, D.S., Suhonen, J.: Microscopic anharmonic vibrator approach for beta decays. Nucl. Phys. A **781**, 88–103 (2007)
46. Mohr, P.: α-Nucleus potentials, α-decayhalf-lives, and shell closures for superheavy nuclei. Phys. Rev. C **73**, 031301(R)/1–5 (2006)
47. Lotti, P., Maglione, E., Catara, F., Insolia, A.: Surface clustering and two-nucleon pick-up in Samarium Isotopes. Europhys. Lett. **6**, 125–129 (1988)
48. Hagino, K.: Role of dynamical particle-core vibration coupling in reconciliation of the $d_{3/2}$ puzzle for spherical proton emitters. Phys. Rev. C **64**, 041304(R)/1–5 (2001)
49. Davids, C.N., Esbensen, H.: Particle-vibration coupling in proton decay of near-spherical nuclei. Phys. Rev. C **64**, 034317/1–7 (2001)
50. Semmens, P.B.: Proton emission from spherical and near-spherical nuclei. Nucl. Phys. A **682**, 239c–246c (2001)
51. Esbensen, H., Davids, C.N.: Coupled-channels treatment of deformed proton emitters. Phys. Rev. C **63**, 014315/1–13 (2000)
52. Ferreira, L.S., Maglione, E.: ^{151}Lu: Spherical or deformed? Phys. Rev. C **61**, 021304(R)/1–3 (2000)
53. Ter-Akopian, G.M., et al.: Neutron multiplicities and yields of correlated Zr–Ce and Mo–Ba fragment pairs in spontaneous fission of ^{252}Cf. Phys. Rev. Lett. **73**, 1477–1480 (1994)
54. Săndulescu, A., Florescu, A., Carstoiu, F., Greiner, W., Hamilton, J.H., Ramayya, A.V., Babu, B.R.S.: Isotopic yields in the cold fission of ^{252}Cf. Phys. Rev. C **54**, 258–265 (1996)
55. Wu, S.-C., et al.: New determination of the Ba–Mo yield matrix for ^{252}Cf. Phys. Rev. C **62**, 041601(R)/1–4 (2000)
56. Ramayya, A.V., et al.: Binary and ternary fission studies with ^{252}Cf. Progr. Part. Nucl. Phys. **46**, 221–229 (2001)
57. Jandel, M., et al.: Gamma-ray multiplicity distribution in ternary fission of ^{252}Cf. J. Phys. G **28**, 2893–2905 (2002)
58. Hamilton, J.H., et al.: Yad. Fiz. **65**, 677 (2002)
59. Hamilton, J.H., et al.: Phys. At. Nucl. **65**, 695 (2002)
60. Săndulescu, A., Greiner, W.: J. Phys. G **3**, L189 (1977)
61. Săndulescu, A., Greiner, W.: Cluster decays. Rep. Prog. Phys. **55**, 1423–1481 (1992)

62. Möller, P., Nix, J.R.: Macroscopic poetential-energy surfaces for symmetric fission and heavy-ion reactions. Nucl. Phys. A **272**, 502–532 (1976)
63. Cârjan, N., Sierk, A.J., Nix, J.R.: Effect of dissipation on ternary fission in very heavy nuclear systems. Nucl. Phys. A **452**, 381–397 (1986)
64. Mignen, J., Royer, G.: A geometric model for ternary fission. J. Phys. G **16** L227–L232 (1990)
65. Gönnenwein, F., Börsig, B.: The model of cold fission. Nucl. Phys. A **530**, 27–57 (1991)
66. Knitter, H.-H., Hambsch, F.-J., Buditz-Jorgensen, C.: Nuclear mass and charge distribution in the cold region of the spontaneous fission of ^{252}Cf. Nucl. Phys. A **536**, 221–259 (1992)
67. Royer, G., Mignen, J.: Binary and ternary fission of hot rotating nuclei. J. Phys. G **18**, 1781–1792 (1992)
68. Royer, G., Haddad, F., Mignen, J.: On nuclear ternary fission. J. Phys. G **18**, 2015–2026 (1992)
69. Săndulescu, A., Mişicu, S., Carstoiu, F., Greiner, W.: Cold fission modes in ^{252}Cf. Fiz. Elem. Chastits Az. Yadra **30**, 908–953 (1999)
70. Săndulescu, A., Mişicu, S., Carstoiu, F., Greiner, W.: Phys. Part. Nucl. **30**, 386 (1999)
71. Săndulescu, A., Carstoiu, F., Bulboacă, I., Greiner, W.: Cluster decsription of cold (neutronless) α ternary fission of ^{252}Cf. Phys. Rev. C **60**, 044613/1–13 (1999)
72. Carstoiu, F., Bulboacă, I., Săndulescu, A., Greiner, W.: Half-lives of trinuclear molecules. Phys. Rev. C **61**, 044606/1–7 (2000)
73. Săndulescu, A., Mişicu, Ş., Carstoiu, F., Florescu, A., Greiner, W.: Role of the higher static deformations of fragments in the cold binary fission of ^{252}Cf. Phys. Rev. C **58**, 2321–2328 (1998)
74. Delion, D.S., Săndulescu, A., Greiner W.: Probing mean field of neutron rich nuclei by cold fission. Phys. Rev. C **68**, 041303(R)/1–5 (2003)
75. Mişicu, Ş., Săndulescu, A., Greiner, W.: Coupling between fragment radial motion and the transversal degrees of freedom in cold fission. Phys. Rev. C **64**, 044610/1–10 (2001)
76. Delion, D.S., Săndulescu, A., Mişicu, Ş., Cârstoiu, F., Greiner, W.: Double fine structure in binary cold fission. J. Phys. G **28**, 289–306 (2009)
77. Specht, H.J.: Nuclear fission. Rev. Mod. Phys. **46**, 773–787 (1974)
78. Mirea, M., Delion, D.S., Săndulescu, A.: Microscopic cold fission yields of ^{252}Cf. Phys. Rev. C **81**, 044317/1–4 (2010)
79. Mizutori, S., Dobaczewski, J., Lalazissis, G.A., Nazarewicz, W., Reinhard, P.-G.: Nuclear skins and halos in the mean field theory. Phys. Rev. C **61**, 044326/1–14 (2000)

Chapter 7
Ternary Emission Processes

7.1 Coupled Channels Equations for Two Proton Emission

Binary decay processes are possible when the final fragments are more bound than the initial nucleus. It turns out that, energetically, it is possible that not only two, but also three or more fragments can be simultaneously emitted from the ground state of the parent nucleus. Such process is called ternary emission and it is of the following form:

$$P(J_iM_i) \to D_1(J_1M_1) + D_2(J_2M_2) + D_3(J_3M_3). \qquad (7.1)$$

As in the case of binary processes, it is allowed if the Q-value is positive, i.e.,

$$Q = B_i - B_1 - B_2 - B_3 > 0. \qquad (7.2)$$

There are mainly two kinds of ternary emission processes, namely (1) two proton emission and (2) cold ternary fission.

Firstly, we will analyze the two proton emission.

Two proton emission is a process in which the third fragment (3) is the heaviest one, while the system of the other two protons (1, 2) is the lightest one. For this reason, this process has a true three body character at variance with the ternary fission, where the Born–Oppenheimer adiabatic approach is a proper framework to treat the problem.

Decay process in which two protons are emitted is a very exotic event, but it is energetically possible for some nuclei. In the earlier 1960s Goldanski [1] proposed two extreme mechanisms, in which the particles are emitted either simultaneously or sequentially. Due to different Coulomb barriers the penetrability in the two cases leads to decay widths differing by several orders of magnitude. A systematic theoretical analysis of the processes involving the three body problem was performed in Ref. [2] for the first time and later on in [3], but the interest in this subject increased especially in the recent years.

In a series of references [4–8] it was shown that the analysis of this exotic decay can reveal details about two-body correlations on the nuclear surface. By using the

D. S. Delion, *Theory of Particle and Cluster Emission*, Lecture Notes in Physics, 819,
DOI: 10.1007/978-3-642-14406-6_7, © Springer-Verlag Berlin Heidelberg 2010

simple idea of Ref. [2] that the sequential two proton decay width is a superposition of one proton width multiplied by the density of the two proton distribution, the emission from ^{12}O [9] was investigated within the R-matrix approach in a series of papers [10, 11]. By using this method, the two proton emission from ^{18}Ne [12] was analyzed in Ref. [13] and from ^{45}Fe in Ref. [14].

In a series of recent papers a more sophisticated technique based on the Feshbach reaction theory and the shell model in continuum was used to analyze different nuclei emitting two protons [15, 16] and the results were systematized in the review paper [17].

The old dilemma about proton emission: is it a simultaneous diproton or a sequential mechanism was not clarified yet. In Ref. [5] it is given a comparison between these two mechanisms of two proton emission, in terms of the so-called T and Y Jacobi systems of coordinates as in Fig. 7.1. They are defined in terms of hyperspherical coordinates [18] for the partition $core + p + p$, as follows:

$$\begin{aligned}
\theta_T &= \arctan\left[\sqrt{\frac{A_{\text{core}}+2}{4A_{\text{core}}}}\frac{X_T}{Y_T}\right],\\
\rho^2 &= \frac{1}{2}X_T^2 + \frac{2A_{\text{core}}}{A_{\text{core}}+2}Y_T^2,\\
\theta_Y &= \arctan\left[\sqrt{\frac{A_{\text{core}}(A_{\text{core}}+2)}{(A_{\text{core}}+1)^2}}\frac{X_Y}{Y_Y}\right],\\
\rho^2 &= \frac{A_{\text{core}}}{A_{\text{core}}+1}X_Y^2 + \frac{A_{\text{core}}+1}{A_{\text{core}}+2}Y_Y^2,
\end{aligned} \tag{7.3}$$

where $\theta_{T,Y}$ is the hyperangle and ρ the permutationally invariant hyper-radius (notice that it plays the role of radius and not reduced radius).

The T system of coordinates describes emission of a correlated di-proton object and this idea is connected with pairing correlations which occur between two protons at the nuclear surface. The Y system of coordinates corresponds to the emission of one proton from the core-proton correlated system, i.e., to the sequential decay. We expand the wave function in terms of hyperharmonics:

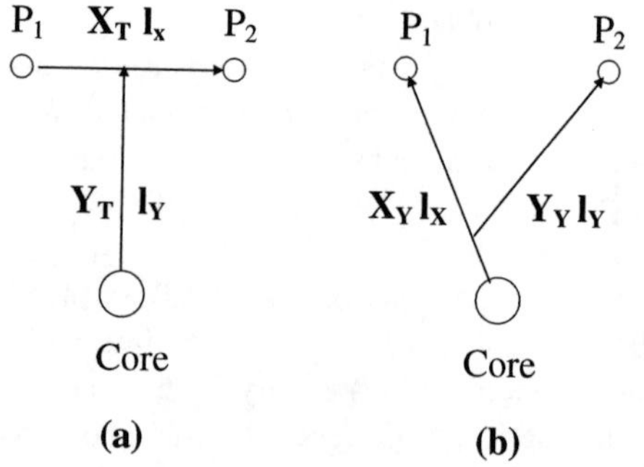

Fig. 7.1 **a** T system of coordinates, **b** Y system of coordinates

$$\Psi(\mathbf{X},\mathbf{Y}) = \Psi(\rho,\Omega_\rho) = \frac{1}{\rho^{5/2}}\sum_{K\gamma} f_{K\gamma}(\rho)\mathscr{I}_{K\gamma}(\Omega_\rho), \tag{7.4}$$

where $\mathscr{I}_{K\gamma}(\Omega_\rho)$ is a set of ortho-normal angular functions on the hypersphere $\Omega_\rho = (\theta, \Omega_X, \Omega_Y)$. The hypermomentum is denoted by K, while the other quantum numbers by γ. By projecting out a given hyper-channel, one obtains the system of coupled equations for hyper-radial components as follows:

$$\begin{aligned} \frac{d^2}{d\rho^2} - \frac{\mathscr{L}(\mathscr{L}+1)}{\rho^2} + \frac{2M}{\hbar^2}[E - \langle K\gamma|V(\rho)|K\gamma\rangle]f_{K\gamma} \\ = \frac{2M}{\hbar^2}\sum_{K'\gamma'}\langle K\gamma|V(\rho)|K'\gamma'\rangle f_{K'\gamma'}, \end{aligned} \tag{7.5}$$

where M is the scaling nucleonic mass, $\mathscr{L} = K + 3/2$ and

$$\langle K\gamma|V(\rho)|K'\gamma'\rangle \equiv V_{K\gamma,K'\gamma'} = \int d\Omega_\rho \mathscr{I}^\dagger_{K\gamma}(\Omega_\rho)\sum_{i>j}\hat{V}(\mathbf{r}_{ij})\mathscr{I}_{K'\gamma'}(\Omega_\rho). \tag{7.6}$$

The decay width can be computed by using a relation similar to the distorted wave one (4.42), i.e.,

$$\Gamma = \frac{4}{\hbar v}\sum_{K\gamma}\left|\sum_{K'\gamma'}\sum_{K''\gamma''}\int_0^{\rho_{\max}} d\rho f^{(R)}_{K\gamma,K'\gamma'}(\rho)\left[V_{K'\gamma',K''\gamma''} - V^{(s)}_{K'\gamma',K''\gamma''}\right]f^{(\mathrm{box})}_{K''\gamma''}(\rho)\right|^2, \tag{7.7}$$

where $f^{(\mathrm{box})}{}_{K\gamma}(\rho)$ are the radial hyperspherical components of the so-called "box" wave function which is a good approximation of the resonant wave function in the interval $[0, \rho_{\max}]$, i.e.,

$$\left(\hat{H} - E\right)\Psi_{\mathrm{box}}(\rho,\Omega_\rho) = 0, \tag{7.8}$$

calculated with zero boundary conditions on both sides. A good approximation of $f^{(R)}$ and $V^{(s)}$ is given by the diagonal approximation of the three-body Coulomb field, i.e.,

$$\begin{aligned} f^{(R)}_{K\gamma,K'\gamma'} &= \delta_{K\gamma,K'\gamma'}F_{\mathscr{L}}(\chi_{K\gamma},\kappa\rho), \quad \mathscr{L} = K + 3/2, \\ V^{(s)}_{K\gamma,K'\gamma'} &= \frac{\rho_{\max}}{\rho}V^{(\mathrm{Coul})}_{K\gamma,K\gamma}(\rho_{\max}). \end{aligned} \tag{7.9}$$

In Refs. [7, 8], it is shown that the decay width of the three-body problem, in analogy with the two-body case, can be expressed in an equivalent form, derived from the continuity equation (2.77) for the hypersphere of the radius $\rho_{\max}$, i.e.,

$$\Gamma = \frac{\hbar\oint d\Omega_\rho \mathscr{J}(\rho_{\max},\Omega_\rho)}{\int_0^{\rho_{\max}}\rho^5 d\rho\oint d\Omega_\rho \mathscr{P}(\rho,\Omega_\rho)}, \tag{7.10}$$

where the probability flux and total probability are respectively given by:

$$\begin{aligned}\mathcal{J}(\rho,\Omega_\rho) &= \frac{\hbar}{M}\Psi^*(\rho,\Omega_\rho)\rho^{5/2}\frac{\partial}{\partial\rho}\rho^{5/2}\Psi(\rho,\Omega_\rho)\\ \mathcal{P}(\rho,\Omega_\rho) &= |\Psi(\rho,\Omega_\rho)|^2.\end{aligned} \tag{7.11}$$

The first estimate of Goldanski [1], gives the angular distribution of two independent protons as a product of two penetrabilities, i.e.,

$$\Gamma(\theta) = 2\frac{6}{\pi r_c M_{\rm red}^{3/2} E_T^{1/2}}\sin\theta_k\cos\theta_k P_l(E_T\sin^2\theta_k, Z_{\rm core}, r_c)P_l(E_T\cos^2\theta_k, Z_{\rm core}, r_c), \tag{7.12}$$

where $\theta_k = \arctan[\sqrt{E_x/E_y}]$ is the hyperangle in momentum space and $M_{\rm red}$ the reduced mass of the core-proton sub-system. It interesting to point out that this very simple estimate gives results in qualitative agreement with the above three-body exact theory [19].

7.2 Coupled Channels Equations for Ternary Fission

Now we will analyze the ternary fission process. Several experimental groups evidenced the ternary fission of ^{252}Cf, where the third emitted light cluster is an α-particle [20], ^{10}Be or ^{14}C [21, 22]. An interesting experiment was performed in Dubna [23], allowing the measurement of the averaged energy and angular distribution of light clusters, emitted in the ternary fission of ^{252}Cf.

In Refs. [24, 25], it is proposed a microscopic treatment of the ternary fission process, using the superposition of preformation factors, calculated for each heavy fragment separately. The computed energy of the α-particle is close to the experimentally observed averaged value. The spread of the energy spectrum can be easily understood in terms of the three-body dynamics.

We will describe the ternary emission process according to Ref. [26]. Let us consider the ternary system of two heavy fragments (1, 2) plus a light cluster (3). We denote the distance between the centers of the heavy fragments by $R = R_1 + R_2$, as in Fig. 7.2.

The light cluster motion is described in terms of the radius $\mathbf{r}$ from the center of mass (cm) of the heavy system. We approximate the cm of the whole ternary system by the cm of the binary 1–2 sub-system. Therefore, the Hamiltonian of the ternary fission process, involving the transition to ground states of the heavy fragments can be written as follows [27]:

$$H = -\frac{\hbar^2}{2\mu_{12}}\nabla_R^2 - \frac{\hbar^2}{2\mu_3}\nabla_r^2 + V^{(12)}(R) + V^{(3)}(R,\mathbf{r}). \tag{7.13}$$

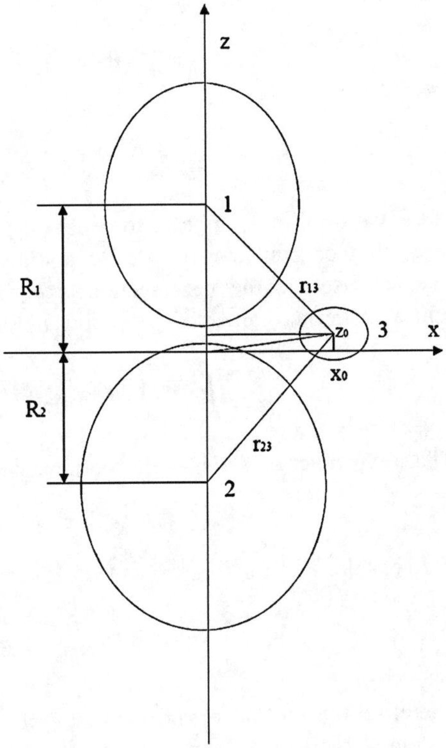

Fig. 7.2 The geometry of the cold ternary fission process

We suppose that the emission of the ternary particle takes place in the final stage, corresponding to the ground state deformations of the heavy fragments, because we investigate the cold fission channel. The strong equatorial emission of light fragments is an experimental argument supporting this hypothesis. Obviously some other shape configurations can be important for a detailed description of the angular distribution.

We also suppose that the system preserves its axial symmetry during the decay process. The total wave function has a vanishing projection of the angular momentum on the symmetry axis of the system. We consider the following "molecular" ansatz, for the wave function:

$$\psi_K(\mathbf{R},\mathbf{r}) = \sum_n \psi_{nK}^{(12)}(\mathbf{R})\psi_{n-K}^{(3)}(R,\mathbf{r}), \tag{7.14}$$

where by n we denote the eigenvalue number and by K the angular momentum projection on the intrinsic axis. The validity of this ansatz is based on the large mass asymmetry between the light cluster and heavy fragments and will be confirmed by our calculations. In the summation we will consider only narrow resonant states, by neglecting the influence of the continuum.

The two factors entering the wave function (7.14) are given respectively by:

$$\begin{aligned}\psi_{nK}^{(12)}(\mathbf{R}) &= \sum_{l} \frac{g_{nlK}^{(12)}(R)}{R} Y_{lK}(\hat{R}),\\ \psi_{nK}^{(3)}(R,\mathbf{r}) &= \sum_{l} \frac{g_{nlK}^{(3)}(R,r)}{r} Y_{lK}(\hat{r}).\end{aligned} \tag{7.15}$$

Here, we used the symbol g to denote the radial wave function in the laboratory system of coordinates in order to avoid any possible confusion with the focal distance used in the next section. For narrow resonances, it makes sense the approximate normalization procedure in the internal region:

$$\sum_{n} \int d\mathbf{R} \int d\mathbf{r} \left[\psi_{nK}^{(12)}(\mathbf{R})\right]^2 \left[\psi_{n-K}^{(3)}(R,\mathbf{r})\right]^2 = 1. \tag{7.16}$$

The wave function satisfies the stationary Schrödinger equation:

$$\mathbf{H}\psi_K(\mathbf{R},\mathbf{r}) = E_K \psi_K(\mathbf{R},\mathbf{r}). \tag{7.17}$$

We consider that for some distance R the motion of the light particle is given by:

$$\left[-\frac{\hbar^2}{2\mu_3}\nabla_r^2 + V^{(3)}(R,\mathbf{r}) - E_{nK}^{(3)}(R)\right]\psi_{nK}^{(3)}(R,\mathbf{r}) = 0. \tag{7.18}$$

Therefore the motion of the two heavy fragments is governed by the following system of equations:

$$\left[-\frac{\hbar^2}{2\mu_{12}}\nabla_R^2 + V^{(12)}(R) + E_{nK}^{(3)}(R) - E_K\right]\psi_{nK}^{(12)}(\mathbf{R}) = \sum_{n'} \mathscr{A}_{n-K,n'-K}(R)\psi_{n'K}^{(12)}(\mathbf{R}), \tag{7.19}$$

where

$$\begin{aligned}\mathscr{A}_{nK,n'K'}(R) &= \frac{\hbar^2}{2\mu_{12}} \int d\mathbf{r}[\psi_{nK}^{(3)*}(R,\mathbf{r})\nabla_R^2\psi_{n'K'}^{(3)}(R,\mathbf{r})\\ &\quad + 2\nabla_R\psi_{nK}^{(3)*}(R,\mathbf{r})\nabla_R\psi_{n'K'}^{(3)}(R,\mathbf{r})]\\ &= \delta_{KK'}\frac{\hbar^2}{2\mu_{12}}\sum_{l}\int_0^{r_m} dr\left[g_{nlK}^{(3)}\frac{d^2}{dR^2}g_{n'lK}^{(3)} + 2\frac{d}{dR}g_{nlK}^{(3)}\frac{d}{dR}g_{n'lK}^{(3)}\right].\end{aligned} \tag{7.20}$$

is called the adiabatic matrix. For a small norm of this matrix one can consider that the motion has an adiabatic character. This approximation is also called Born–Oppenheimer method. We stress on the fact that in (7.19) the potential of the binary system $V^{(12)}(R)$ is modified by the energy of the light particle $E_{nK}^{(3)}(R)$.

By performing standard manipulations one obtains the following system of coupled equations for radial components:

$$\begin{aligned}
&\left[-\frac{\hbar^2}{2\mu_3}\left(\frac{d^2}{dr^2}-\frac{l(l+1)}{r^2}\right)-E^{(3)}_{nK}(R)\right]g^{(3)}_{nlK}(r)+\sum_{l'}V^{(3)}_{lK,l'K}(R,r)g^{(3)}_{nl'K}(R,r)=0,\\
&\left[-\frac{\hbar^2}{2\mu_{12}}\left(\frac{d^2}{dR^2}-\frac{l(l+1)}{R^2}\right)+V^{(12)}(R)+E^{(3)}_{nK}(R)-E_K\right]g^{(12)}_{nlK}(R)\\
&=\sum_{n'}\mathscr{A}_{n-K,n'-K}(R)g^{(12)}_{n'lK}(R),
\end{aligned} \tag{7.21}$$

where

$$\begin{aligned}
V^{(3)}_{lK,l'K}(R,r)&=\langle Y_{lK}(\hat{r})|V^{(3)}(R,\mathbf{r})|Y_{l'K}(\hat{r})\rangle\\
&=\sum_L V^{(3)}_L(R,r)\langle Y_{lK}(\hat{r})|Y_{L0}(\hat{r})|Y_{l'K}(\hat{r})\rangle.
\end{aligned} \tag{7.22}$$

Thus, the Q-value is distributed continuously between the light cluster energy $E^{(3)}_{nK}$ and the energy of heavy fragments: $E^{(12)}_{nK}=E_K-E^{(3)}_{nK}$.

This system of equations is written in spherical coordinates. The light cluster moves in a hyperdeformed field of heavy fragments. The description of this potential can be done in a better way by using the prolate spheroidal system of coordinates.The necessary details are given in Appendix (14.7).

For the adiabatic case the wave functions of the heavy fragments and light particle channels can be normalized independently:

$$\begin{aligned}
&\sum_l\int_0^{R_m}\left[g^{(12)}_{nlK}(R)\right]^2 dR=1,\\
&\sum_l\int_0^{r_m}\left[g^{(3)}_{nlK}(R,r)\right]^2 dr=1.
\end{aligned} \tag{7.23}$$

7.3 Ternary Potential

We suppose that the potential $V^{(12)}$ acting between heavy clusters is given by a double folding procedure using Coulomb and M3Y nuclear interaction. The position of the first resonant state is adjusted by a repulsive core. The details of calculations are given in Ref. [28].

The potential acting on the light cluster is a sum of the Coulomb and nuclear parts, i.e.,

$$V^{(3)}(R,\mathbf{r})=V^{(3)}_C(R,\mathbf{r})+V^{(3)}_N(\mathbf{r}). \tag{7.24}$$

In order to obtain the multipole expansion of the Coulomb potential, one uses its expression in terms of the individual interactions (see Fig. 7.2):

$$V_C^{(3)}(R,\mathbf{r}) = V_C^{(13)}(R,\mathbf{r}_{13}) + V_C^{(23)}(R,\mathbf{r}_{23}), \tag{7.25}$$

where

$$V_C(R,\mathbf{r}_{k3}) \equiv V_C(R,r_{k3},\theta_{k3}) = \sum_{\lambda=0,2,4} V_{0\lambda\lambda}^{(k3)}(r_{k3})Y_{\lambda 0}(\theta_{k3}), \quad k=1,2. \tag{7.26}$$

Here $V_{0\lambda\lambda}(r_{k3})$ are the multipole form factors of each cluster-heavy fragment potential and:

$$\begin{aligned} r_{13}^2 &= r^2 + R_1^2 - 2rR_1\cos\theta, \quad r_{23}^2 = r^2 + R_2^2 + 2rR_2\cos\theta, \\ \cos\theta_{13} &= \frac{r\cos\theta - R_1}{r_{13}}, \quad \cos\theta_{23} = \frac{r\cos\theta + R_2}{r_{23}}, \\ R_1 &= \frac{RA_2}{A_1+A_2}, \quad R_2 = \frac{RA_1}{A_1+A_2}. \end{aligned} \tag{7.27}$$

By multiplying (7.24) with $2\pi Y_{L0}(\theta)$ and using numerical integration over angle θ one obtains the multipole coefficients $V^{(3)}{}_L(R, r)$.

7.4 Angular Distribution of the Light Particle

The decay width of a resonant state, defined by quantum numbers (n, K), can be derived from the Schrödinger equation with complex energy, as it is done in Sect. 2.5 in Chap. 2. For the light cluster it is given by:

$$\Gamma_{nK}^{(3)}(R) = \hbar v_{nK} \lim_{r\to\infty} \int d\hat{r} |\psi_{nK}^{(3)}(R,\mathbf{r})|^2, \tag{7.28}$$

where v_{nK} is the velocity of the light particle at infinity. At large distances the wave function has the form of a decaying (outgoing) Gamow state:

$$\psi_{nK}^{(3)}(R,\mathbf{r}) \to_{r\to\infty} \sum_l N_{nlK}^{(3)}(R) \frac{H_l^{(+)}(\kappa_{nK} r)}{r} Y_{lK}(\hat{r}), \tag{7.29}$$

with the wave number $\kappa_{nK} = \sqrt{2\mu_3 E_{nK}^{(3)}}/\hbar$. The coefficients $N_{nlK}^{(3)}(R)$ are determined from the matching condition between internal and external components and their derivatives, as described below. One gets the following result:

$$\Gamma_{nK}^{(3)}(R) = \hbar v_{nK} \sum_l |N_{nlK}^{(3)}(R)|^2. \tag{7.30}$$

The probability that the light particle is emitted in some direction θ is given by a similar to (7.28) relation, but without integration on angles:

$$
\begin{aligned}
\Gamma^{(3)}_{nK}(R,\theta) &= \hbar v_{nK} \lim_{r\to\infty} |r\psi^{(3)}_{nK}(R,\mathbf{r})|^2 \\
&\equiv \Gamma^{(3)}_{nK}(R) W^{(3)}_{nK}(R,\theta),
\end{aligned}
\tag{7.31}
$$

where we introduced the following function defining angular distribution:

$$
\begin{aligned}
W^{(3)}_{nK}(R,\theta) &= \sum_L A_{nLK} Y_{L0}(\theta), \\
A_{nLK} &= \left[\sum_l |N^{(3)}_{nlK}(R)|^2\right]^{-1} \sum_{ll'} N^{(3)}_{nlK}(R) N^{(3)}_{nl'K}(R) \\
&\times \sqrt{\frac{(2ll+1)(2l'+1)}{4\pi(2L+1)}} \langle l0; l'0|L0\rangle \langle lk; l'-K|L0\rangle \cos\frac{l(l+1)-l'(l'+1)}{\chi}, \\
\chi &= \frac{2Z_{12}Z_3e^2}{\hbar v_{nK}}.
\end{aligned}
\tag{7.32}
$$

We would like to mention that this quantum description, based on the molecular ansatz (7.14), gives the general angular distribution shape, at variance with semiclassical approaches, where a Monte-Carlo procedure, generating randomly the trajectories is necessary. In addition, our formalism takes into account the influence of the heavy fragment motion.

Let us analyze the following three ternary splittings of ^{252}Cf:

$$
\begin{aligned}
&(A) \quad {}^{252}\mathrm{Cf} \to {}^{142}\mathrm{Xe} + {}^{106}\mathrm{Mo} + {}^{4}\mathrm{He}, \\
&(B) \quad {}^{252}\mathrm{Cf} \to {}^{146}\mathrm{Ba} + {}^{96}\mathrm{Sr} + {}^{10}\mathrm{Be}, \\
&(C) \quad {}^{252}\mathrm{Cf} \to {}^{142}\mathrm{Xe} + {}^{96}\mathrm{Sr} + {}^{14}\mathrm{C}.
\end{aligned}
\tag{7.33}
$$

We consider only the true cold (neutronless) fission process, i.e., the fragments are emitted in their ground states. The ground state quadrupole deformations of heavy fragments are given in Table 7.1.

The energy and angular distributions of the light cluster were measured. The results are given in Table 7.2 [23]. The energy spectra of emitted light fragments are centered around the values given in the second column. They have gaussian shapes and their widths ΔE are given in the third column. The angular distributions are centered around angles given in the fourth column.

Table 7.1 The quadrupole deformation parameters β_2 [29] for the heavy fragments involved in the cold fission of ^{252}Cf

Nucleus	Z	A	β_2
^{142}Xe	54	142	0.145
^{106}Mo	42	106	0.361
^{146}Ba	56	146	0.199
^{96}Sr	38	96	0.338

Table 7.2 The averaged values of energy spectra for the light fragment in the processes (7.33) (given in the second column)

Process	E (MeV)	ΔE (MeV)	θ_0 (deg.)	Q (MeV)	TKE (MeV)
A	16	6	84 ± 1	215.350	173
B	18	7	84 ± 2	202.063	167
C	24	8	83 ± 2	221.132	166

The widths are given in the third column and the angular distributions are centered around angles given in the forth column. In the last two columns the Q-value and total kinetic energy of fragments (TKE), respectively [23] are given

From the last columns, which give Q-values and total kinetic energies of heavy fragments (TKE), respectively, one sees that these results include not only cold, but also excited heavy fragments and emitted neutrons.

From this table one can see that the averaged energies of emitted clusters are close to each other. It is also known that ^{10}Be yields are by two orders of magnitudes less than for ^{4}He [23]. This situation is in obvious contrast with the cluster decay phenomenon because Coulomb barriers in the polar direction are very different. For the process B the barrier is almost twice higher than for A. This cannot explain similar yields at similar cluster energies, given by so different barriers.

A simple explanation can be given as follows. We have already shown in Refs. [24, 30] that the emission probability of the light cluster is very much enhanced in the neck region. The cluster emission takes place only from the neck region, because it is forbidden from neutron rich heavy fragments. We simulate the neck emission by a nuclear potential, generating the cluster. Thus, we add to the Coulomb interaction a nuclear potential, having its minimum at $x_0 = 6$ fm. The maximum probability of this process, corresponding to the minimum of this potential, was predicted at this distance by a microscopic estimate of Ref. [30].

The resulting potential has the following ansatz:

$$V_N^{(3)}(x,z) = -V_0 + v\frac{\hbar^2}{2\mu_3}\beta^2\left[d_x(x-x_0)^2 + d_z(z-z_0)^2\right], \tag{7.34}$$

where we introduced the deformation factors, as follows:

$$d_x = 1 + \frac{4}{3}\delta, \quad d_z = 1 - \frac{2}{3}\delta. \tag{7.35}$$

The harmonic oscillator (ho) parameter of the emitted light cluster is defined in a standard way:

$$\beta = \frac{\mu_3\omega_0}{\hbar}. \tag{7.36}$$

The size parameter of the ho potential should be related to the measured shape of the free cluster. The parameter β is determined from scattering experiments by using a Gaussian density distribution

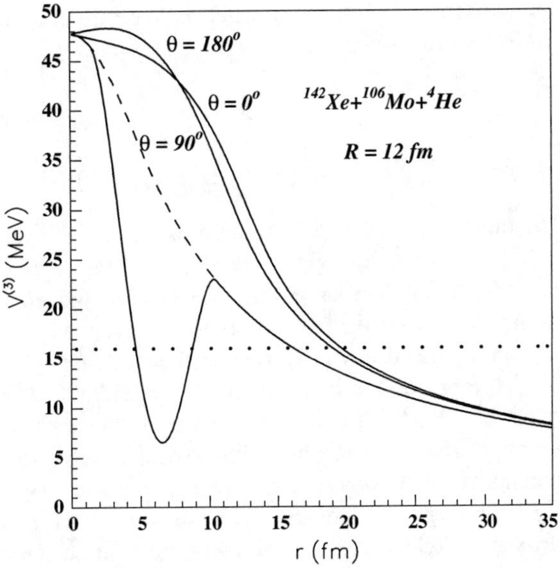

Fig. 7.3 The cold fission potential of the α-particle

$$\rho_\tau(r_\tau) = N_\tau[\beta/\pi]^{3/2} e^{-\beta r_\tau^2}, \quad \tau = p, n. \tag{7.37}$$

Sometimes one uses the equivalent parameter $b = 1/\beta^{1/2}$ [31]. This potential is shown at various angles in Fig. 7.3 for the emission of α-particle and in Fig. 7.4 for ^{10}Be.

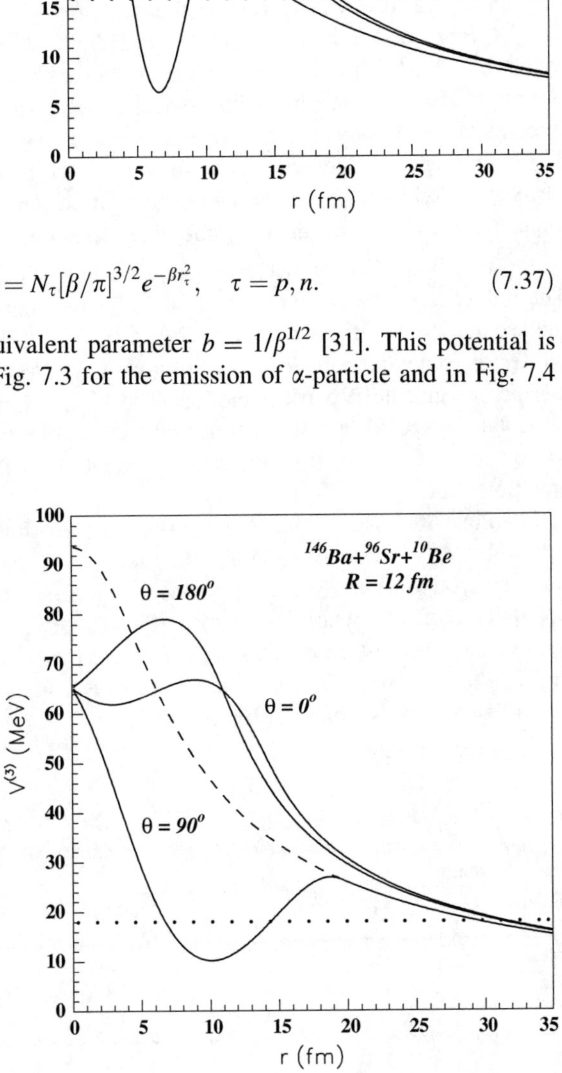

Fig. 7.4 The cold fission potential of ^{10}Be

The coordinate x_0, measured from the symmetry axis, is called eccentricity and the coordinate z_0 equatorial distance. We estimate z_0 as the half distance between nuclear surfaces at $x_0 = 0$ (see Fig. 7.2), i.e.,

$$z_0 = \frac{1}{2}\left[R\frac{A_2 - A_1}{A_2 + A_1} + R_{m2} - R_{m1}\right],$$
$$R_{mk} = R_0[1 + \beta_2 Y_{20}(0^o)], \quad k = 1,2. \tag{7.38}$$

The factor v in (7.34) will be used as a scaling parameter.

The barrier is strongly lowered in case (B) and becomes comparable with the case (A), as can be seen in Figs. 7.3 and 7.4. Indeed, in Table 7.3 one can see that the ho strength is five times smaller in the second case. In this way, we are able to explain similar decay properties in the two cases.

The depth of the ho potential, $V_0 = 27.7$ MeV, reproduces the experimental value $E^{(3)} = 16$ MeV in the middle of this interval, i.e., for $R = 15$ fm. The variation of the eigenvalue with respect to R is about 6 MeV. In this way, the dynamical three-body effect gives the width of the energy distribution.

Our numerical calculations have shown that the adiabatic matrix is by one order of magnitude less than the energy of the light cluster resonant states and therefore we can neglect it. Thus, the only coupling effect consists in a strong enhancement of the two-body Coulomb barrier between heavy fragments $V^{(12)}(R) + E^{(3)}_{nK}(R)$. The influence of the cluster motion on the heavy fragments dynamics is given by the second system of equations in (7.21). In Fig. 7.5, by the lower solid line we plotted the potential between heavy fragments $V^{(12)}(R)$ versus the distance R in the absence of any third particle motion. The higher solid line represents the same potential corrected by the energy of the emitted particle $V^{(12)}(R) + E^{(3)}(R)$ (= 16 MeV). This quantity enters in the equation of fragment motion (7.21) as a new potential.

In Fig. 7.6, we plotted the two wave functions for $E^{(3)} = 0$, and $E^{(3)} = 16$ MeV, respectively. One remarks that the motion of heavy fragments is centered around a given distance, inside a large Coulomb barrier and the predicted two-body half life would be very large. The two-body wave function corresponding to the first resonant state has a sharp maximum, as can be seen from Fig. 7.6. It corresponds to an already preformed α-particle, because we add $E^{(3)}$ to the initial two-body potential for all radii.

Table 7.3 The harmonic oscillator parameter $b = 1/\beta^{1/2}$ for different emitted light fragments [31] (the second column), $\beta = \mu_3\omega_0/\hbar$ (the third column) and the harmonic oscillator strength (the last column)

Nucleus	b (fm)	β (fm^{-2})	$\frac{\hbar^2}{2\mu_3}\beta^2$ (MeV fm^{-2})
^{4}He	1.19	0.706	2.603
^{10}Be	1.44	0.480	0.481
^{14}C	1.65	0.367	0.201

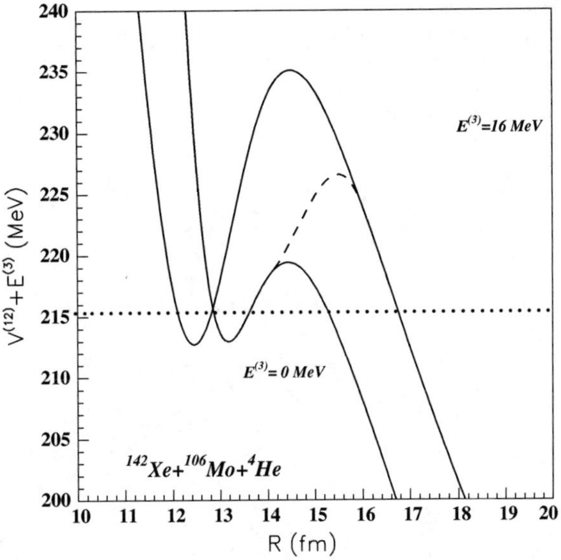

Fig. 7.5 The potential between heavy fragments corresponding to $E^{(3)} = 0$ (*the lower solid line*) and 16 MeV for the process (A) (*the upper solid line*). The *Q*-value is given by a *dotted line*. The potential describing an intermediate creation of the α-particle is plotted by a *dashed line*

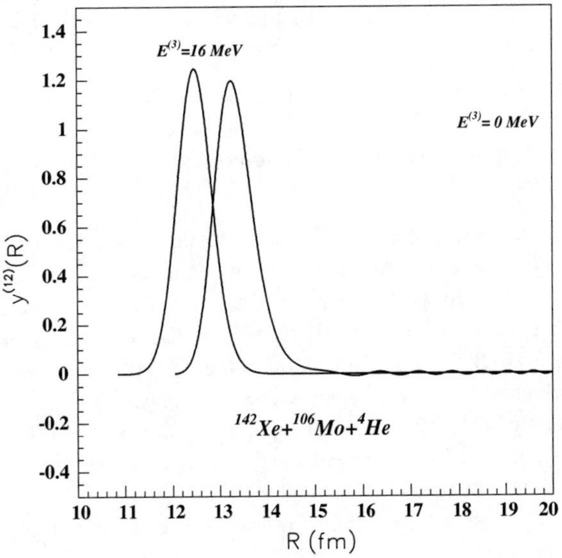

Fig. 7.6 The wave functions corresponding to the potentials plotted by *solid lines* in Fig. 7.5

It is also possible to consider a more general scenario in which the third light cluster is born at some intermediate distance. This situation is depicted by a dashed line in Fig. 7.5. In this case, the barrier is lower and heavy fragments are less localized. In this case the half life of the binary system is still much larger than of the light particle resonant state and qualitatively the picture does not change.

In Fig. 7.7 we see that a small deviation from the equatorial parameter z_0, given by (7.38), strongly changes the position of the maximum in the angular

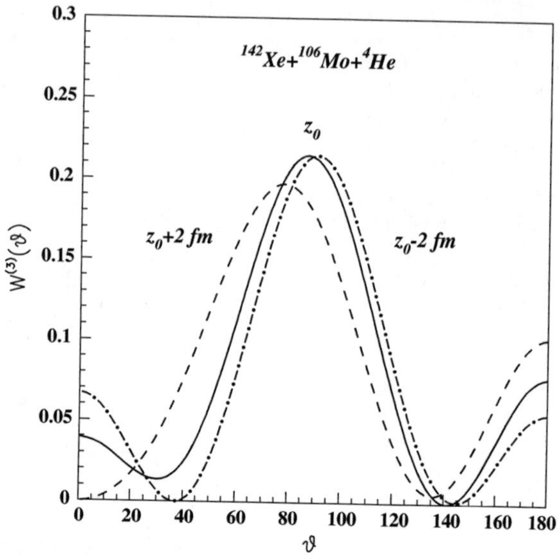

Fig. 7.7 The normalized angular distribution of the process A for z_0 (*solid line*), $z_0 + 2$ fm (*dashed line*) and $z_0 - 2$ fm (*dot-dashed line*). The parameters of the nuclear potential are $V_0 = 27.7$ MeV, $R = 15$ fm, $x_0 = 6$ fm

distribution. We stress on the fact that z_0, corresponding to the solid line, gives a very result close the experimental value in Table 7.2.

Thus, we answered the two main questions raised by the experiment, namely:

1. Why the energies and half lives of the emitted light clusters are close to each other, in spite of the different Coulomb barriers in the equatorial direction?
2. Why the angular distribution of the light clusters are centered around the same angle?

To answer the first question, we assumed that the light cluster is born in the neck region. The process is simulated using an ho well with a size parameter given by its known nuclear density. The strength of the potential is five times smaller for ^{10}Be than for ^{4}He and 13 times smaller for ^{14}C. Thus, Coulomb barriers in the equatorial direction become similar in the three cases, giving comparable eigenvalues of the first resonant states. We have chosen the minimum of the ho potential at an equal distance with respect to surfaces of heavy fragments. The values of the eccentricity with respect to the symmetry axis is taken from the microscopic estimate of the total preformation amplitude.

To answer the second question, we analyzed the influence of all potential parameters on the maximum of the light particle angular distribution. It turns out that the only sensitive parameters are the deviation from the equator and the cluster deformation. From the position of the maximum in the angular distribution, we conclude that the light cluster deformation should be born in the neck region indeed. For ^{10}Be we found out that a significant deformation improves the agreement with the experiment.

The experimental facts suggest that half lives for ^{10}Be are shorter than 10^{-13} s. Thus, we conclude that the ho potential in which the cluster deformation is born should be flatter than its measured size and the intrinsic deformation should be significant.

References

1. Goldansky, V.I.: On neutron deficient isotopes of light nuclei and the phenomena of proton and two-proton radioactivity. Nucl. Phys. **19**, 482–495 (1960)
2. Swan, P.: Sequential decay theory for the $B^{11}(p, 3\alpha)$ reaction. Rev. Mod. Phys. **37**, 336–345 (1965)
3. Bernstein, M.A., Friedman, W.A., Lynch, W.G.: Emission of particle unstable resonances from compound nuclei. Phys. Rev. C **29**, 132–138 (1984)
4. Mukha, I.G., Schrieder, G.: Two-proton radioactivity as a genuine three-body decay: the ^{19}Mg probe. Nucl. Phys. A **690**, 280c–283c (2001)
5. Grigorenko, L.V., Johnson, R.C., Mukha, I.G., Thomson, I.J., Zhukov, M.V.: Two-proton radioactivity and three-body decay: general problems and theoretical approach. Phys. Rev. C **64**, 054002/1–12 (2001)
6. Grigorenko, L.V., Zhukov, M.V.: Two-proton radioactivity and threebody decay II. Exploratory studies of lifetimes and correlations. Phys. Rev. C **68**, 054005/1–15 (2003)
7. Grigorenko, L.V., Zhukov, M.V.: Two-proton radioactivity and three-body decay. III. Integral formulas for decay widths in a simplified semianalytical approach. Phys. C **76**, 014008/1–17 (2007)
8. Grigorenko, L.V., Zhukov, M.V.: Two-proton radioactivity and three-body decay. IV. Connection to quasiclassical formulation. Phys. C **76**, 014009/1–9 (2007)
9. Kryger, R.A., et. al.: Two-proton emission from the ground state of ^{12}O. Phys. Rev. Lett. **74**, 860–863 (1995)
10. Barker, F.C.: Width of the ^{12}O ground state. Phys. Rev. C **59**, 535–538 (1999)
11. Barker, F.C.: ^{12}O Ground-state decay by ^{2}He emission. Phys. Rev. C **63**, 047303/1–2 (2001)
12. Goméz del Campo, J., et. al.: Decay of a resonance in ^{18}Ne by the simultaneous emission of two protons. Phys. Rev. Lett. **86**, 43–46 (2001)
13. Brown, B.A., Barker, F.C., Millener, D.J.: Di-proton decay of the 6.15 MeV 1^- state in ^{18}Ne. Phys. Rev. C **61**, 051309(R)/1–3 (2002)
14. Brown, B.A., Barker, F.C., Millener, D.J.: Di-proton decay of the 6.15 MeV 1^- states in ^{18}Ne. Phys. Rev. C **67**, 041304(R)/1–3 (2003)
15. Bennaceur, K., Nowacki, F., Okolowicz, J., Ploszajczak, M.: Analysis of the $^{16}O(p\gamma)^{17}F$ capture reaction using the shell model embedded in the continuum. Nucl. Phys. A **671**, 203–232 (2000)
16. Okolowicz, J., Ploszajczak, M., Rotter, I.: Dynamics of quantum systems emebeded in continuum. Phys. Rep. **374**, 271-383 (2003)
17. Rotter, I., Okolowicz, J., Ploszajczak, M.: Microscopic theory of the two-proton radioactivity. Phys. Rev. Lett. **95**, 042503/1–4 (2005)
18. Zhukov, M.V., et.al.: Bound state properties of borromean hallo nuclei: ^{6}He and ^{11}Li. Phys. Rep. **231**, 153–199 (1993)
19. Grigorenko, L.V., Johnson, R.C., Mukha, I.G., Thomson, I.J., Zhukov, M.V.: Theory of two-proton radioactivity with application to ^{19}Mg and ^{48}Ni. Phys. Rev. Lett. **85**, 22–25 (2000)
20. Ramayya, A.V., et al.: Cold (neutronless) α ternary fission of ^{252}Cf. Phys. Rev. C **57**, 2370–2374 (1998)
21. Hamilton, J.H., et al.: New cold and ultra hot binary and cold ternary spontaneous fission mopdes for ^{252}Cf and new band structures with gammasphere. Prog. Part. Nucl. Phys. **38**, 273-287 (1997)
22. Ramayya, A.V., et al.: Observation of ^{10}Be emission in the cold ternary spontaneous fission of ^{252}Cf. Phys. Rev. Lett. **81**, 947–950 (1998)

23. Ter-Akopian, G.M.: Angular Distribution in the Cold Fission of ^{252}Cf. Seminar, 6 May 2002, Physics Department, Frankfurt/Main University (2002)
24. Florescu, A., Săndulescu, A., Delion, D.S., Hamilton, J.H., Ramayya, A., Greiner, W.: Preformation probabilities for light ternary particles in cold (neutronless) fission of ^{252}Cf. Phys. Rev. C **61**, 051602(R)/1–4 (2000)
25. Delion, D.S., Florescu, A., Săndulescu, A.: Microscopic description of the cold α and ^{10}Be ternary fission yields of ^{252}Cf in spheroidal coordinates. Phys. Rev. C **63**, 044312/1–10 (2001)
26. Delion, D.S., Săndulescu, A., Greiner, W.: Anisotropy in the ternary cold fision. J. Phys. G **29**, 317–336 (2003)
27. Delion, D.S., Săndulescu, A., Greiner, W.: Self-consistent description of the ternary cold fission: tri-rotor mode. J. Phys. G **28**, 2921–2938 (2002)
28. Delion, D.S., Săndulescu, A., Mişicu, S., Cârstoiu, F., Greiner, W.: Double fine structure in binary cold fission. J. Phys. G **28**, 289–306 (2009)
29. Möller, P., Nix, R.J., Myers, W.D., Swiatecki, W.: Nuclear ground-state masses and deformations. At. Data Nucl. Data Tables **59**, 185–381 (1995)
30. Delion, D.S., Săndulescu, A., Mişicu, S., Cârstoiu, F., Greiner, W.: Quasimolecular resonances in the binary cold fission of ^{252}Cf. Phys. Rev. C **64**, 041303/1–5 (2001)
31. Dao Khoa, T.: α-nucleus potential in the double-folding model. Phys. Rev. C **63**, 034007/1–15 (2001)
32. Thomas, R.G.: A formulation of the theory of alpha-particle decay from time-independent equations. Prog. Theor. Phys. **12**, 253–264 (1954)

Part II
Microscopic Description of Emission Processes

The most general framework to describe emission processes is, of course, the time dependent approach. In this part of the book, devoted to microscopic approaches, we will derive the general expression of the decay width, known as the Fermi golden rule, by using time-dependent Schrödinger equation. Then, we will show that this relation can be recovered within various approaches like reaction theory or R-matrix approach. The R-matrix theory [1, 2] makes a step forward with respect to the Gamow theory [3], by expressing the decay width as a product between the particle preformation probability and the penetration through the barrier [4–9]. This relation is similar to its phenomenological counterpart (2.87), but the role of the wave function at the matching point $f_c(R)/R$ is played by the preformation amplitude, already introduced by Eq. 2.143 and defined as the overlap between the initial wave function and the product of the daughter and α-particle wave functions. This approach takes into account the nuclear structure details, by expressing the cluster wave function in terms of two proton and two neutron orbitals in some mean field, interacting with each other via two-body residual forces [5, 8]. We will describe the general procedure to estimate the preformation amplitude within the so-called Multi-step Shell Model technique (MSM) [10, 11]. Due to the antisymmetrization effects between the α-particle and daughter wave functions, the interaction becomes non-local in the internal region [12, 13].

It was shown that the usual shell model space using $N = 6$–8 major shells underestimates the experimental decay width by several orders of magnitude [14, 15], due to the exponential decrease of bound single particle wave functions [16]. An answer to the problem would be the inclusion of the sp narrow resonances lying in continuum [17–19], i.e. Gamow states. In spite of the fact that the true asymptotic behaviour of the wave functions is achieved, the value of the half life is still not reproduced [20]. Only the background components in continuum can describe the right order of magnitude of experimental decay widths [21–25]. The inclusion of the background contribution becomes important because an important part of the α-clustering process proceeds through such states.

The problem of considering the continuum part of the spectrum in microscopic calculations is rather involved, but very important especially for drip line nuclei [26]. The idea to replace the integration over the real spectrum in continuum by sp Gamow resonances plus an integration along a contour in the complex plane including these resonances was considered by Berggren in Ref. [27]. The calculation is very much simplified if one considers that in some physical processes only the narrow resonances are relevant and the integration, giving the background, can be neglected [28, 29]. This was shown to be an adequate approach, for instance in giant resonances [30] and in the nucleon decay processes [31]. To estimate the decay width, the states in continuum can be taken into account effectively by including a cluster component [32], or by considering a sp basis with a larger harmonic oscillator (ho) parameter for states in continuum [25].

The α-decaying state can be described as a sp resonance, namely by using the matching between logarithmic derivatives of the preformation amplitude and external Coulomb function. The derivative of the α-particle preformation factor, estimated within the shell model is almost constant along any neutron chain and therefore is not consistent with the decreasing behaviour of Q-values along such chains [33, 34]. We will show that the slope of the preformation amplitude can be corrected by changing the ho parameter of sp components. These components are related to an α-cluster term, not predicted by the standard shell model [32]. Indeed, the even–odd pair staggering of binding energies found along the α-lines with $N-Z=$ const., can be explained in terms of a "pairing" in the isospin space between proton and neutron pairs, considered as bosons [35, 36]. The generalization of this approach in terms many-body Greens's functions was performed in Ref. [37, 38]. This suggest that α-particles are already preformed at least in the low density region of the nuclear surface and they can explain the above inconsistency. We will discuss all these points within the so-called selfconsistent emission theory.

Finally we will describe the α-decay fine structure to vibrational states within the Quasiparticle Random Phase Approximation (QRPA) and the generalization of the preformation amplitude to heavy cluster decays. A brief description of fission-like theory within the Two Center Shell Model is also given.

References

1. Thomas, R.G.: A formulation of the theory of alpha-particle decay from time-independent equations. Prog. Theor. Phys. **12**, 253–264 (1954)
2. Lane, A.M., Thomas, R.G.: R-Matrix theory of nuclear reactions. Rev. Mod. Phys. **30**, 257–353 (1958)
3. Gamow, G.: Zur Quantentheorie des Atomkernes. Z. Phys. **51**, 204–212 (1928)
4. Mang, H.J.: Calculation of α-transition probabilities. Phys. Rev. **119**, 1069–1075 (1960)
5. Săndulescu, A.: Reduced widths for favoured alpha transitions. Nucl. Phys. A **37**, 332–343 (1962)
6. Soloviev, V.G.: Effect of pairing correlations on the alpha decay rates. Phys. Lett. **1**, 202–205 (1962)

7. Mang, H.J., Rasmussen, J.O.: Mat. Fys. Skr. Dan. Vid. Selsk. **2**, 3 (1962)
8. Mang, H.J.: Alpha decay. Ann. Rev. Nucl. Sci. **14**, 1–28 (1964)
9. Poggenburg, J.K., Mang, H.J., Rasmussen, J.O.: Theoretical alpha-decay rates for the actinide region. Phys. Rev. **181**, 1697–1719 (1969)
10. Liotta, R.J., Pomar, C.: Multi-step shell-model treatment of sux-particle system. Nucl. Phys. A **362**, 137–162 (1981)
11. Liotta, R.J., Pomar, C.: A graphical procedure to evaluate the many-body shell-model equations. Nucl. Phys. A **382**, 1–19 (1982)
12. Fliessbach, T.: The reduced width amplitude in the reaction theory for composite particles. Z. Phys. A **272**, 39–46 (1975)
13. Lovas, R.G., Liotta, R.J., Insolia, A., Varga, K., Delion, D.S.: Microscopic theory of cluster radioactivity. Phys. Rep. **294**, 265–362 (1998)
14. Fliessbach, T., Mang, H.J., Rasmussen, J.O.: The reduced width amplitude in the reaction theory for composite particles. Phys. Rev. C **13**, 1318–1323 (1976)
15. Tonozuka, I., Arima, A.: Surface α-clustering and α-Decays of ^{211}Po. Nucl. Phys. A **323**, 45–60 (1979)
16. Fliessbach, T., Okabe, S.: Surface alpha-clustering in the Lead region. Z. Phys. A **320**, 289–294 (1985)
17. Janouch, F.A., Liotta, R.: Influence of α-cluster formation on α decay. Phys. Rev. C **27**, 896–898 (1983)
18. Dodig-Crnkovic, G., Janouch, F.A., Liotta, R.J.: The continuum and the alpha-particle formation. Phys. Scr. **37**, 523–525 (1988)
19. Lenzi, S.M., Dragun, O., Maqueda, E.E., Liotta, R.J., Vertse, T.: Description of alpha clustering including continuum configurations. Phys. Rev. C **48**, 1463–1465 (1993)
20. Delion, D.S., Suhonen, J.: Microscopic description of α-like resonances. Phys. Rev. C **61**, 024304/1–12 (2000)
21. Insolia, A., Curutchet, P., Liotta, R.J., Delion, D.S.: Microscopic description of alpha decay of deformed nuclei. Phys. Rev. C **44**, 545–547 (1991)
22. Delion, D.S., Insolia, A., Liotta, R.J.: Anisotropy in alpha decay of odd-mass deformed nuclei. Phys. Rev. C **46**, 884–888 (1992)
23. Delion, D.S., Insolia, A., Liotta, R.J.: Alpha widths in deformed nuclei: microscopic approach. Phys. Rev. C **46**, 1346–1354 (1992)
24. Delion, D.S., Insolia, A., Liotta, R.J.: Microscopic description of the anisotropy in alpha decay. Phys. Rev. C **49**, 3024–3028 (1994)
25. Delion, D.S., Insolia, A., Liotta, R.J.: New single particle basis for microscopic description of decay processes. Phys. Rev. C **54**, 292–301 (1996)
26. Dobaczewski, J., Nazarewicz, W., Werner, T.R., Berger, J.F., Chinn, C.R., Decharge, J.: Mean-field description of ground-state properties of drip-line nuclei: pairing and continuum effects. Phys. Rev. C **53**, 2809–2840 (1996)
27. Berggren, T.: The use of resonant states in Eigenfunction expansions of scattering and reaction amplitudes. Nucl. Phys. A **109**, 265–287 (1968)
28. Vertse, T., Curuchet, P., Civitarese, O., Ferreira, L.S., Liotta, R.J.: Application of Gamow resonances to continuum nuclear spectra. Phys. Rev. C **37**, 876–879 (1988)
29. Curuchet, P., Vertse, T., Liotta, R.J.: Resonant random phase approximation. Phys. Rev. C **39**, 1020–1031 (1989)
30. Vertse, T., Liotta, R.J., Maglione, E.: Exact and approximate calculation of giant resonances. Nucl. Phys. A **584**, 13–34 (1995)
31. Maglione, E., Ferreira, L.S., Liotta, R.J.: Nucleon decay from deformed nuclei. Phys. Rev. Lett. **81**, 538–541 (1998)
32. Varga, K., Lovas, R.G., Liotta, R.J.: Cluster-configuration shell model for alpha decay. Nucl. Phys. A **550**, 421–452 (1992)
33. Rasmussen, J.O.: Alpha-decay barrier penetrabilities with an exponential nuclear potential: even-even nuclei. Phys. Rev. **113**, 1593–1598 (1959)

34. Akovali, Y.A.: Review of alpha-decay data from double-even nuclei. Nucl. Data Sheets **84**, 1–114 (1998)
35. Dussel, G., Caurier, E., Zuker, A.P.: Mass predictions based on α-line systematics. At. Data Nucl. Data Tables, **39**, 205–211 (1988)
36. Gambhir, Y.K., Ring, P., Schuck, P.: Nuclei: a superfluid condensate of α particles? A study within the Interacting-Boson Model. Phys. Rev. Lett. **51**, 1235–1238 (1983)
37. Dussel, G.G., Fendrik, A.J., Pomar, C.: Microscopic description of four-body excitations in heavy nuclei. Phys. Rev. C **34**, 1969–1973 (1986)
38. Dussel, G.G., Fendrik.: Microscopic description of four body excitations in heavy nuclei. Phys. Rev. C **34**, 1097–1109 (1986)

Chapter 8
Microscopic Emission Theories

8.1 Time Dependent Approach

The most general approach of the emission problem, describing the binary splitting $P \to D_1 + D_2$, is given by the time-dependent Schrödinger equation (2.3 in Chap. 2)

$$i\hbar\frac{\partial}{\partial t}|\Phi(t)\rangle = \mathbf{H}|\Phi(t)\rangle. \tag{8.1}$$

The wave function at a certain time t can be written in terms of the evolution operator

$$\Phi(t) = \mathbf{U}(t, t_0)\Phi(t_0), \tag{8.2}$$

satisfying a similar to (8.1) equation

$$i\hbar\frac{\partial \mathbf{U}}{\partial t} = \mathbf{HU}. \tag{8.3}$$

By using the initial condition $\mathbf{U}(t_0, t_0) = \mathbf{1}$ the solution is given by

$$\mathbf{U}(t, t_0) = \mathbf{1} - \int_{t_0}^{t} \mathbf{HU}(t', t_0)dt'. \tag{8.4}$$

8.1.1 Integral Formula. Fermi "Golden Rule"

Let us split the Hamiltonian into two parts

$$\mathbf{H} = \mathbf{H}_0 + \mathbf{W}, \tag{8.5}$$

D. S. Delion, *Theory of Particle and Cluster Emission*, Lecture Notes in Physics, 819,
DOI: 10.1007/978-3-642-14406-6_8,

where the first part $\mathbf{H}_0$ describes the parent bound state and the second one $\mathbf{W}$ the perturbation, similar to that given in Fig. 2.7 in Chap. 2. The time dependent solution can be expanded in terms of eigenstates ψ_n of the not perturbed Hamiltonian [1, 2]

$$\Phi(t,\mathbf{x}) = \sum_n c_n(t)\psi_n(\mathbf{x}) \equiv \sum_n a_n(t)e^{-\frac{i}{\hbar}E_n t}\psi_n(\mathbf{x}), \tag{8.6}$$

where E_n are the eigenvalues of $\mathbf{H}_0$. The new coefficients $a_n(t)$ are nothing else than the Heisenberg representation on the wave function, because the Schrödinger equation (8.1) acquires the following form

$$i\hbar\frac{da_m(t)}{dt} = \sum_n \mathbf{W}_{mn}e^{\frac{i}{\hbar}(E_m-E_n)t}a_n(t), \tag{8.7}$$

where

$$\mathbf{W}_{mn} = \langle\psi_m|\mathbf{W}|\psi_n\rangle. \tag{8.8}$$

Let us now write this system of equations for an initial bound state and several final states in continuum, i.e.

$$\begin{aligned} i\hbar\frac{da_f(t)}{dt} &= \mathbf{W}_{fi}e^{\frac{i}{\hbar}(E_f-E_i)t}a_i(t), \quad a_f(0)=0 \\ i\hbar\frac{da_i(t)}{dt} &= \int dE_f \mathbf{W}_{if}e^{\frac{i}{\hbar}(E_i-E_f)t}a_f(t), \quad a_i(0)=1, \end{aligned} \tag{8.9}$$

where we understood the integration on energy as $dE_f \to \rho(E_f)dE_f$, in terms of the density of states $\rho(E_f)$. The above sytem can be integrated as follows

$$\begin{aligned} a_f(t) &= -\frac{i}{\hbar}\int_0^t dt' \mathbf{W}_{fi}e^{\frac{i}{\hbar}(E_f-E_i)t'}a_i(t') \\ a_i(t) &= 1-\frac{i}{\hbar}\int dE_f \int_0^t dt'' \mathbf{W}_{if}e^{\frac{i}{\hbar}(E_i-E_f)t''}a_f(t''). \end{aligned} \tag{8.10}$$

Let us now consider a time independent perturbation. By replacing the first line into the second one, and by exchanging the order of integration, one gets

$$a_i(t) = 1+\frac{i}{\hbar}\int dE_f|\mathbf{W}_{fi}|^2 \int_0^t dt' a(t')\frac{1-e^{\frac{i}{\hbar}(E_f-E_i)(t-t')}}{E_f-E_i}. \tag{8.11}$$

For a large time interval, we use the following limit

$$\frac{1-e^{\frac{i}{\hbar}(E_f-E_i)(t-t')}}{E_f-E_i} \to_{t\to\infty} i\pi\delta(E_f-E_i)+\mathscr{P}\frac{1}{E_f-E_i}. \tag{8.12}$$

Then, the coefficient of the initial state satisfies the following equation

$$a_i(t) = 1 - \int_0^t a_i(t')dt', \tag{8.13}$$

which has an exponential decreasing solution

$$a_i(t) = e^{-\lambda t}. \tag{8.14}$$

The coefficient λ is given by

$$\lambda = \frac{1}{\hbar}\left(\frac{\Gamma}{2} - i\Delta E_i\right), \tag{8.15}$$

in terms of the decay width and energy shift, respectively

$$\begin{aligned} \Gamma &= 2\pi \int dE_f |\mathbf{W}_{fi}|^2 \delta(E_f - E_i) \\ \Delta E_i &= \mathcal{P} \int dE_f \frac{|\mathbf{W}_{fi}|^2}{E_f - E_i}. \end{aligned} \tag{8.16}$$

The first line of the above relations is called Fermi "golden rule" for decay processes. The coefficient of the final state can be easily derived by replacing (8.14) in the first line of Eq. 8.10

$$a_f(t) = -\mathbf{W}_{fi} \frac{e^{\frac{i}{\hbar}(E_f - E_i)} e^{-\lambda t} - 1}{E_f - E_i + i\hbar\lambda} \rightarrow_{t\rightarrow\infty} \frac{\mathbf{W}_{fi}}{E_f - E_i - \Delta E_i + \frac{i}{2}\Gamma}. \tag{8.17}$$

Thus, one can explain from "first principles" the exponential low of the particle density in decay processes (2.72 in Chap. 2) and the Lorentzian distribution of the wave function, already used in Eq. 2.5 in Chap. 2 to describe narrow resonant states. Let us now write down these relations for the binary decay process. As an initial condition we suppose the parent wave function, i.e.

$$|\Phi\rangle_{t=0} = |\Psi^{(P)}\rangle. \tag{8.18}$$

The general solution at a certain moment t can be represented as a sum of the parent and several channel wave functions (8.6)

$$|\Phi\rangle = c(t)|\Psi^{(P)}\rangle + \sum_c \int b_c(E', t)|\Psi^c_{E'}\rangle dE'. \tag{8.19}$$

The basis channel wave functions $|\Phi^c_E\rangle$ should be orthogonal, i.e.

$$\begin{aligned} \langle \Psi^c_E | \Psi^{c'}_{E'} \rangle &= \delta_{cc'} \delta(E - E') \\ \langle \Psi^c_E | \mathbf{H} | \Psi^{c'}_{E'} \rangle &= E \delta_{cc'} \delta(E - E'). \end{aligned} \tag{8.20}$$

The set $|\Psi^c_E\rangle$ does not contain the eigenstates of the Hamiltonian $\mathbf{H}$, but diagonalizes it in a subspace of the Hilbert space. The Fermi "golden rule" for a transition to the level E_0 can be rewritten as follows [2]

$$\begin{aligned}\Gamma &= \sum_c \Gamma_c = 2\pi \sum_c |\langle \Psi^c_{E_0}|(\mathbf{H} - E^c_0)|\Psi^{(P)}\rangle|^2, \\ E^c_0 &= \langle \Psi^{(P)}|\mathbf{H}_0|\Psi^{(P)}\rangle = \langle \Psi^{(P)}|\mathbf{H}|\Psi^{(P)}\rangle.\end{aligned} \tag{8.21}$$

We considered that the initial wave function $\Psi^{(P)}$ has non-vanishing values inside the nucleus, where the transition operator does not contribute. These relations hold under the condition that $\Gamma_c \ll E^c{}_0$, which is very well satisfied for most emission processes.

8.1.2 Surface Formula. Preformation Amplitude

Each channel wave function (8.19) has a similar form with the usual cluster-like expansion (2.141 in Chap. 2), but containing the antisymmetrized product between the relative and fragment wave functions, i.e.

$$|\Psi^c_E\rangle = |\mathcal{A}\left[U^c_E \Psi^c_1 \Psi^c_2\right]\rangle. \tag{8.22}$$

By considering the α-decay process to a daughter nucleus with $A = Z + N$ nucleons, the antisymmetrization operator $\mathcal{A}$ is defined as follows

$$\mathcal{A} \equiv \left[\binom{Z}{2}\binom{N}{2}\right]^{-1/2} \sum_P (-)^P \hat{P}, \tag{8.23}$$

where by $\hat{P}$ we denote the permutation operator.

By using a similar to (4.45 in Chap. 4) relation the matrix element defining the partial decay width can be brought in the form of a surface integral [2, 3], as given by Eq. 4.46 in Chap. 4, i.e.

$$\langle \Psi^{(P)}|(\mathbf{H} - E^c_0)|\Psi_{E^c_0}\rangle = \frac{\hbar^2 R^2}{2\mu} \int \left(U^c_E \frac{\partial \mathcal{F}_c}{\partial r} - \mathcal{F}_c \frac{\partial U^c_E}{\partial r} \right)_{r=R} d\hat{r}, \tag{8.24}$$

where we introduced the channel preformation amplitude by the following overlap integral

$$\mathcal{F}_c(\mathbf{r}) = \langle \mathcal{A}\left[\Psi^c_1 \Psi^c_2\right]|\Psi^{(P)}\rangle. \tag{8.25}$$

The antisymmetrization becomes less important in the region beyond the nuclear surface.

8.1.3 Time-Dependent Approach for Proton Emitters

We illustrate the time-dependent formalism using a simple example, namely proton emission. Proton emission is a deep subbarrier process and therefore the stationary assumption is a very good approximation. However, the time-dependence of the wave function inside and beyond the barrier can bring useful informations concerning the transient and asymptotic regimes of this process. In order to analyze transitions between ground states, one solves the time-dependent Schrödinger equation

$$i\hbar\frac{\partial\Phi(t,\mathbf{r})}{\partial t}=\mathbf{H}\Phi(t,\mathbf{r}), \tag{8.26}$$

where the wave function contains only the relative component. The simplest case is the adiabatic process with spherical symmetry, where the Hamiltonian is given by the kinetic term plus the potential $V_0(r, \mathbf{s})$, defined by Eq. 2.133 in Chap. 2. The above Schrödinger equation is a one-dimensional parabolic equation in spatial coordinates, which can be solved numerically by using standard techniques, like for instance the Krank–Nicholson scheme [4, 5]. As an initial condition, one uses the wave function

$$\Phi(0,r)=\Psi^{(0)}_{ljm}(r), \tag{8.27}$$

which is an eigenstate of the modified Hamiltonian

$$\mathbf{H}_0(r,\mathbf{s})=-\frac{\hbar^2}{2\mu}\Delta^2+V_0(r,\mathbf{s})+W(r), \tag{8.28}$$

where the transition operator is given by

$$\begin{aligned} W(r)\ &=0, \quad r\leq r_{\max},\\ &=V_0(r_{\max})+(r-r_{\max})\tan\theta-V_0(r), \quad r>r_{\max}. \end{aligned} \tag{8.29}$$

The eigenstates of $\mathbf{H}_0$ are bound states. The tunneling probability is given by the probability in finding the proton beyond a certain point r_B

$$P(t,r_B)=\int_{r_B}^{\infty}|\Phi(t,r)|^2r^2dr. \tag{8.30}$$

The decay rate is calculated according to the relation

$$\Gamma(t,r_B)=\frac{\hbar}{1-P(t,r_B)}\frac{dP(t,r_B)}{dt}. \tag{8.31}$$

This quantity can be compared with the decay width given by the stationary approach for $\Gamma(\infty,r_B)$, for radius r_B beyond the nuclear surface.

The Q-value of the quasistationary process is defined as follows

$$E_0 = \int_0^{r_B} \Phi(t,r)^* \mathbf{H} \Phi(t,r) r^2 dr \left[\int_0^{r_B} |\Phi(t,r)|^2 r^2 dr \right]^{-1} . \qquad (8.32)$$

In Ref. [6] a rather academic example of the proton emission from ^{208}Pb was investigated with the aim of revealing the main feature of the transient process. It was shown that the decay width (8.31), as a function of time, has an oscillatory evolution during the motion of the wave packet inside the Coulomb barrier and reaches a quasiconstant value when the proton moves in the asymptotic region. The evolution of the average position, defined by

$$r_{av}(t, r_B) = \int_0^{r_B} r |\Phi(t,r)|^2 r^2 dr \left[\int_0^{r_B} |\Phi(t,r)|^2 r^2 dr \right]^{-1} , \qquad (8.33)$$

shows an increase during the transient regime, while small-amplitude fluctuations, of about a limiting value, appear in the asymptotic region. This corresponds to the last maximum of the wave packet as a function of the radius. This behaviour is expected because the wave packet does not significantly change its shape during the time evolution. The quasistationary wave packet penetrates through the barrier by performing these oscillations, instead of going through the barrier in a continuous fashion. Thus, the inverse period of these oscillations can be associated with the assault frequency λ_0, which is a quantity appearing as a multiplicative factor in front of the WKB penetration integral (5.1 in Chap. 5). The asymptotic value of the decay widths is rather close to the WKB estimate, as it is shown in Table 8.1 for $r_B = 25$ fm.

A similar analysis was performed in [7] for a realistic proton emitter, namely ^{109}I. It was shown that the asymptotic behaviour of the decay width does not depend upon the initial condition, i.e. upon different values of the angles θ in (8.29). Later, in [8], these authors analyze other proton emission processes from spherical nuclei and compare the asymptotic half lives with the corresponding DWA and WKB values, obtaining the results presented in Table 8.2

In Ref. [9], the authors extend this analysis to axially deformed nuclei. They investigate the angular distribution of emitted protons, defined as

Table 8.1 Asymptotic values of the decay widths as compared to the corresponding WKB values at energies E_0

E_0 (MeV)	$\Gamma(\infty)$ (s^{-1})	$\Gamma_{\rm WKB}$ (s^{-1})
7.19	1.90×10^{19}	2.33×10^{19}
7.71	1.25×10^{20}	1.74×10^{20}
12.45	2.23×10^{20}	2.59×10^{20}
16.55	2.79×10^{19}	2.52×10^{19}

Table 8.2 Asymptotic values of the decay widths as compared with the distorted wave and WKB values for different nuclei at energies E_0

Nucleus	E_0 (keV)	$T(\infty)$	T_{DW}	T_{WKB}
^{105}Sb	491	(19.6 ± 0.1) s	20 s	24 s
^{113}Cs	977	(534 ± 0.1) ns	540 ns	640 ns
^{147}Tm	1132	(206.80 ± 0.05) μs	210 μs	260 μs
^{151}Lu	1125	(58.4 ± 0.1) ms	60 ms	90 ms
^{157}Ta	947	(227 ± 2) ms	220 ms	170 ms
^{185}Bi	1611	(3.17 ± 0.01) μs	3.2 μs	2.5 μs

$$\Gamma(t,\theta) = \int_{R(\theta)}^{\infty} |\Phi(t,r,\theta)|^2 r^2 dr, \tag{8.34}$$

where $R(\theta)$ denotes the nuclear surface. The mean value of the angular momentum was taken as

$$\langle \mathbf{L}^2 \rangle = \int_0^{\pi} \sin\theta d\theta \int_{R(\theta)}^{\infty} |\Phi(t,r,\theta)|^2 \mathbf{L}^2 r^2 dr \times \left[\int_0^{\pi} \sin\theta d\theta \int_{R(\theta)}^{\infty} |\Phi(t,r,\theta)|^2 r^2 dr \right]^{-1}. \tag{8.35}$$

This formalism was applied to the hypothetical emitter ^{208}Pb. The initial quasistationary proton states were obtained from a deformed single particle shell model. It turns out that the results display a picture which is rather complex and does not always agree with that of a particle escaping along the path of minimum Coulomb barrier, as introduced by Hill and Wheeler in α-decay. The tunneling path is given not by the shape of the potential, but by the spatial distribution of the initial wave function, which essentially depends upon the quantum numbers. A crucial role in the time evolution of the wave packet is played by the distribution of the residual angular momentum and its time variation. Another interesting feature shows that not all states have an asymptotic decay width. For these states, no exponential decay was found until the end of the calculated time, although more than half of the wave function already escaped.

A special attention was paid to the numerical reflection on boundaries. In Ref. [10], artificial boundary conditions had to be imposed on numerical solution of the time-dependent Schrödinger equation, in order to eliminate the reflections of the wave packet at the numerical boundaries. The impact of reflections on the time-dependent rate could be reduced by replacing, according to the continuity equation, the time derivative of the tunneling probability in proton emission by the flux at the outer turning point.

Finally, we mention that several papers were devoted to the time-dependent approach to the α-decay. In Ref. [11], the Schrödinger equation with a time dependent fissioning potential was solved in order to compute the angular distribution of the emitted α-particles.

8.2 Resonating Group Method (RGM)

The Resonating Group Method (RGM) is the most general many-body approach to describe scattering processes involving composite objects, like for instance α-scattering on nuclei. Outgoing Gamow-like solutions can be extracted in a standard way, in order to evaluate decay widths of the emission process. RGM is closely related to the Generator Coordinate Method (GCM) by an unitary transformation between the corresponding basis vectors and operators [12].

We will give basic relation concerning RGM [13, 14]. The wave function of the system is given by Eq. 8.22, i.e.

$$\Psi(\mathbf{R}_{12}, \mathbf{x}_1, \mathbf{x}_2) = \mathscr{A}[u(\mathbf{R}_{12})\Phi_1(\mathbf{x}_1)\Phi_2(\mathbf{x}_2)], \tag{8.36}$$

where $\mathbf{R}_{12}$ is the distance between the cm of the two fragments and Φ_k are the shell model wave functions of the fragments. By introducing the basis states in the coordinate representation, corresponding to the separation distance $\mathbf{R}$ between fragments

$$\langle \mathbf{r}_1, \ldots, \mathbf{r}_N | \mathbf{R} \rangle = \mathscr{A}[\delta(\mathbf{R} - \mathbf{R}_{12})\Phi_1\Phi_2], \tag{8.37}$$

the wave function (8.36) can be written as follows

$$|\Psi\rangle = \int d\mathbf{R} u(\mathbf{R})|\mathbf{R}\rangle. \tag{8.38}$$

By projecting out from the stationary Schrödinger equation

$$\mathbf{H}|\Psi\rangle = E|\Psi\rangle, \tag{8.39}$$

a given component of the basis (8.37), one obtains the RGM integral equation

$$\int d\mathbf{R}'[H(\mathbf{R}, \mathbf{R}') - EN(\mathbf{R}, \mathbf{R}')]u(\mathbf{R}') = 0, \tag{8.40}$$

where the corresponding kernels are given by

$$\begin{aligned} N(\mathbf{R}, \mathbf{R}') &\equiv \langle \mathbf{R} | \mathbf{R}' \rangle = \delta(\mathbf{R} - \mathbf{R}') - K(\mathbf{R}, \mathbf{R}') \\ H(\mathbf{R}, \mathbf{R}') &\equiv \langle \mathbf{R} | (\mathbf{H} - E_1 - E_2) | \mathbf{R}' \rangle. \end{aligned} \tag{8.41}$$

The RGM equation (8.40) has a similar form with the CGM equation [12], by replacing delta function with a Gaussian distribution in the basis state (8.37), i.e.

$$\langle \mathbf{r}_1, \ldots, \mathbf{r}_N | \mathbf{S} \rangle = \mathscr{A}[\Gamma(\mathbf{S}, \mathbf{R}_{12})\Phi_1\Phi_2], \tag{8.42}$$

where

$$\Gamma(\mathbf{S}, \mathbf{R}_{12}) = \left(\frac{\pi}{\beta}\right)^{3/2} exp\left[-(\mathbf{S} - \mathbf{R}_{12})^2/(2\beta^2)\right], \tag{8.43}$$

so that (8.38) becomes

$$|\Psi^{\mathrm{CGM}}\rangle = \int d\mathbf{S} u^{\mathrm{CGM}}(\mathbf{S})|\mathbf{S}\rangle. \tag{8.44}$$

The CGM basis (8.42) has practical advantages, due to the finite Gaussian distribution of the kernel (8.43).

The RGM equation (8.40) can be symbolically written as follows

$$\left[\hat{H} - E(1 - \hat{K})\right]|u\rangle = 0. \tag{8.45}$$

By excluding the vanishing eigenvalues $\lambda_n = 0$ of the metric matrix

$$\left(1 - \hat{K}\right)|\psi_n\rangle = \lambda_n|\psi_n\rangle, \tag{8.46}$$

one obtains an equivalent standard Schrödinger equation, i.e.

$$\left(\hat{H}_\Omega - E\right)|\Omega\rangle = 0, \tag{8.47}$$

where we introduced a new basis vector

$$|\Omega\rangle \equiv \left(1 - \hat{K}\right)^{\frac{1}{2}}|u\rangle, \tag{8.48}$$

and the transformed operator in this new basis is given by

$$\hat{H}_\Omega \equiv \left(1 - \hat{K}\right)^{\frac{1}{2}}\hat{H}\left(1 - \hat{K}\right)^{-\frac{1}{2}}. \tag{8.49}$$

This procedure is equivalent to the diagonalization in a non-orthogonal basis, described in Appendix (14.9). The advantage to work with the new basis vectors (8.48) is that they are properly normalized, i.e.

$$\langle\Omega_E|\Omega_{E'}\rangle = \delta(E - E'). \tag{8.50}$$

This property leads to the idea of replacing in the matrix element (8.24), defining the decay width, the wave function $U(\mathbf{R})$ by the properly normalized function $\Omega(\mathbf{R})$ [13]. In this way, in order to keep constant the product

$$U(\mathbf{R})\mathscr{F}(R) = \Omega(\mathbf{R})\mathscr{G}(R), \tag{8.51}$$

one has to replace the preformation amplitude $\mathscr{F}(R)$ (8.25) by a new one, defined as follows

$$\mathscr{G}(R) = \left(1 - \hat{K}\right)^{-\frac{1}{2}}\mathscr{F}(R). \tag{8.52}$$

Anyway, the overlap kernel $1-\hat{K}$ strongly depends upon the radius inside the nuclear volume [13], where the exchange effects are important. For this reason, in this case it is more appropriate to speak about α-correlations than about an α-particle. For distances beyond the geometrical touching point, where Pauli effects diminish, the exchange kernel $\hat{K}$ practically vanishes and the two amplitudes coincide $\mathscr{G} \approx \mathscr{F}$.

8.3 Feshbach Reaction Theory

A similar to (8.21) relation for the decay width can be derived within a general approach of open systems, by using the Feshbach theory of nuclear reactions [15], applied for the α-decay from Po isotopes in [16]. This theory estimates the decay probability from some parent wave function Ψ_k in a bound state k to a scattering state Ψ_E^c given by Eq. 2.59 in Chap. 2, due to the penetration through the Coulomb barrier. Thus, the total wave function can be splitted into two orthogonal components

$$|\Psi\rangle = Q|\Psi\rangle + P|\Psi\rangle, \tag{8.53}$$

where the projection operators are written in a standard way

$$\begin{aligned} Q &= \textstyle\sum_k |\Phi_k\rangle\langle\Phi_k| \\ P &= \textstyle\sum_c \int dE|\Psi_E^c\rangle\langle\Psi_E^c|. \end{aligned} \tag{8.54}$$

From the stationary Schrödinger equation $(E-H)|\Psi\rangle = 0$, by using the orthogonality of the projection operators, one derives the following equations

$$\begin{aligned} (E-\mathbf{H}_{PP})P|\Psi\rangle &= \mathbf{H}_{PQ}Q|\Psi\rangle \\ (E-\mathbf{H}_{QQ})Q|\Psi\rangle &= \mathbf{H}_{QP}P|\Psi\rangle, \end{aligned} \tag{8.55}$$

where $\mathbf{H}_{PP} = P\mathbf{H}P$, $\mathbf{H}_{QQ} = Q\mathbf{H}Q$, $\mathbf{H}_{PQ} = P\mathbf{H}Q$. Then, by replacing the outgoing solution of the first equation in terms of the Green operator, i.e.

$$P|\Psi\rangle = PG^{(+)}P\mathbf{H}_{PQ}|\Psi\rangle \equiv P\frac{1}{E+i\varepsilon-\mathbf{H}}P\mathbf{H}_{PQ}|\Psi\rangle, \tag{8.56}$$

into the second one, we obtain the following relation

$$\left(E-\mathbf{H}_{QQ}-\mathbf{H}_{QP}P\frac{1}{E+i\varepsilon-\mathbf{H}}P\mathbf{H}_{PQ}\right)Q|\Psi\rangle = 0. \tag{8.57}$$

By using the expression of the projecting operators (8.54) and the well known decomposition of the Green operator spectral components

$$\frac{1}{E-E'+i\varepsilon} = \mathscr{P}\frac{1}{E-E'} + i\pi\delta(E-E'), \tag{8.58}$$

one obtains

$$\sum_{k'}\left[(E-E_{k'})-\Delta E_{kk'}+\frac{i}{2}\Gamma_{kk'}\right]\langle\Phi_{k'}|\Psi\rangle=0, \tag{8.59}$$

where the energy shift is given by the principal value of the integral

$$\Delta E_{kk'}=\sum_{c'}\mathscr{P}\int dE'\frac{D_{kc'}(E')D^*_{k'c'}(E')}{E-E'}, \tag{8.60}$$

while the decay width is the residue of the integral, i.e.

$$\Gamma_{kk'}=2\pi\sum_{c}D_{kc}(E)D^*_{k'c}(E), \tag{8.61}$$

where we introduced the following notation

$$D_{kc}(E)\equiv\langle\Phi_k|\mathbf{H}|\Psi^c_E\rangle. \tag{8.62}$$

In most decay processes the width is much smaller than the decay energy and the above matrices are practically diagonal. In this case, the above relations are similar to Eq. 8.16, derived within the time dependent approach. There is one difference, namely the transition operator $\mathbf{W}$ is replaced here by the full Hamiltonian $\mathbf{H}=\mathbf{H_0}+\mathbf{W}$. On the other hand, Φ_k is an eigenstate of the non-perturbed Hamiltonian $\mathbf{H}_0$. Due to the orthogonality between the two sets of states, $\langle\Phi_k|\Psi^c_E\rangle=0$, one obtains the equivalence between the time dependent and reaction theory.

The bound state can be obtained by a direct diagonalization of the shell model Hamiltonian in the internal region, while in order to find scattering state, one has to solve the following integro-differential equation

$$(E-H)|\Psi^c_E\rangle=\sum_k D_{kc}(E)|\Phi_k\rangle. \tag{8.63}$$

By using the auxiliary functions satisfying the equations

$$\begin{aligned}(E-\mathbf{H})|\Psi^c_{E0}\rangle&=0\\(E-\mathbf{H})|\Psi^c_{Ek}\rangle&=|\Phi_k\rangle,\end{aligned} \tag{8.64}$$

from the orthogonality condition $\langle\Phi_k|\Psi^c_E\rangle=0$ one obtains the following algebraic system of equations for unknowns $D_{kc}(E)$

$$\sum_{k'}D_{ck'}\langle\Phi_k|\Psi_{Ek'}\rangle=\langle\Phi_k|\Psi^c_{E0}\rangle. \tag{8.65}$$

By expanding the auxiliary functions in the core-angular multipole basis (2.60 in Chap. 2), one gets a system of differential equations for radial components

$$\sum_{c'}\left\{\left[\frac{\hbar^2}{2\mu}\left(\frac{d^2}{dr^2}-\frac{l_c(l_c+1)}{r^2}\right)-V_0(r)+E_c\right]\delta_{cc'}-\langle c|V_d(\mathbf{r})|c'\rangle\right\}$$
$$\times\begin{bmatrix} g_{c'0}(r) \\ g_{c'k}(r) \end{bmatrix}=\begin{bmatrix} 0 \\ r\mathscr{F}_{ck}(r) \end{bmatrix}, \tag{8.66}$$

where the inhomogeneous part is the overlap integral between the internal part and external core-angular function (2.143 in Chap. 2), i.e. the preformation amplitude

$$\mathscr{F}_{ck}(r)=\langle c|\Phi_k\rangle. \tag{8.67}$$

8.4 R-Matrix Approach

In the R-matrix theory, one divides the space into an internal and an external region at a radius R. The fundamental ingredient of the theory is the logarithmic derivative of the internal wave function at R, i.e. $\beta_c^{(\mathrm{int})}(E, R)$. In the standard version of the R-matrix theory [17–19] this quantity is taken as a free parameter because the microscopic evaluation of the internal solution was considered impossible in the processes thought to be of interest at that time. Nowadays, one can perform that calculation in various processes. Thus, for instance, concerning proton emission, in [20] the quantities β were evaluated by using a diagonalization procedure in a spherical harmonic oscillator basis. On the other hand, the preformation amplitude (8.25) in the internal region can be estimated for α and other cluster decays. Due to the fact that R-matrix theory played a central role in the microscopic theory of the α-decay and heavy custer emission, we will present it in detail.

8.4.1 Spherical Emitters

Let us first present the R-matrix theory for spherical boson emitters, like even-even α-emitters. Here, the channel index is identified with the angular momentum of the relative motion $c = l$. Within the R-matrix theory, one expands the internal wave function by using a *different basis* $u_l(r)$ with respect to the standard diagonalization basis, defined only for the internal region $r \in [0, R]$. This basis is obtained by imposing the boundary condition

$$\beta_l^{(\mathrm{int})}(E,R)=\left[\frac{r}{f_{nl}(r)}\frac{df_{nl}(r)}{dr}\right]_{r=R}. \tag{8.68}$$

for all values of n.

The system of functions $f_{nl}(r)$ is orthonormal in the *finite interval* $[0, R]$, i.e.

$$\int_0^R f_{nl}^*(r) f_{n'l} dr = \delta_{nn'}. \tag{8.69}$$

which differs from the orthonormality condition on the infinite interval of Eq. 4.32 in Chap. 4. In this basis the internal wave function reads

$$f_l^{(\mathrm{int})}(E,r) = \sum_n a_n(E) f_{nl}(r). \tag{8.70}$$

The difference between this expansion and the corresponding one in a harmonic oscillator basis is very small inside the nuclear volume.

By using again Eq. 2.42 in Chap. 2 for two wave functions with energies E and E_n one readily obtains the expansion coefficient as

$$\begin{aligned} a_n(E) &= \int_0^R f_{nl}^*(r) f_l^{(\mathrm{int})}(E,r) dr \\ &= \frac{\hbar^2}{2\mu R} \frac{f_{nl}(R)}{E_n - E} \left[R \frac{df_l^{(\mathrm{int})}(E,R)}{dr} - f_l^{(\mathrm{int})}(E,R) \beta_l^{(\mathrm{int})}(E_n,R) \right]. \end{aligned} \tag{8.71}$$

Then the internal wave function can be expressed as follows

$$f_l^{(\mathrm{int})}(E,r) = \mathcal{G}(r,R) \left[R \frac{df_l^{(\mathrm{int})}(E,R)}{dr} - f_l^{(\mathrm{int})}(E,R) \beta_l^{(\mathrm{int})}(E_n,R) \right], \tag{8.72}$$

in terms of the Green's function

$$\mathcal{G}(r,R) = \sum_n \frac{\gamma_{nl}(r)\gamma_{nl}(R)}{E_n - E}, \tag{8.73}$$

where $\gamma_{nl}(R)$ is the reduced width (2.43 in Chap. 2). This relation allows us to express the logarithmic derivative of the internal wave function at the radius R

$$\begin{aligned} \left[\frac{R}{f_l^{(\mathrm{int})}(E,R)} \frac{df_l^{(\mathrm{int})}(E,R)}{dr} - \beta_l^{(\mathrm{int})}(E_n,R) \right]^{-1} &= \sum_n \frac{\gamma_{nl}^2(R)}{E_n - E} \\ &\equiv \mathcal{R}(R), \end{aligned} \tag{8.74}$$

where $\mathcal{R}$ denotes the R-matrix, which turns out to be the Green's function for equal arguments $\mathcal{G}(R,R)$.

The external logarithmic derivative is given in terms of the Coulomb–Hankel function

$$\beta_l^{(\mathrm{ext})}(E,R) \equiv \frac{\kappa R}{H_l^{(+)}(\chi,\kappa R)} \frac{dH_l^{(+)}(\chi,\kappa R)}{d\rho} \equiv S_l(E,R) + iP_l(E,R). \tag{8.75}$$

where we introduced the so-called shift factor [19] and penetrability (2.37 in Chap. 2), in terms of regular and irregular Coulomb functions as follows

$$S_l(E,R) \equiv \kappa R \frac{F_l'(\chi,\kappa R)F_l(\chi,\kappa R) + G_l'(\chi,\kappa R)G_l(\chi,\kappa R)}{F_l^2(\chi,\kappa R) + G_l^2(\chi,\kappa R)},$$
$$P_l(E,R) \equiv \frac{\kappa R}{F_l^2(\chi,\kappa R) + G_l^2(\chi,\kappa R)}. \tag{8.76}$$

If the energy is close to an eigenvalue E_n one obtains from (8.74)

$$\left[\frac{r}{f_l^{(\mathrm{int})}(E,r)}\frac{df_l^{(\mathrm{int})}(E,r)}{dr}\right]_{r=R} \approx \beta_l^{(\mathrm{int})}(E_n,R) - \gamma_{nl}^{-2}(R)(E-E_n), \tag{8.77}$$

which we already used in Eq. 2.38 in Chap. 2. By using the equality between the internal and external logarithmic derivative, as given by Eq. 8.75, the S-matrix becomes

$$\mathcal{S}_l \approx \overline{\mathcal{S}}_l(E)\left[1 - \frac{i\Gamma_{nl}(E,R)}{E - E_n - \Delta_{nl}(E,R) + \frac{i}{2}\Gamma_{nl}(E,R)}\right], \tag{8.78}$$

where $\overline{\mathcal{S}}_l$ is the background contribution (2.37 in Chap. 2) and the width is given by (2.44 in Chap. 2). The energy shift is defined by

$$\Delta_{nl}(E,R) = -[S_l(E,R) - \beta_l^{(\mathrm{int})}(E,R)]\gamma_{nl}^2(R). \tag{8.79}$$

At the pole of the S-matrix one gets the resonant energy $E_r = E^{(0)} - \frac{i}{2}\Gamma_r$ which should, therefore, satisfy the equation

$$E_r - E_n - \Delta_{nl}(E_r,R) + \frac{i}{2}\Gamma_{nl}(E_r,R) = 0, \tag{8.80}$$

which is a nonlinear equation for the real ($E^{(0)}$) and imaginary (Γ_r) part of the energy of the Gamow state.

This procedure ensures the independence of the decay width upon the matching radius and restores the properties of the resonant states at large distances, only if the dynamics is described by a potential defined for external, as well as internal regions, as it is the case of proton emission. If the internal wave function is computed independently from the potential describing the external wave function, as is the case of the preformation amplitude, then this condition is not anymore fulfilled. This point will be discussed later, within the selfconsistent decay theory.

8.4.2 Deformed Emitters

Next we will describe the generalization of the R-matrix formalism to deformed nuclei [19]. The internal parent wave function can be expanded in terms of some eigenstates

$$\Psi^{(P)} = \sum_n a_n \Psi_n^{(P)}, \tag{8.81}$$

where

$$\begin{aligned} \mathbf{H}\Psi_n^{(P)} &= E_n \Psi_n^{(P)} \\ \langle \Psi_n^{(P)} | \Psi_{n'}^{(P)} \rangle &= \delta_{nn'}. \end{aligned} \tag{8.82}$$

The coefficients a_n in Eq. 8.81 are given by

$$a_n = \langle \Psi_n^{(P)} | \Psi^{(P)} \rangle. \tag{8.83}$$

The parent wave function and its nth component have a channel representation in terms of core-angular harmonics similar to Eq. 2.59 in Chap. 2, i.e.

$$\begin{aligned} \Psi^{(P)} &\to \sum_c \frac{f_c^{(\mathrm{int})}(R)}{R} \mathscr{Y}^{(c)} \\ \Psi_n^{(P)} &\to \sum_c \frac{f_{nc}(R)}{R} \mathscr{Y}^{(c)}. \end{aligned} \tag{8.84}$$

Thus, the radial components can be projected out by using the core-angular hamonics (2.60 in Chap. 2) and they can be identified with the preformation amplitude (8.25), i.e

$$\begin{aligned} \frac{f_c^{(\mathrm{int})}(R)}{R} &\to \mathscr{F}_c(R) = \langle Y^{(c)} | \Psi^{(P)} \rangle \\ \frac{f_{nc}(R)}{R} &\to \mathscr{F}_{nc}(R) = \langle \mathscr{Y}^{(c)} | \Psi_n^{(P)} \rangle. \end{aligned} \tag{8.85}$$

For each channel one introduces the reduced width and its gradient

$$\begin{aligned} \gamma_{nc}(R_c) &= \sqrt{\frac{\hbar^2}{2\mu_c R_c}} f_{nc}(R_c) \\ \delta_{nc}(R_c) &= \sqrt{\frac{\hbar^2 R_c}{2\mu_c}} \left[\frac{df_{nc}(r)}{dr} \right]_{r=R_c}. \end{aligned} \tag{8.86}$$

One has similar definitions for $\gamma_c(R_c)$, $\delta_c(R_c)$ involving parent function $\Psi^{(P)}$. The usual condition within the R-matrix theory is that the logarithmic derivative of the internal wavefunction component

$$\frac{\delta_{nc}(R_c)}{\gamma_{nc}(R_c)} = \beta_c^{(\mathrm{int})}(R_c), \tag{8.87}$$

has the same value for all eigenstates n. The generalization of the R-function (8.74) to the deformed case is straightforward. By using Schrödinger equation for two different energies

$$\mathbf{H}\Psi_1 = E_1\Psi_1, \quad \mathbf{H}\Psi_2 = E_2\Psi_2, \tag{8.88}$$

we multiply each equality on the left with the other wave function conjugate, substract the equalities and integrate over the coordinates of the fragments and the internal volume. For a Hermitian potential, one obtains the Green's theorem, i.e.

$$\begin{aligned}(E_2 - E_1)\int \Psi_1^*\Psi_2 d\mathbf{r} &= \frac{\hbar^2}{2\mu_c}\int_S (\Psi_2^*\nabla_S\Psi_1 - \Psi_1\nabla_S\Psi_2^*)dS \\ &= \sum_c (\gamma_{2c}^*\delta_{1c} - \gamma_{1c}\delta_{2c}^*),\end{aligned} \tag{8.89}$$

where by the integration over the surface S, we understand an integration over fragment coordinates and relative angular variable. From (8.83), with $\Psi_1 \to \Psi^{(P)}(E,r)$ and $\Psi_2 \to \Psi_n^{(P)}(r)$, one obtains the expression of the expansion coefficients similar to Eq. 8.71, i.e.

$$a_n = \sum_c \frac{\gamma_{nc}}{E_n - E}\left[\delta_c(E) - \gamma_c(E)\beta_c^{(\mathrm{int})}(E_n)\right], \tag{8.90}$$

where we used the expansions (8.85) and the orthogonality of core-angular harmonics. We also pointed out the energy as the argument in the above relation, except for the functions with the lower index "n", which implicitlly have as argument E_n. The wave function is given by

$$\Psi^{(P)}(E,r) = \sum_c\left[\sum_n \frac{\Psi_n^{(P)}(r)\gamma_{nc}(R_c)}{E_n - E}\right]\left[\delta_c(E,R_c) - \gamma_c(E,R_c)\beta_c^{(\mathrm{int})}(E_n,R_c)\right], \tag{8.91}$$

where the function within the first brackets plays the role of the Green's function. We evidenced here the radial argument, because this relation expresses the wave function in any point r in terms of the values and derivatives at the channel surface R_c. By multiplying to left with the core-angular harmonics $\mathscr{Y}^{(c')}$ and integrating one obtains the channel reduced width

$$\gamma_{c'}(E) = \sum_c \mathscr{R}_{c'c}\left[\delta_c(E) - \gamma_c(E)\beta_c^{(\mathrm{int})}(E_n)\right], \tag{8.92}$$

where the R-matrix is defined as follows [19]

$$\mathscr{R}_{cc'} = \sum_n \frac{\gamma_{nc'}\gamma_{nc}}{E_n - E}. \tag{8.93}$$

The above relation (8.92) is the analog of Eq. 8.74. One obtains the S-matrix in the one-level approximation by using the equality between internal and external logarithmic derivatives with the channel angular momentum l_c in (8.76), i.e

$$\mathcal{S}_{cc'} \approx \overline{\mathcal{S}}_{cc'}(E)\left[\delta_{cc'} - \frac{i\Gamma_{nc}^{1/2}(E,R)\Gamma_{nc'}^{1/2}(E,R)}{E - E_n - \Delta_n(E,R) + \frac{i}{2}\Gamma_n(E,R)}\right], \tag{8.94}$$

where the total width and energy shift are

$$\begin{aligned} \Gamma_n(E,R) &= \sum_c \Gamma_{nc}(E,R), \\ \Delta_n(E,R) &= \sum_c \Delta_{nc}(E,R), \end{aligned} \tag{8.95}$$

with the partial quantities defined as follows

$$\begin{aligned} \Gamma_{nc}(E,R) &\equiv 2P_{l_c}(E,R)\gamma_{nc}^2(R), \\ \Delta_{nc}(E,R) &\equiv -[S_{l_c}(E,R) - \beta_c^{(\mathrm{int})}(E,R)]\gamma_{nc}^2(R). \end{aligned} \tag{8.96}$$

The above expression of the S-matrix (8.94) is the generalization of the spherical version, given by Eq. 2.46 in Chap. 2. The channel reduced width has the already known expression in (8.86). We introduced the partial channel shift and penetrability as a generalization of (8.75) and (8.76)

$$\frac{\kappa R}{H_{l_c}^{(+)}(\chi,\kappa R)}\frac{dH_{l_c}^{(+)}(\chi,\kappa R)}{d\rho} \equiv S_{l_c}(E,R) + iP_{l_c}(E,R). \tag{8.97}$$

As we already mentioned, in order to make the theory selfconsistent it is necessary that the product between the reduced width and penetrability in the decay width (8.96) should not depend upon the matching radius R. This is not at all a trivial operation and we will discus this point later in Sect. 10.4 in Chap. 10.

In order to obtain the complex resonant energy, $E_r = E^{(0)} - \frac{i}{2}\Gamma_r$ one finds the pole of the S-matrix (8.94), i.e.

$$E_r - E_n - \Delta_n(E_r,R) + \frac{i}{2}\Gamma_n(E_r,R) = 0, \tag{8.98}$$

By inserting the energy dependence (8.97), one obtains a nonlinear equation, giving the real $E^{(0)}$ and imaginary part Γ_r of the Gamow state energy.

This procedure to estimate the proton decay width from triaxial nuclei was used in [20]. Here, there were first computed the eigenvalue E_n^{HO} and decay width Γ_n^{HO}, by using the diagonalization procedure with an harmonic-oscillator basis. Then a linear ansatz for the energy shift was used, i.e.

$$\Delta_{nc}(E) \approx -\left[S_{l_c}(E_\lambda) - \beta_c^{(\mathrm{int})} + \frac{dS_{l_c}(E_n)}{dE}(E - E_n)\right]\gamma_{nc}^2, \tag{8.99}$$

giving the following nth solutions of Eq. 8.98 for the resonant energy and decay width, respectively

$$
\begin{aligned}
E^{(0)} &= \left[E_n^{HO} + \sum_c \left[\beta_c^{(\text{int})} - S_{l_c}(E_n) \right] \gamma_{nc}^2 \right] \left[1 + \sum_c \frac{dS_{l_c}(E_n)}{dE} \gamma_{nc}^2 \right]^{-1}, \\
\Gamma_r &= \Gamma_n^{HO}(E_n) \left[1 + \sum_c \frac{dS_{l_c}(E_n)}{dE} \gamma_{nc}^2 \right]^{-1},
\end{aligned}
\tag{8.100}
$$

where the core+harmonic oscillator channel is indicated by $c \to (J_f Knlj)$.

8.4.3 R-Matrix Approach for Two Proton Emission

The sequential emission of two protons can be described within the R-matrix approach by considering one proton emission from the core-proton correlated system, like for instance the sequential emission from ^{12}O through the tail of ^{11}N [21, 22]. The density of the probability is given as follows

$$
N(E, U) \sim \frac{\Gamma_1(E, U)}{[E - Q_{2p} - \Delta_{\text{tot}}(E)]^2 + \Gamma_{\text{tot}}^2(E)/4}, \tag{8.101}
$$

in terms of the total decay energy E, the relative energy of the second decay U and the two-proton decay energy Q_{2p}. The total width and energy shift are given by integration over the energy of the second decay, i.e.

$$
\begin{aligned}
\Gamma_{\text{tot}}(E) &= \int_0^E \Gamma_1(E, U) dU. \\
\Delta_{\text{tot}}(E) &= \int_0^E \Delta_1(E, U) dU,
\end{aligned}
\tag{8.102}
$$

Here, the width and energy shift under the integral sign are written in the standard R-matrix factorized form as follows

$$
\begin{aligned}
\Gamma_1(E, U) &= 2\gamma_1^2 P_{1l}(E - U)\rho(U), \\
\Delta_1(E, U) &= -\gamma_1^2 \left[S_{1l}(E - U) - S_{1l}(Q_{2p} - U) \right] \rho(E, U),
\end{aligned}
\tag{8.103}
$$

with the density of states given by

$$
\begin{aligned}
\rho(U) &= c \frac{\Gamma_2(U)}{[U - Q_{1p} - \Delta_2(U)]^2 + \Gamma_2^2(U)/4}, \\
\int_0^\infty \rho(U) dU &= 1,
\end{aligned}
\tag{8.104}
$$

where

$$\begin{aligned}\Gamma_2(U) &= 2\gamma_2^2 P_{2l}(U),\\ \Delta_2(U) &= -\gamma_1^2\left[S_{2l}(U) - S_{2l}(Q_{2p})\right].\end{aligned} \tag{8.105}$$

A very good approximation is, as usually, the Thomas ansatz, i.e. the linear dependence of energy shift factors S_{1l} and S_{2l} upon the energy, as discussed in [22].

References

1. Harada, K., Rauscher, E.A.: Unified theory of alpha decay. Phys. Rev. **169**, 818–824 (1968)
2. Fliessbach, T., Mang, H.J.: On absolute values of α-decay rates. Nucl. Phys. A **263**, 75–85 (1976)
3. Wildermuth, K., Fernandez, F., Kanellopoulos, E.J., Sünkel, W.: J. Phys. G **6**, 603–617 (1980)
4. Fox, L.: Numerical Solution of Ordinary and Partial Differential Equations. Pergamon Press, New York (1962)
5. Ixaru, L.: Numerical Methods for Differential Equations and Applications. Reidel, Boston (1984)
6. Mişicu, S., Cârjan, N.: Proton decay from excited states in sopherical nuclei. J. Phys. **G24**, 1745–1755 (1998)
7. Talou, P., Strottman, D., Cârjan, N.: Exact calculation of proton decay rates from excited states in spherical nuclei. Phys. Rev. C **60**, 054318/1–7 (1999)
8. Talou, P., Cârjan, N., Negrevergne, C., Strottman, D.: Exact dynamical approach to spherical ground-state proton emission. Phys. Rev. C **62**, 014609/1–4 (2000)
9. Talou, P., Cârjan, N., Strottman, D.: Time-dependent approach to bidimensional quantum tunneling: application to the proton emission from deformed nuclei. Nucl. Phys. A **647**, 21–46 (1999)
10. Cârjan, N., Rizea, M., Strottman, D.: Improved boundary conditions for the decay of low lying metastable proton states in a time-dependent approach. Comput. Phys. Commun. **173**, 41–60 (2005)
11. Tanimura, O., Fliessbach, T.: Dynamic model for alpha particle emission during fission. Z. Phys. A **328**, 475–486 (1987)
12. Canto, F., Brink, D.M.: Microscopic description of the collision between nuclei. Nucl. Phys. A **279**, 85–96 (1977)
13. Fliessbach, T.: The reduced width amplitude in the reaction theory for composite particles. Z. Phys. A **272**, 39–46 (1975)
14. Fliessbach, T., Walliser, H.: The structure of the resonanting group equation. Nucl. Phys. A **377**, 84–104 (1982)
15. Feshbach, H.: Unified theory of nuclear reactions. Ann. Phys. (NY) **5**, 357–390 (1958)
16. Săndulescu, A., Silişteanu, I., Wünsch, R.: Alpha decay within Feshbach theory of nuclear reactions. Nucl. Phys. A **305**, 205–212 (1978)
17. Teichmann, T., Wigner, E.P.: Sum rules in the dispersion theory of nuclear reactions. Phys. Rev. **87**, 123–135 (1952)
18. Thomas, R.G.: A formulation of the theory of alpha-particle decay from time-independent equations. Prog. Theor. Phys. **12**, 253–264 (1954)
19. Lane, A.M., Thomas, R.G.: R-Matrix theory of nuclear reactions. Rev. Mod. Phys. **30**, 257–353 (1958)

20. Kruppa, A.T., Nazarewicz, W.: Gamow and R-Matrix approach to proton emitting nuclei. Phys. Rev. C **69**, 054311/1–11 (2004)
21. Kryger, R.A., et al.: Two-proton emission from the ground state of ^{12}O. Phys. Rev. Lett. **74**, 860–863 (1995)
22. Barker, F.C.: Width of the ^{12}O ground state. Phys. Rev. C **59**, 535–538 (1999)

Chapter 9
Preformation Amplitude

9.1 Definition

The main building blocks of emitted clusters are single particle (sp) wave functions, computed as eigenstates of the nuclear plus Coulomb mean field. For heavier clusters emission, like for instance cold fission, the α-like description becomes not tractable and the so called fission-like theories are more suitable. In such theories the building blocks should also be sp states, but estimated as eigenstates in the two-center representation of the mean field, as described in the last chapter.

We will consider the α-decay as typical cluster emission process

$$P \to D + \alpha. \tag{9.1}$$

The amplitude of this process, called the α-particle preformation amplitude, is given as the following overlap integral [1, 2]

$$\mathscr{F}(\mathbf{R}_\alpha) = \langle \alpha D | P \rangle = \int d\mathbf{x}_\alpha d\mathbf{x}_D \left[\psi_\alpha^{(\beta_\alpha)}(\mathbf{x}_\alpha) \Psi^{(D)}(\mathbf{x}_D) \right]^* \Psi^{(P)}(\mathbf{x}_P) \,, \tag{9.2}$$

where by $\mathbf{x}$ we denoted the internal coordinates of the fragments. Therefore, the final result should depend upon the relative radius between emitted fragments. The antisymmetrization effects are neglected, because we are interested in distances beyond the geometrical touching point, where Pauli principle diminishes.

9.2 α-Particle Wave Function

The first ingredient of the preformation amplitude (9.2) is given by the α-particle wave function. In order to understand the structure of the α-particle function let us transform the product of two proton and two neutron Gaussians to the center of mass (cm) system

D. S. Delion, *Theory of Particle and Cluster Emission*, Lecture Notes in Physics, 819,
DOI: 10.1007/978-3-642-14406-6_9, © Springer-Verlag Berlin Heidelberg 2010

$$e^{-\beta_\alpha r_1^2/2} e^{-\beta_\alpha r_2^2/2} e^{-\beta_\alpha r_3^2/2} e^{-\beta_\alpha r_4^2/2} = e^{-\beta_\alpha (r_\pi^2 + r_\nu^2 + r_\alpha^2)/2} e^{-\beta_\alpha R_\alpha^2/2}$$
$$\sim \psi_{\rm rel}^{(\beta_\alpha)}(r_\pi, r_\nu, r_\alpha) \psi_{\rm cm}^{(\beta_\alpha)}(R_\alpha), \tag{9.3}$$

where we have labelled by 1, 2 the proton and by 3, 4 the neutron coordinates. The relative and cm normalized wave functions are respectively given by

$$\begin{aligned} \psi_{\rm rel}^{(\beta_\alpha)}(r_\pi, r_\nu, r_\alpha) &= \phi_{00}^{(\beta_\alpha)}(r_\pi) \phi_{00}^{(\beta_\alpha)}(r_\nu) \phi_{00}^{(\beta_\alpha)}(r_\alpha), \\ \psi_{\rm cm}^{(\beta_\alpha)}(R_\alpha) &= \phi_{00}^{(\beta_\alpha)}(R_\alpha). \end{aligned} \tag{9.4}$$

Here $\phi_{nl}^{(\beta_\alpha)}$ is the radial harmonic oscillator (ho) wave function with the parameter $\beta_\alpha \approx 0.5$ fm^{-2} [3]. These wave functions are written in terms of the relative and cm Moshinsky coordinates

$$\begin{pmatrix} \mathbf{r}_\pi \\ \mathbf{R}_\pi \end{pmatrix} = \frac{\mathbf{r}_1 \mp \mathbf{r}_2}{\sqrt{2}}, \quad \begin{pmatrix} \mathbf{r}_\nu \\ \mathbf{R}_\nu \end{pmatrix} = \frac{\mathbf{r}_3 \mp \mathbf{r}_4}{\sqrt{2}}, \quad \begin{pmatrix} \mathbf{r}_\alpha \\ \mathbf{R}_\alpha \end{pmatrix} = \frac{\mathbf{R}_\pi \mp \mathbf{R}_\nu}{\sqrt{2}}, \tag{9.5}$$

Therefore the internal α-particle wave function is written as follows

$$\psi_\alpha^{(\beta_\alpha)}(\mathbf{x}_\alpha) = \psi_{\rm rel}^{(\beta_\alpha)}(r_\pi, r_\nu, r_\alpha) \chi_{00}^{(\pi)}(s_1, s_2) \chi_{00}^{(\nu)}(s_3, s_4), \tag{9.6}$$

where we introduced spin singlet wave function

$$\chi_{00}^{(\tau)} = \left[\chi_{\frac{1}{2}}^{(\tau)} \otimes \chi_{\frac{1}{2}}^{(\tau)} \right]_{00}. \tag{9.7}$$

The absolute and relative volume elements are connected by the following relation

$$d\mathbf{r}_1 d\mathbf{r}_2 d\mathbf{r}_3 d\mathbf{r}_4 = 8 d\mathbf{r}_\pi d\mathbf{r}_\nu d\mathbf{r}_\alpha d\mathbf{R} \equiv d\mathbf{x}_\alpha d\mathbf{R}_\alpha, \quad \mathbf{R} = \mathbf{R}_\alpha/2. \tag{9.8}$$

9.3 Multi-step Shell Model (MSM)

The other ingredients entering the preformation amplitude (9.2) are parent and daughter wave functions. Some emitters have a simple α-clustering structure, like in the case of ^{212}Po. The antisymmetrization effects are small beyond the touching configuration, where we will estimate the decay width, and the wave function of the parent nucleus (P) can be approximately factorized as a product between a four-body state and the daughter (D) wave function, i.e.

$$\Psi^{(P)} = \Psi_\alpha \Psi^{(D)}. \tag{9.9}$$

In general, this factorization is valid in all α-decays for inter-fragment distances corresponding to a weak overlap of nuclear densities. Taking into account this factorization, the integration over daughter coordinates $\mathbf{x}_D$ can be carried out and the overlap integral becomes

$$\mathscr{F}_\alpha(\mathbf{R}_\alpha) = \int d\mathbf{x}_\alpha \psi^*_\alpha(\mathbf{x}_\alpha)\Psi_\alpha. \tag{9.10}$$

The four-body wave function Ψ_α can be written in the Fock space as follows

$$\Psi_\alpha|D\rangle = \sum_{L_\alpha} P^\dagger_{L_\alpha M_\alpha}|D\rangle, \tag{9.11}$$

where the short-hand notation $L_\alpha \equiv (L^\pi_\alpha, a)$ denotes the angular momentum, parity and eigenvalue index. In order to built the four-body wave function, we proceed according to the Multi-step Shell Model (MSM) procedure [4, 5]. To this purpose, let us first introduce the correlated two-particle state with the multipolarity L. The two-particle creation operator for correlated pairs defines an eigenstate of the Tamm–Dankoff approach (TDA), which has actually a two-particle shell model ansatz

$$P^\dagger_{\tau_1\tau_2 LM} = \sum_{j_1 j_2} X_{\tau_1\tau_2}(j_1 j_2; L) C^\dagger_{LM}(\tau_1 j_1 \tau_2 j_2). \tag{9.12}$$

We label sp states by the following short-hand notation $j \equiv (\varepsilon, l, j)$, where ε is sp energy, l angular momentum and j total spin. We also labeled by τ the isospin index defining proton (π) or neutron (ν) degrees of freedom. The pair short-hand index in the operator P and amplitude X has the following meaning $L \equiv (L^\pi, k)$, where L^π is angular momentum with the parity $\pi = \pm$ and k the eigenvalue number. If one considers as daughter the double magic nucleus ^{208}Pb, the $\pi\pi$ correlated pair defines a state in ^{210}Po, the $\nu\nu$ pair in ^{210}Pb, while the $\pi\nu$ pair in ^{210}Bi. The summation is taken over all the indices, denoting spherical shell model quantum numbers, $j_1 j_2$ if $\tau_1 \neq \tau_2$ and over $j_1 \leq j_2$ if $\tau_1 = \tau_2$. The normalized pair operators are defined as follows

$$\begin{aligned} C^\dagger_{LM}(\tau_1 j_1 \tau_2 j_2) &= \frac{1}{\Delta_{j_1 j_2}}\left(a^\dagger_{\tau_1 j_1} \otimes a^\dagger_{\tau_2 j_2}\right)_{LM} \\ \Delta_{j_1 j_2} &\equiv \sqrt{1 + \delta_{\tau_1\tau_2}\delta_{j_1 j_2}}, \end{aligned} \tag{9.13}$$

in terms of sp creation operators $a^\dagger_{\tau jm}$. Here by parentheses we denoted the angular momentum coupling. These operators, as well as the correlated ones defined by Eq. 9.12, satisfy the boson commutation rules rule from which one gets the TDA orthonormality relations for amplitudes

$$\sum_{j_1 j_2} X_{\tau_1\tau_2}(j_1 j_2; L) X_{\tau_1\tau_2}(j_1 j_2; L') = \delta_{LL'}. \tag{9.14}$$

The Hamiltonian for a spherical nucleus is written as a superposition of different multipoles in the particle–particle coupling scheme:

$$H = \sum_{\tau j} \varepsilon_{\tau j} N_{\tau j} + \frac{1}{2} \sum_{L} \sum_{\tau_1 \tau_2} \sum_{i_1 i_2} \sum_{j_1 j_2} \langle \tau_1 i_1 \tau_2 i_2 | V | \tau_1 j_2 \tau_2 j_2 \rangle_L \times \sum_{M} C^{\dagger}_{LM}(\tau_1 i_1 \tau_2 i_2) C_{LM}(\tau_1 j_1 \tau_2 j_2). \tag{9.15}$$

The equation of motion

$$\left[H, P^{\dagger}_{\tau_1 \tau_2 LM}\right] = E_{\tau_1 \tau_2 L} P^{\dagger}_{\tau_1 \tau_2 LM}, \tag{9.16}$$

where the brackets denote the commutator, leads to the two-particle TDA system of equations

$$\sum_{j_1' j_2'} \langle \tau_1 j_1 \tau_2 j_2 | V | \tau_1 j_1' \tau_2 j_2' \rangle_L X_{\tau_1 \tau_2}(j_1' j_2'; L) = (E_{\tau_1 \tau_2 L} - \varepsilon_{\tau_1 j_1} - \varepsilon_{\tau_2 j_2}) X_{\tau_1 \tau_2}(j_1 j_2; L), \tag{9.17}$$

where $\varepsilon_{\tau j}$ is the sp energy of the jth spherical level.

In the second step, one defines the four-body wave function Ψ_α by the correlated four-particle creation operator, which has two components

$$P^{\dagger}_{L_\alpha M_\alpha} = \sum_{L_1 L_2} X_{\pi\pi\nu\nu}(L_1 L_2; L_\alpha) \left(P^{\dagger}_{\pi\pi L_1} \otimes P^{\dagger}_{\nu\nu L_2}\right)_{L_\alpha M_\alpha} + \sum_{L_1 L_2} X_{\pi\nu\pi\nu}(L_1 L_2; L_\alpha) \left(P^{\dagger}_{\pi\nu L_1} \otimes P^{\dagger}_{\pi\nu L_2}\right)_{L_\alpha M_\alpha}, \tag{9.18}$$

where we introduced a similar notation $L_\alpha \equiv (L_\alpha^\pi, k_\alpha)$ for the quartet index. For a core state $|D\rangle = |^{208}\mathrm{Pb}\rangle$ these two terms correspond to the couplings $(^{210}\mathrm{Po} \otimes {}^{210}\mathrm{Pb})_{L_\alpha M_\alpha}$ and $(^{210}\mathrm{Bi} \otimes {}^{210}\mathrm{Bi})_{L_\alpha M_\alpha}$, respectively.

By using the short-hand notations $J \equiv (\tau_1, \tau_2, L)$, the above relation can be rewritten as follows

$$P^{\dagger}_{L_\alpha M_\alpha} = \sum_{J_1 J_2} X(J_1 J_2; L_\alpha) \left(P^{\dagger}_{J_1} \otimes P^{\dagger}_{J_2}\right)_{L_\alpha M_\alpha}. \tag{9.19}$$

Using the TDA equation of motion for the four-particle system

$$[H, P^{\dagger}_{L_\alpha M_\alpha}] = E_{L_\alpha} P^{\dagger}_{L_\alpha M_\alpha}, \tag{9.20}$$

and the analogous two-particle equation (9.16) one obtains the following system of equations by using the symmetrized double commutator

$$\sum_{J_1' J_2'} \mathscr{H}_{L_\alpha}(J_1 J_2; J_1' J_2') X(J_1' J_2'; L_\alpha) = E_{L_\alpha} \sum_{J_1' J_2'} \mathscr{I}_{L_\alpha}(J_1 J_2; J_1' J_2') X(J_1' J_2'; L_\alpha), \tag{9.21}$$

where the metric matrix is defined by the following overlap integral

$$\mathscr{I}_{L_\alpha}(J_1J_2;J_1'J_2') \equiv \langle 0|\left(P_{J_1}^\dagger \otimes P_{J_2}^\dagger\right)_{L_\alpha}^\dagger \left(P_{J_1'}^\dagger \otimes P_{J_2'}^\dagger\right)_{L_\alpha}|0\rangle. \tag{9.22}$$

The Hamiltonian matrix is proportional to the metric matrix

$$\begin{aligned} H_{L_\alpha}(J_1J_2;J_1'J_2') &\equiv \frac{1}{2}\langle 0|\left[\left(P_{J_1}^\dagger \otimes P_{J_2}^\dagger\right)_{L_\alpha}^\dagger, H, \left(P_{J_1'}^\dagger \otimes P_{J_2'}^\dagger\right)_{L_\alpha}\right]|0\rangle \\ &= \frac{1}{2}(E_{J_1} + E_{J_2} + E_{J_1'} + E_{J_2'})\mathscr{I}_{L_\alpha}(J_1J_2;J_1'J_2'). \end{aligned} \tag{9.23}$$

The MSM system of equations (9.21), derived in Ref. [6], is formally different from Ref. [4, 5] which uses a non-symmetric Hamiltonian matrix, but it can be shown that they are equivalent. The metric matrix elements are computed using the standard angular momentum recoupling procedure, and the final result is given in Appendix (14.10). In such a way, the interaction is involved only in the first two-particle step. In the second four-particle step, only the two-particle energies enter the equation of motion. The system of equations (9.21) can be solved using as an orthonormal basis the eigenstates of the metric matrix, as described in Appendix (14.9).

9.3.1 Two-Body Correlations

In order to determine the function $\mathscr{F}_\alpha(\mathbf{R}_\alpha)$ (9.10) describing the amount of the α-clustering given by sp shell model constituents, we have to perform an integration over the relative coordinates. We stress on the fact that the four-body wave function Ψ_α given by Eqs 9.11 and 9.19 depends upon the *absolute* coordinates of two protons and two neutrons. The configuration counterpart of the pair-creation operators entering these quartets has the following form

$$\begin{aligned} \left[a_{j_1}^\dagger \otimes a_{j_2}^\dagger\right]_L &\to \mathscr{A}\{\psi_{j_1}(\mathbf{x}_1) \otimes \psi_{j_2}(\mathbf{x}_2)\}_L \\ &\equiv \frac{1}{\sqrt{2}}\left\{\left[\psi_{j_1}(\mathbf{x}_1) \otimes \psi_{j_2}(\mathbf{x}_2)\right]_L - \left[\psi_{j_1}(\mathbf{x}_2) \otimes \psi_{j_2}(\mathbf{x}_1)\right]_L\right\}. \end{aligned} \tag{9.24}$$

To estimate the overlap integral (9.10), when the above mentioned radial sp wave functions are given as numerical solutions of the Schrödinger equation, is a difficult task, due to the change from absolute to the cm and relative coordinates.

The main difficulty consists in the non-separability between the radial and angular coordinates in the new system. It is, therefore, necessary to perform an angular integration before integrating over the radial coordinates. There is one case in which this operation can be done analytically, namely the spherical ho basis, by using Talmi–Moshinsky recoupling coefficients. This technique was used for the first time in Ref. [2]. It is possible to have a very good representation of the sp wave functions $\psi_j(\mathbf{x})$ in terms of the ho basis

$$\psi_{\tau\varepsilon ljm}(\mathbf{x}) \equiv \sum_n c_n(\tau\varepsilon lj)\varphi^{(\beta)}_{nljm}(\mathbf{x}), \tag{9.25}$$

where the following standard definitions were used

$$\begin{aligned}\varphi^{(\beta)}_{nljm}(\mathbf{x}) &\equiv \left[\phi^{(\beta)}_{nl}(\mathbf{r}) \otimes \chi_{\frac{1}{2}}(s)\right]_{jm},\\ \phi^{(\beta)}_{nlm}(\mathbf{r}) &\equiv \mathscr{R}^{(\beta)}_{nl}(r)Y_{lm}(\hat{r}).\end{aligned} \tag{9.26}$$

Here, ε is the eigenvalue index $c_n(\tau\varepsilon lj)$ the expansion coefficients in the spherical ho basis $\mathscr{R}^{(\beta)}_{nl}(r)$ and $\beta \equiv M_N\omega/\hbar$ is the ho parameter.

In the structure of the quartet operator (9.18) enter $\pi\pi$–$\nu\nu$ components. The pair $\pi\pi$ building block is given by the product of the sp wave functions entering the ho basis. Due to the fact that the α-particle wave function is antisymmetric with respect to 12 coordinates, one obtains that the antisymmetric product of the sp wave functions can be written as

$$\begin{aligned}\mathscr{A}\{\psi_{j_1}(\mathbf{x}_1) \otimes \psi_{j_2}(\mathbf{x}_2)\}_{L_\pi M_\pi} &\rightarrow \sqrt{2}\left[\psi_{j_1}(\mathbf{x}_1) \otimes \psi_{j_2}(\mathbf{x}_2)\right]_{L_\pi M_\pi}\\ &= \sqrt{2}\sum_{n_1 n_2} c_{n_1}(j_1)c_{n_2}(j_2)\left[\varphi^{(\beta)}_{n_1 l_1 j_1}(\mathbf{x}_1) \otimes \varphi^{(\beta)}_{n_2 l_2 j_2}(\mathbf{x}_2)\right]_{L_\pi M_\pi}.\end{aligned} \tag{9.27}$$

By changing the j–j to the L–S coupling scheme only the singlet component gives contribution in the overlap with the α-particle wave function, i.e.

$$\begin{aligned}&\left[\varphi^{(\beta)}_{n_1 l_1 j_1}(\mathbf{x}_1) \otimes \varphi^{(\beta)}_{n_2 l_2 j_2}(\mathbf{x}_2)\right]_{L_\pi M_\pi}\\ &\equiv \left\{\left[\phi^{(\beta)}_{n_1 l_1}(\mathbf{r}_1) \otimes \chi_{\frac{1}{2}}(\mathbf{s}_1)\right]_{j_1} \otimes \left[\phi^{(\beta)}_{n_2 l_2}(\mathbf{r}_2) \otimes \chi_{\frac{1}{2}}(\mathbf{s}_2)\right]_{j_2}\right\}_{L_\pi M_\pi}\\ &\rightarrow \langle (l_1 l_2)L_\pi(\tfrac{1}{2}\tfrac{1}{2})0; L_\pi|(l_1\tfrac{1}{2})j_1(l_2\tfrac{1}{2})j_2; L_\pi\rangle\\ &\quad\times \left\{\left[\phi^{(\beta)}_{n_1 l_1}(\mathbf{r}_1) \otimes \phi^{(\beta)}_{n_2 l_2}(\mathbf{r}_2)\right]_{L_\pi} \otimes \left[\chi_{\frac{1}{2}}(\mathbf{s}_1) \otimes \chi_{\frac{1}{2}}(\mathbf{s}_2)\right]_0\right\}_{L_\pi M_\pi}.\end{aligned} \tag{9.28}$$

Here, the angular brackets denote the recoupling jj–LS coefficient. The scalar product of the spin part with the corresponding part of the α-particle gives unity. Then, the configuration part can be be written in the relative + cm system using the Talmi–Moshinsky transformation. Due to the fact that the α-particle wave function contains only the monopole radial component, one obtains

$$\begin{aligned}\left[\phi^{(\beta)}_{n_1 l_1}(\mathbf{r}_1) \otimes \phi^{(\beta)}_{n_2 l_2}(\mathbf{r}_2)\right]_{L_\pi M_\pi} &\rightarrow \sum_{n_\pi N_\pi} \langle n_\pi 0 N_\pi L_\pi; L_\pi|n_1 l_1 n_2 l_2; L_\pi\rangle\\ &\quad\times \left[\phi^{(\beta)}_{n_\pi 0}(\mathbf{r}_\pi) \otimes \phi^{(\beta)}_{N_\pi L_\pi}(\mathbf{R}_\pi)\right]_{L_\pi M_\pi}.\end{aligned} \tag{9.29}$$

For the proton part of the overlap integral one obtains an expansion in terms of the ho wave functions depending on the relative proton coordinate,. The overlap of the $\pi\pi$ wave function with the two-proton part of the α-particle wave function (9.6) is given by the following integral

$$
\begin{aligned}
\mathscr{F}_\pi(\mathbf{R}_\pi) &= \int d\mathbf{x}_\pi [\phi_{00}^{(\beta_\alpha)}(\mathbf{r}_\pi)\chi_{00}^{(\pi)}(12)]^* \\
&\times \sum_{j_1 \le j_2} \frac{1}{\Delta_{j_1 j_2}} X_{\pi\pi}(j_1 j_2; L_\pi) \mathscr{A}\{\psi_{j_1}(\mathbf{x}_1) \otimes \psi_{j_2}(\mathbf{x}_2)\}_{L_\pi M_\pi} \\
&= \sum_{N_\pi} G_\pi(N_\pi L_\pi) \phi_{N_\pi L_\pi M_\pi}^{(\beta)}(\mathbf{R}_\pi),
\end{aligned}
\tag{9.30}
$$

where we have introduced the proton G-coefficients as follows

$$
\begin{aligned}
G_\pi(N_\pi L_\pi) &= \sum_{12} B_\pi(n_1 l_1 j_1 n_2 l_2 j_2; L_\pi) \langle (l_1 l_2) L_\pi (\tfrac{1}{2}\tfrac{1}{2})0; L_\pi | (l_1 \tfrac{1}{2}) j_1 (l_2 \tfrac{1}{2}) j_2; L_\pi \rangle \\
&\times \sum_{n_\pi} \langle n_\pi 0 N_\pi L_\pi; L_\pi | n_1 l_1 n_2 l_2; L_\pi \rangle \mathscr{I}_{n_\pi 0}^{(\beta\beta_\alpha)}.
\end{aligned}
\tag{9.31}
$$

The overlap integral between spherical ho radial functions is defined as

$$
\mathscr{I}_{nm}^{(\beta\beta_\alpha)} \equiv \langle \mathscr{R}_{n0}^{(\beta)} | \mathscr{R}_{m0}^{(\beta_\alpha)} \rangle.
\tag{9.32}
$$

The B-coefficient contains the information on sp and two-body wave functions

$$
\begin{aligned}
B_\pi(n_1 l_1 j_1 n_2 l_2 j_2; L_\pi) \equiv & \sqrt{2} \sum_{\varepsilon_1 \le \varepsilon_2} \frac{1}{\Delta_{j_1 j_2}} X_{\pi\pi}(j_1 j_2; L_\pi) \\
& \times c_{n_1}(\pi\varepsilon_1 l_1 j_1) c_{n_2}(\pi\varepsilon_2 l_2 j_2),
\end{aligned}
\tag{9.33}
$$

were we used for clarity the full sp notation. The introduction of this coefficient reduces the computer time, because it is necessary to evaluate the recoupling coefficients in (9.31) only once for all sp states entering the summation with the same quantum numbers in (9.33). One also obtains a similar $\nu\nu$ expansion.

9.3.2 Four-Body Correlations

Next, we perform the transformation of the $\pi\pi$–$\nu\nu$ product to the relative + cm coordinates of the α-particle. Due to the fact that the α-particle wave function of the relative coordinate contains only the monopole component, one obtains

$$
\begin{aligned}
\left[\phi_{N_\pi L_\pi}^{(\beta)}(\mathbf{R}_\pi) \otimes \phi_{N_\nu L_\nu}^{(\beta)}(\mathbf{R}_\nu)\right]_{L_\alpha M_\alpha} \to & \sum_{n_\alpha N_\alpha} \langle n_\alpha 0 N_\alpha L_\alpha; L_\alpha | N_\pi L_\pi N_\nu L_\nu; L_\alpha \rangle \\
& \times \left[\phi_{n_\alpha 0}^{(\beta)}(\mathbf{r}_\alpha) \otimes \phi_{N_\alpha L_\alpha}^{(\beta)}(\mathbf{R}_\alpha)\right]_{L_\alpha M_\alpha}.
\end{aligned}
\tag{9.34}
$$

Finally, we obtain the overlap integral, defining the $\pi\pi$–$\nu\nu$ part of the preformation amplitude. It is written as a superposition of the ho functions in the cm coordinate of the α-cluster

$$\mathscr{F}_\alpha(\mathbf{R}_\alpha) = \sum_{L_\alpha} \mathscr{F}_{L_\alpha}^{(\alpha)}(\mathbf{R}_\alpha) = \sum_{L_\alpha} \sum_{N_\alpha} W(N_\alpha L_\alpha) \phi_{N_\alpha L_\alpha M_\alpha}^{(\beta)}(\mathbf{R}_\alpha), \tag{9.35}$$

where

$$\begin{aligned} W(N_\alpha L_\alpha) \equiv 8 & \sum_{N_\pi N_\nu} \sum_{L_\pi L_\nu} G_\pi(N_\pi L_\pi) G_\nu(N_\nu L_\nu) X_{\pi\pi\nu\nu}(L_\pi L_\nu; L_\alpha) \\ & \times \sum_{n_\alpha} \langle n_\alpha 0 N_\alpha L_\alpha; L_\alpha | N_\pi L_\pi N_\nu L_\nu; L_\alpha \rangle \mathscr{I}_{n_\alpha 0}^{(\beta\beta_\alpha)} . \end{aligned} \tag{9.36}$$

It can also be written in terms of the true cm radius $\mathbf{R} = \mathbf{R}_\alpha/2$, but for the radial ho wave functions depending upon 4β, i.e.

$$\mathscr{F}_\alpha(\mathbf{R}) = \sum_{L_\alpha} \mathscr{F}_{L_\alpha}^{(\alpha)}(\mathbf{R}) = \sum_{L_\alpha} \sum_{N_\alpha} W(N_\alpha L_\alpha) \phi_{N_\alpha L_\alpha M_\alpha}^{(4\beta)}(\mathbf{R}). \tag{9.37}$$

The second part of the overlap integral is given by the $\pi\nu$–$\pi\nu$ four-particle component. It is computed in a similar way, by replacing the four-particle amplitude and G-coefficients with the corresponding $\pi\nu$ terms. To do this one uses the equivalent $\pi\nu$ representation of the α-particle wave function (9.6), by exchanging the "2" and "3" coordinates. The corresponding overlap integral gives a small contribution, because the proton and neutron active orbitals are different for α-emitters with $Z > 50$.

9.4 α-Decay from ^{212}Po

9.4.1 Single-Particle Basis

Narrow sp Gamow resonances can become a useful tool in analyzing the high-lying α-like four-particle states. They were observed long time ago as resonances in the α-particle anomalous large angle scattering (ALAS) and were connected with the so called "quasimolecular states". Such states were mainly observed and analyzed in the α-scattering on light nuclei like ^{16}O [7], ^{40}Ca [8] or ^{28}Si [9].

The nucleus ^{212}Po exhibits a nice α-like structure on top of the double-magic nucleus ^{208}Pb. This structure allows us to apply the above described procedure to estimate the α-decay preformation amplitude. In order to compute it, we used a sp spectrum provided by the numerical integration of the radial Schrödinger equation. As a mean field we used the Woods–Saxon (WS) potential with the so called universal parametrization [10]. The bound states and resonances in continuum were computed using the code GAMOW [11] in a revised version. The results for

Table 9.1 Single-particle energies for ^{208}Pb used in the two-particle basis (in MeV) for protons and neutrons

Protons					Neutrons				
k	l	$2j$	Real (ε_k)	Imag (ε_k)	k	l	$2j$	Real (ε_k)	Imag (ε_k)
1	5	9	−3.571	0.000	1	4	9	−3.690	0.000
2	3	7	−3.333	0.000	2	6	11	−2.538	0.000
3	6	13	−1.605	0.000	3	2	5	−1.876	0.000
4	1	3	−0.485	0.000	4	7	15	−1.601	0.000
5	3	5	−0.306	0.000	5	0	1	−1.283	0.000
6	1	1	0.696	0.000	6	2	3	−0.614	0.000
7	4	9	4.264	0.000	7	4	7	−0.546	0.000
8	6	11	5.689	0.000	8	3	7	2.199	−0.998
9	7	15	6.240	0.000	9	5	11	2.472	−0.038
10	2	5	6.954	−0.003	10	8	17	5.348	−0.002
11	0	1	8.017	−0.047	11	5	9	5.594	−0.808
12	4	7	8.316	−0.001	12	7	13	5.701	−0.013
13	2	3	8.717	−0.036					

^{208}Pb are given in Table 9.1 for protons and neutrons, respectively. Here, the states are labeled by the index k. The angular momentum l and total spin j are also given. Concerning the resonances in continuum, only those states having energies with Real$(\varepsilon_k) <$ 12.5 MeV and $|\text{Imag}(\varepsilon_k)| < 1$ MeV were selected. One can see that there are more narrow proton resonances, due to the Coulomb barrier. The calculated values are rather close to those obtained in Ref. [12] were a different parametrization of the WS potential was used. In principle, the continuum part of the spectrum, which we have neglected, contains the background given by the integration on a contour around the included resonances.

In order to perform the overlap integral (9.10) defining the preformation amplitude of the α-like state, it is necessary to expand the sp wave functions in terms of the spherical ho basis as in Eq. 9.25. The expansion coefficients are found by a standard fitting procedure for a given ho parameter β. The maximal value of the radial quantum number is taken as $n_{\max} = 9$.

9.4.2 Two-Particle States

The two-particle states were treated by using Eq. 9.17 with the two-body interaction matrix elements obtained from the surface delta force [13] with one adjustable strength scaling parameter for each of the neutron–neutron, proton–proton and proton–neutron interaction channels. These three parameters were adjusted in the nuclei ^{210}Pb, ^{210}Po and ^{210}Bi, respectively, to yield reasonable values for the energies of the lowest states in these nuclei. In addition, the values of the radial single-particle functions on the nuclear surface were evaluated for the surface delta force and not approximated by one single constant, as discussed in [13].

Table 9.2 Experimental and theoretical excitation energies in ^{210}Po (a), ^{210}Pb (b) and ^{210}Bi (c)

(a)			(b)			(c)		
$J^{\pi}_{k_2}$	$E_{\rm exp}$ [MeV]	$E_{\rm th}$ [MeV]	$J^{\pi}_{k_2}$	$E_{\rm exp}$ [MeV]	$E_{\rm th}$ [MeV]	$J^{\pi}_{k_2}$	$E_{\rm exp}$ [MeV]	$E_{\rm th}$ [MeV]
0^+_1	0.000	0.000	0^+_1	0.000	0.000	1^-_1	0.000	0.000
2^+_1	1.181	1.280	2^+_1	0.800	0.903	0^-_1	0.047	0.338
4^+_1	1.427	1.447	4^+_1	1.098	1.027	9^-_1	0.271	0.262
6^+_1	1.473	1.483	6^+_1	1.196	1.076	2^-_1	0.320	0.338
8^+_1	1.557	1.493	8^+_1	1.279	1.103	3^-_1	0.348	0.235
8^+_2	2.188	1.618	(10^+_1)	1.799	2.101	7^-_1	0.433	0.284
2^+_2	2.290	1.650	3^-_1	1.870		5^-_1	0.439	0.274
6^+_2	2.326	1.669	(8^+_2)	2.003	2.191	4^-_1	0.502	0.338
4^+_2	2.383	1.675	(4^+_2)	2.038	2.223	6^-_1	0.550	0.338
3^-_1	2.387					(1^-_2)	0.563	0.310
1^+_1	2.394	1.746				8^-_1	0.583	0.338
5^+_1	2.403	1.746				10^-_1	0.670	1.431
3^+_1	2.414	1.746				8^-_2	0.916	0.416
7^+_1	2.438	1.746				(2^-_2)	0.972	0.422

All experimentally known states are taken below 2.5 MeV for ^{210}Po, below 2.0 MeV for ^{210}Pb, and below 1.0 MeV for ^{210}Bi

In Table 9.2a, b, c we list the available experimental data for the nuclei ^{210}Po, ^{210}Pb and ^{210}Bi, respectively. Also, the corresponding calculated values are given for comparison. One notices that for all of these nuclei, the lowest states of a given multipolarity J^{π} are very well reproduced by the surface delta interaction (for all the multipoles we have used the one and the same interaction-strength parameter mentioned above). For ^{210}Po and ^{210}Bi, the second calculated state of a given multipole falls well below the experimental second state producing a too compressed spectrum around 1.5 MeV in ^{210}Po and around 0.5 MeV in ^{210}Bi. In ^{210}Pb and ^{210}Po the first 3^- state is very collective and most likely of the 3 particle—1 hole character which is outside our configuration space. This is why no counterpart of it is given on the theoretical side in Table 9.2a and b.

9.4.3 *α*-Like States

In order to have confidence in the description of the high-lying four-particle resonances, the known low-lying states in ^{212}Po should be reproduced. They were computed in Ref. [14], using a method similar to the MSM approach and in Ref. [15] within the shell model approach. In the later paper, a very good agreement was obtained for the first 0^+, 2^+, 4^+, 6^+, 8^+ states. As mentioned earlier, our initial MSM basis is non-orthogonal and overcomplete, and a new orthogonal basis obtained using the eigenstates of the metric matrix should be used. The states having small metric eigenvalues are spurious and should be excluded. This is due to the fact that $\pi\nu$–$\pi\nu$ four-particle basis can be expressed in terms of the $\pi\pi$–$\nu\nu$

basis if all the two-particle eigenstates are used. In our calculation we used only the lowest two-particle $\pi\pi$, $\nu\nu$, $\pi\nu$ eigenstates for multipoles up to $J = 9$ and only some components become spurious. The result of calculation is given in Table 9.3. In the second column the experimental energies are given. In the next three columns we give the computed eigenvalues obtained by solving Eq. 9.21. They correspond to different minimal accepted eigenvalues $D_{\min}$ of the metric matrix, written in parentheses.

One can observe that the energies in the third and fourth columns, corresponding to $D_{\min} = 0.05$ and $D_{\min} = 0.2$, respectively, are practically identical and they reproduce the experimental data within 400 keV. The values of the last column, corresponding to $D_{\min} = 0.5$, are worse and, therefore, we fixed the minimal threshold of the metric matrix eigenvalues at $D_{\min} = 0.2$. One can see that, in spite of the fact that only the lowest two-particle states were included in calculation, the agreement with the experimental low-lying part of the spectrum is quite satisfactory, somewhere between Ref. [14] and Ref. [15].

It is now important to analyze the structure of these four-particle states in terms of the two-particle states, containing Gamow sp resonances. In Table 9.4 (second column) the structure of the four-particle states in terms of the most important two-particle pairs is given. One can see that the first six low-lying states have practically one major component. The four-particle amplitude X is given in the third column and it is larger than unity, due to the non-orthogonality of the basis.

We also stress the fact that the lowest monopole two-particle components $0_1^+(\pi\pi)$ and $0_1^+(\nu\nu)$ entering in the four-particle wave function are the only ones having Gamow resonances in continuum in their structure. This can be seen in Tables 9.5a and 9.6a, where we give the structure of these states in terms of the sp states labeled in Table 9.1a, b for protons and neutrons, respectively. The other low-lying states in Table 9.3, which are not given in Table 9.4, have no such states in their structure. It is important to mention that the gs structure of ^{212}Po is practically of the form gs(^{210}Po) $\otimes$ gs(^{210}Pb) in agreement with the supposed ansatz of Refs. [17, 18].

Table 9.3 Low-lying excitation energies in ^{212}Po (in MeV) for different thresholds of the metric matrix eigenvalues (in parantheses)

$J^{\pi}_{k_4}$	E_{exp}	E (0.05)	E (0.2)	E (0.5)
0_1^+	0.000	0.000	0.000	0.000
2_1^+	0.727	0.949	0.952	0.952
4_1^+	1.132	1.087	1.090	1.081
6_1^+	1.355	1.081	1.144	1.135
8_1^+	1.476	1.131	1.174	1.166
2_2^+	1.513	1.203	1.203	2.398
1_1^+	1.621	1.907	1.909	1.901
2_3^+	1.679	1.783	1.995	2.423
0_2^+	1.801	2.080	2.091	2.081
2_4^+	1.806	2.248	2.253	2.607

The experimental data are taken from [16]

Table 9.4 Quartet structure of the low-lying α-like states in terms of the two-particle pairs (the second column)

$J^{\pi}_{k_4}$	$J(\pi\pi)$, $J(\nu\nu)$	X	S_1	S_2	HF$_1$	HF$_2$
0^+_1	$0^+_1(\pi\pi), 0^+_1(\nu\nu)$	1.084	1.19 (−2)	8.49 (−7)	1.00	1.00
2^+_1	$0^+_1(\pi\pi), 2^+_1(\nu\nu)$	1.091	1.38 (−3)	5.16 (−8)	0.12	0.06
4^+_1	$0^+_1(\pi\pi), 4^+_1(\nu\nu)$	1.092	7.49 (−4)	1.39 (−9)	0.07	0.00
6^+_1	$0^+_1(\pi\pi), 6^+_1(\nu\nu)$	1.092	4.44 (−4)	1.02 (−8)	0.04	0.01
8^+_1	$0^+_1(\pi\pi), 8^+_1(\nu\nu)$	1.092	3.33 (−4)	5.36 (−9)	0.02	0.00
2^+_2	$2^+_1(\pi\pi), 0^+_1(\nu\nu)$	1.092	1.54 (−3)	6.44 (−8)	0.04	0.73

The amplitude is given in the third column. The spectroscopic factors of the $\pi\pi$–$\nu\nu$ and $\pi\nu$–$\pi\nu$ four-particle terms in the quartet operator defined by Eq. 9.18 are given in the fourth and fifth column. In the sixth and seventh columns the hindrance factors of the $J^{\pi}_{k_4}$ states defined by Eq. 6.2 in Chap. 6 are given

Table 9.5 The structure of the two-particle states $0^+_{k_2}(\pi\pi)$ in ^{210}Po containing proton single-particle Gamow resonances

No.	(a)			(b)			(c)		
	$0^+_1(\pi\pi)$ $E = -8.651$ MeV			$0^+_4(\pi\pi)$ $E = -1.547$ MeV			$0^+_6(\pi\pi)$ $E = 1.219$ MeV		
	X	$k_1(\pi)$	$k_2(\pi)$	X	$k_1(\pi)$	$k_2(\pi)$	X	$k_1(\pi)$	$k_2(\pi)$
1	0.460	1	1	−0.111	2	2	0.111	4	4
2	0.771	2	2	0.379	3	3	0.132	5	5
3	−0.313	3	3	0.769	4	4	−0.977	6	6
4	0.156	4	4	0.475	5	5			
5	0.149	5	5	0.106	6	6			
6	−0.124	7	7						
7	0.115	9	9						

The amplitude is given in the second column. In the third and fourth columns the involved single-particle states labeled according to the first column of Table 9.1a are given

This microscopic structure of the four-particle states is strongly connected with the formation amplitude (9.35) computed as the overlap integral between these states and the α-particle wave function, i.e.

$$\mathscr{F}_J(R) = \int [\psi_\alpha(\mathbf{x}_\alpha)\Psi^{(D)}(\mathbf{x}_D)Y_J(\hat{R})]^*\Psi^{(P)}(\mathbf{x}_P)d\mathbf{x}_\alpha d\mathbf{x}_D d\hat{R}, \qquad (9.38)$$

where $\psi_\alpha(\mathbf{x}_\alpha)$ is the gaussian α-particle wave function (9.6) [3]. It is a function of the distance between the daughter and the cm of the emitted cluster. The most important α-decay of ^{212}Po is the transition from the gs 0^+_1. In Fig. 9.1 we plot the formation amplitude for this transition versus the mentioned distance. One can observe that this function is peaked on the nuclear surface $R_0 = 1.2A^{1/3} = 7.2$ fm. It is of interest to see which would be the equivalent local potential, if one considers the α-particle formation amplitude as a wave function satisfying the Schrödinger equation

Table 9.6 The structure of the two-particle states $0^+_{k_2}(\nu\nu)$ in ^{210}Pb containing neutron single-particle Gamow resonances

No.	(a)			(b)		
	$0^+_1(\nu\nu)$ $E =$ −8.510 MeV			$0^+_6(\nu\nu)$ $E =$ −1.424 MeV		
	X	$k_1(\nu)$	$k_2(\nu)$	X	$k_1(\nu)$	$k_2(\nu)$
1	−0.896	1	1	−0.137	3	3
2	−0.305	2	2	0.926	6	6
3	−0.240	3	3	0.321	7	7
4	0.106	4	4			
5	−0.110	10	10			

The amplitude is given in the second column. In the third and fourth columns the single-particle states labeled according to the first column of Table 9.1 (b) are given

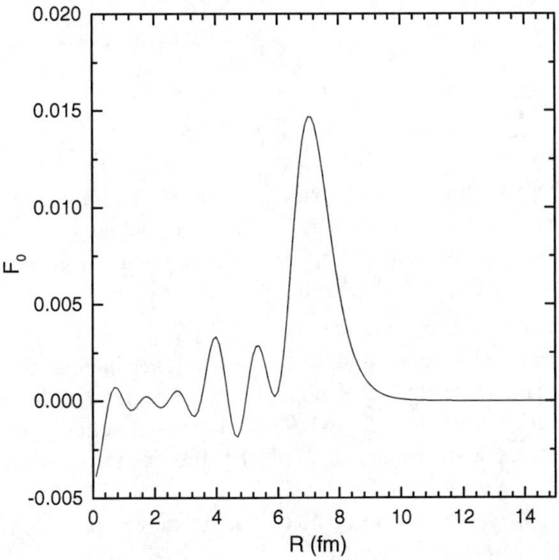

Fig. 9.1 The α-particle formation amplitude given by Eq. 9.35 for the ground state-to-ground state transition versus the distance R of the center of mass of the α-particle from the daughter nucleus ^{208}Pb

$$-\frac{\hbar^2}{2M_\alpha}\nabla^2\mathcal{F}_J(R) + V_J(R)\mathcal{F}_J(R) = E_\alpha\mathcal{F}_J(R), \tag{9.39}$$

where we have denoted the α-particle cm radius by $R \equiv \frac{R_\alpha}{2}$, the angular momentum by J and dropped the eigenvalue index.

By taking into account the expansion (9.35) one obtains for the unknown potential the relation

$$V_J(R) = E_\alpha + \hbar\omega\frac{4\beta}{\beta_\alpha\mathcal{F}_J(R)}\sum_N W(NJ)\left[2\beta R^2 - 2N - J - \frac{3}{2}\right]\mathcal{R}^{(4\beta)}_{NJ}(R). \tag{9.40}$$

In Fig. 9.2, we depict this equivalent potential for $J = 0$ (solid curve) corresponding to the formation amplitude in Fig. 9.1. One clearly sees the molecular

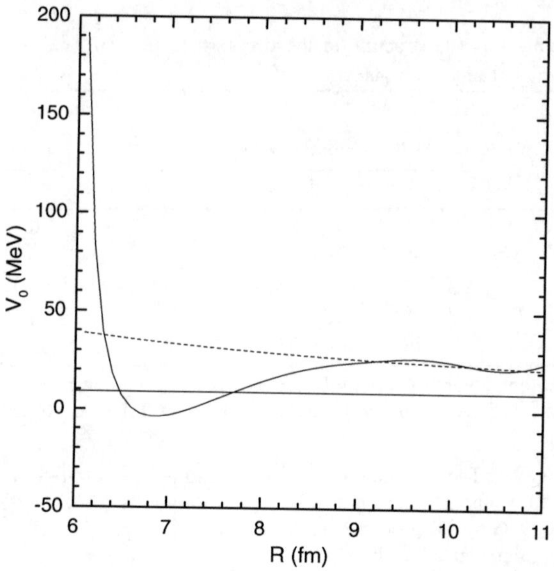

Fig. 9.2 The local equivalent potential computed using Eq. 9.40 corresponding to the α-particle wave function in Fig. 9.1. By a *dashed line* the α-daughter Coulomb potential is plotted. *Horizontal solid line* shows the energy of the emitted α-particle

shape of this local equivalent potential. This kind of local potential was used in some previous works as a phenomenological interaction to reproduce quasi-molecular resonances in the α-particle scattering data [9]. By a dashed line we indicate the pure Coulomb potential between the daughter and α-particle. After the geometrical touching radius $R_c \equiv 1.2(A_A^{1/3} + A_\alpha^{1/3}) = 9$ fm the equivalent potential bears a close resemblance to the Coulomb interaction. At this point the Coulomb barrier is around 25 MeV. The solid horizontal line denotes the energy of the emitted α-particle $E_\alpha = 8.95$ MeV, which actually is the gs energy of the ^{212}Po nucleus with respect to ^{208}Pb. All the excitation energies in Table 9.3 are given relative to this energy.

The following integral of the formation probability defines the microscopic spectroscopic factor (SF)

$$S_{nJ_{k_4}} \equiv \int_0^\infty |\mathscr{F}_{nJ_{k_4}}(R)|^2 R^2 dR, \tag{9.41}$$

where $n = 1$ corresponds to the $\pi\pi$–$\nu\nu$ component and $n = 2$ to the $\pi\nu$–$\pi\nu$ component in the four-particle wave function (9.18). The spectroscopic factors for the $\pi\pi$–$\nu\nu$ and $\pi\nu$–$\pi\nu$ four-particle components of the wave function are given in the fourth and fifth columns of Table 9.4. First of all, one can observe that the total gs to gs SF has the right order of magnitude [19, 20] and is practically given by the $\pi\pi$–$\nu\nu$ component which is much larger that the $\pi\nu$–$\pi\nu$ one. This means that the $\pi\nu$ overlap is much smaller in comparison with the $\pi\pi$ and $\nu\nu$ overlaps because proton and neutron shells above ^{208}Pb are different. The other states have smaller SF's.

In the sixth and seventh columns there are given the so called hindrance factors HF_n defined as the following ratio of the mean values on the interval [21, 22] fm

$$HF_{nJ_{k_4}} \equiv \frac{\langle R\mathscr{F}^2_{nJ_{k_4}}(R)\rangle}{\langle R\mathscr{F}^2_{n0_1}(R)\rangle}. \tag{9.42}$$

Notice that this definition is inverse with respect to Eq. 6.2. The states given in Table 9.4 have $HF_1 > 0.01$. The other states in Table 9.3, not given in Table 9.4, have much smaller HF and have no sp Gamow resonances in their structure. Therefore, there is a straightforward connection between the sp Gamow states and the magnitude of the α-particle overlap integral.

In spite of the strong decrease of the formation amplitude in the region of the geometrical touching radius R_c the decay width, is practically constant, as can be seen from Fig. 9.3 (solid line), proving the validity of our calculation. This is the reason why we used the interval [21, 22] fm to estimate the HF.

According to Eq. 5.10 in Chap. 5 the decay width is a product of two terms which have an opposite behaviour: the formation probability F_0 (dashed line in Fig. 9.3) is decreasing and the penetrability P_0 (dot-dashed line in Fig. 9.3) is increasing. It is important to observe from the same figure that the decay width is underestimated by two orders of magnitude. This means that the inclusion of narrow Gamow resonances is not enough in order to reproduce the absolute value of the decay width. The role of the "background", given by the integration in the complex plane on a contour including the considered resonances, is also very important because the formation probability of an α-cluster on the nuclear surface is proportional to the density of the sp components in the continuum.

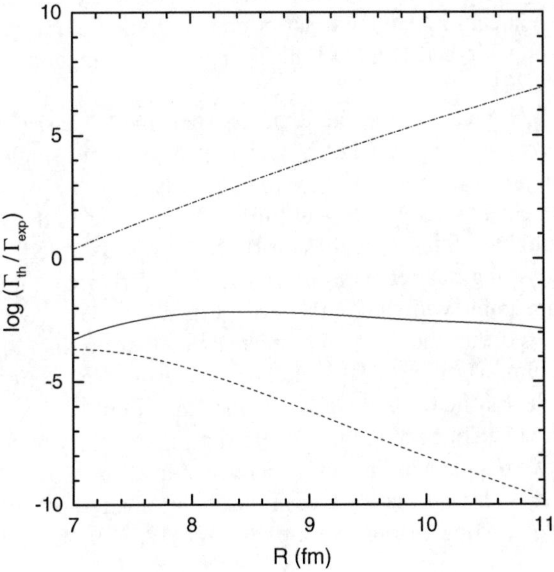

Fig. 9.3 The quantities $\log(\Gamma_{th}/\Gamma_{exp})$ (*solid line*), $\log(F_0)$ (*dashed line*) and $\log(P_0/\Gamma_{exp})$ (*dot-dashed line*) as functions of the distance R of the center of mass of the α-particle from the daughter nucleus ^{208}Pb. F_0 and P_0 are defined by Eq. 5.10

But, as it was pointed out, the narrow sp Gamow resonances play an important role in some low-lying states having a strong overlap with the α-particle wave function and which are called α-like states. They should also play an important role in some high-lying states seen as resonances in the scattering of α-particles on the daughter nucleus, especially above the Coulomb barrier. The states above the barrier have energies larger than 16 (= 25−9) MeV. In Tables 9.5 and 9.6 we list the structure of the two-particle states entering the four-particle eigenstates and containing sp Gamow states for protons and neutrons, respectively. The excitation energies are relative to the gs energy in ^{212}Po. One can see that all of them have a monopole character.

9.5 α-Like Resonances in ^{40}Ca

The α-like states, analyzed in the previous Section, are related with the so called "quasimolecular states", observed long time ago as resonances in the α-particle scattering. For a recent review, see for instance [23], and the references therein. They are connected with the anomalous large angle scattering (ALAS) phenomenon seen in the angular distribution studies. Such states were mainly observed and analyzed in the α-scattering on light nuclei like ^{16}O [7], ^{40}Ca [8] or ^{28}Si [9].

Theoretical interpretation of quasimolecular states is based on the simple picture of a long-living di-nuclear rotating system. Thus, all phenomenological treatments suppose a pocket-like potential between the two nuclei. The resonance levels of this potential provide an explanation of the rotational bands seen in the excitation function. On this line, a recent phenomenological description of the α + ^{40}Ca system was given in Ref. [24].

A lot of work was also done by using different microscopic approaches. We only mention here the description within the Folding Model [25], the Resonanting Group Method (RGM) [26, 27] and the Orthogonality Condition Model (OCM) [28–30].

We stress on the fact that the shell model basis involved in all the previous microscopic approaches was restricted to bound single-particle configurations. In the previous Section, based on results of Ref. [6], we obtained a very good agreement with experiment for low-lying states in the α + ^{208}Pb system within the Multi-step Shell-Model (MSM) [4, 5]. We also predicted in [6] that some of the high-lying quartet states have a strong overlap with the α-particle wave function, comparable with that of the ground state in ^{212}Po. We called them α-like states and showed that their main components consist of single-particle resonances in continuum. Until now, quasimolecular resonances were evidenced only for light nuclear systems as for instance the Ca region. In this section, we apply the MSM formalism of Sect. 9.3 in order to show that high-lying α-like states exist in ^{44}Ti and they coincide with the resonances seen in the α-particles scattering on ^{40}Ca [32].

It is already known that in describing many-body resonances, only the narrow single particle resonances are relevant [12, 33]. This was shown to be an adequate approach for instance for giant resonances [33], but also for the nucleon decay processes [35]. We will show that single-particle resonances are an important ingredient in the structure of quasimolecular states. Moreover, our approach is able to predict the positions of these states in terms of single-particle resonances, without involving additional parameters.

Basically, the α + ^{40}Ca system is very similar to the α + ^{208}Pb one. As mentioned, we have described the resonance structure of the latter system in terms of quartet excitations built on top of the lowest collective pair states. We mention that the inclusion of these resonances in the single particle spectrum to study a system like α + ^{40}Ca is a new feature with respect to the previous microscopic calculations.

The four-body eigenstates resulting from the diagonalization of the system (9.21) should be compared with the α-particle wave function. As in the previous Section, let us consider the α-particle formation amplitude defined by Eq. 9.38. An eigenstate with a large overlap integral should be seen as a maximum of the excitation function describing the α-particle scattering on ^{40}Ca.

The single-particle eigenstates were found using both diagonalization procedure and numerical integration in a spherical Woods–Saxon potential with universal parametrization [36]. In the diagonalization basis we used 16 major shells. The result is given in Table 9.7. The proton states are given on the left side of the Table, while neutron states are given on the right. One can see that the energies for both proton and neutron bound states resulting from the diagonalization procedure (the third/seventh columns) are very close to the values obtained by numerical integration (the fourth/eighth columns). Concerning the states in continuum, due to the Coulomb barrier the proton spectrum obtained by the integration method is much closer to the diagonalization result than in the case of the neutron spectrum. One can also notice the similar relative energies of the first five states for the proton and neutron spectra. In the fifth/ninth column the widths of the resonant states are given. They are computed according to the standard relation

Table 9.7 Proton/neutron single-particle energies (in MeV) for ^{40}Ca

k.	Proton state	E_k (diag.)	E_k (int.)	Γ_k	Neutron state	E_k (diag.)	E_k (int.)	Γ_k
1	$\pi f_{7/2}$	−1.416	−1.417	0.000	$\nu f_{7/2}$	−10.988	−10.988	0.000
2	$\pi p_{3/2}$	1.492	1.408	0.033	$\nu p_{3/2}$	−6.903	−6.908	0.000
3	$\pi p_{1/2}$	3.610	3.401	0.438	$\nu p_{1/2}$	−4.405	−4.418	0.000
4	$\pi f_{5/2}$	6.559	6.601	0.014	$\nu f_{5/2}$	−2.585	−2.587	0.000
5	$\pi g_{9/2}$	7.616	7.622	0.001	$\nu g_{9/2}$	−1.144	−1.147	0.000
6					$\nu d_{5/2}$	0.966	0.657	0.023
7					$\nu h_{11/2}$	8.130	8.351	0.000
8					$\nu d_{5/2}$	7.709	8.467	0.743
9					$\nu g_{7/2}$	7.282	8.724	0.004

In the second/sixth columns the spectroscopic assignments are given. In the third/seventh columns the eigenvalues from the diagonalization and in the fourth/eighth columns the corresponding results from the numerical integration are given. In the fifth/ninth columns the resonance widths are given

$$\Gamma_k = -2\left[\frac{\partial \cot\delta_k}{\partial E}\right]^{-1}_{E=E_r}. \tag{9.43}$$

By using the states of Table 9.7 we obtained the pair eigenstates using the TDA method with the surface-delta residual two-body interaction (SDI) [13]. We adjusted the isoscalar and isovector strengths of the SDI to describe the low-energy spectra [16] of the nuclei ^{42}Ti, ^{42}Ca and ^{42}Sc, used to generate the needed $\pi\pi$, $\nu\nu$ and $\pi\nu$ pairs for the quartet states in ^{44}Ti. The quartet states, in turn, were obtained by solving the eigenvalue problem (9.21). To keep the calculations tractable, we considered the first three collective pair eigenstates in the basis. As described in Ref. [6], the metric matrix was first diagonalized and its eigenstates were used to build a new orthonormal basis. Because our initial basis is overcomplete we excluded states having small eigenvalues of the metric matrix. After this the lowest quartet energy in the diagonalization of Eq. 9.21 is close to the α-particle binding energy. We normalise our four-particle spectrum to this value.

Then we estimated the formation amplitude according to Eq. 9.38. We varied the center of mass (cm) radius within the interval $r \in [5, 7]$ fm, beyond the nuclear radius $R_c = 4.1$ fm. and divided the result with the ground-state value. The mean value of this ratio is the hindrance factor (HF) defined by Eq. 9.42 and it has almost a constant value within the chosen interval of the radial coordinate. In this region the Pauli principle is less important and the antisymmetrization operation in Eq. 9.38 can be neglected. This is actually the important region for the α-scattering process. The HF provides a better description of the quartet resonances than the spectroscopic factor, because it takes care only of the external structure of the wave function.

Results for the hindrance factors are shown in Fig. 9.4 for even spins $J = 0, 2, 4, 6$. One can see that, indeed, for some energies one obtains HF values larger than unity. These states, called α-like states, are the quasimolecular states seen as maxima of the α-scattering excitation function. They are also connected with the ALAS phenomenon as described for instance in Refs. [39, 40]. In the previous Section we showed that the equivalent potential corresponding to the α-particle formation amplitude, considered as a wave function, has indeed a pocket-like shape. This behaviour is connected with the peak of the formation amplitude in the nuclear surface region. This means that the nucleons in these quartet states can be found with a larger probability in the touching configuration region, orbiting with some angular momentum.

The microscopic structure of the first high-lying α-like states in terms of collective pairs is given in the second column of Table 9.8 for $J = 0, 2, 4$, respectively. One can observe that these states have a pronounced collective character, with many amplitudes of similar magnitude. In Table 9.9, we give the single-particle structure of the pair states involved in the quartet states of Table 9.8. One can see that all of them have, practically, one dominant component involving only the first proton resonance $\pi p_{3/2}$ and the first neutron bound state $\nu f_{7/2}$. Therefore, the structure of the first high-lying α-like states is very simple.

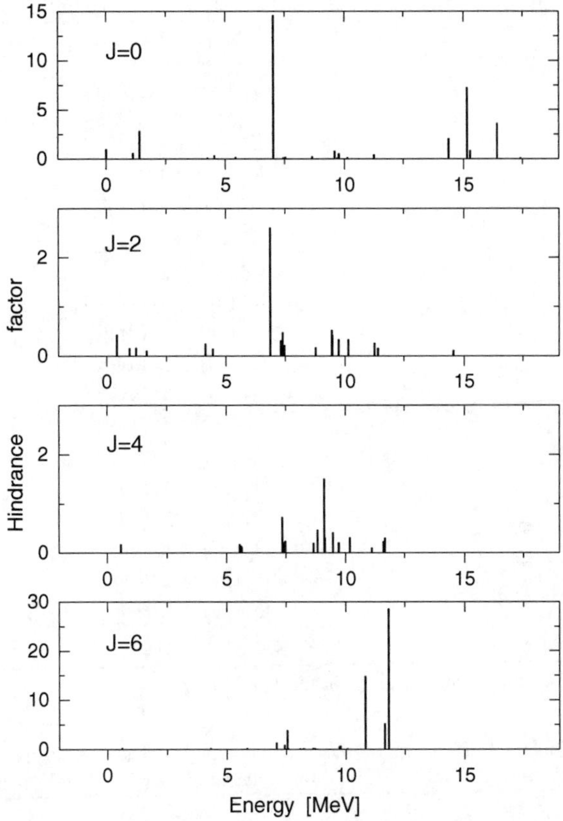

Fig. 9.4 The hindrance factors as functions of the excitation energy for $J = 0, 2, 4, 6$ four-particle eigenstates

The final results of the resonant states are given in Table 9.10. Here we give the computed energies of the first high-lying α-like eigenstates (the third column) for each spin (the first column) as compared to the experimental values (the second column). The experimental energies were obtained using a fitting procedure of the excitation spectrum at large angles [39, 40]. For the $J = 2$ case we considered the mean value of the first two states, which are rather close. In Table 9.10, one can notice the good agreement between our calculation and the experiment. This is especially satisfying when taking into account that these numbers were obtained on top of the pair eigenstates, without any additional parameters. For comparison, in the last column we give the RGM results of Ref. [26], using the Brink–Boecker interaction B1. Concerning these last values, we remark that in Refs. [25, 27] better results were obtained by using a similar formalism, but different two-nucleon forces. Actually, in Ref. [27] it has been clearly shown that the Brink–Boecker interaction B1 fails in reproducing the angular distribution for α + Ca elastic scattering.

We stress on the fact that this structure is stable against variations of the number of pair states included in the quartet basis. By increasing the number of pairs the

Table 9.8 Quartet structure of the first high-lying α-like 0^+_{11}, 2^+_{33}, and 4^+_{43} eigenstates, respectively, in terms of two-particle pairs (the second column)

$J^{\pi}_{k_4}$	$[J^{\pi}_{k_2}(\tau_1\tau_2) \otimes J^{\pi}_{k_2}(\tau'_1\tau'_2)]_{J^{\pi}}$	X
0^+_{11}	$[0^+_2(\pi\pi) \otimes 0^+_1(\nu\nu)]_{0^+}$	0.200
	$[2^+_2(\pi\nu) \otimes 2^+_2(\pi\nu)]_{0^+}$	0.213
	$[3^+_2(\pi\nu) \otimes 3^+_2(\pi\nu)]_{0^+}$	0.155
	$[4^+_2(\pi\nu) \otimes 4^+_2(\pi\nu)]_{0^+}$	0.137
	$[5^+_2(\pi\nu) \otimes 5^+_2(\pi\nu)]_{0^+}$	0.263
2^+_{33}	$[2^+_3(\pi\pi) \otimes 0^+_1(\nu\nu)]_{2^+}$	−0.298
	$[2^+_2(\pi\nu) \otimes 3^+_2(\pi\nu)]_{2^+}$	0.106
	$[3^+_2(\pi\nu) \otimes 3^+_2(\pi\nu)]_{2^+}$	−0.126
	$[2^+_2(\pi\nu) \otimes 4^+_2(\pi\nu)]_{2^+}$	0.218
	$[3^+_2(\pi\nu) \otimes 5^+_2(\pi\nu)]_{2^+}$	0.221
	$[4^+_2(\pi\nu) \otimes 5^+_2(\pi\nu)]_{2^+}$	−0.227
	$[5^+_2(\pi\nu) \otimes 5^+_2(\pi\nu)]_{2^+}$	0.169
4^+_{43}	$[2^+_3(\pi\pi) \otimes 2^+_1(\nu\nu)]_{4^+}$	0.225
	$[2^+_2(\pi\nu) \otimes 3^+_2(\pi\nu)]_{4^+}$	−0.128
	$[2^+_2(\pi\nu) \otimes 2^+_2(\pi\nu)]_{4^+}$	−0.129
	$[3^+_2(\pi\nu) \otimes 4^+_2(\pi\nu)]_{4^+}$	0.112
	$[1^+_2(\pi\nu) \otimes 5^+_2(\pi\nu)]_{4^+}$	0.140
	$[2^+_2(\pi\nu) \otimes 5^+_2(\pi\nu)]_{4^+}$	0.105
	$[5^+_2(\pi\nu) \otimes 5^+_2(\pi\nu)]_{4^+}$	0.325

The amplitudes are given in the third column

Table 9.9 Structure of the low-lying pair states (the second column), entering the quartet structure of Table 9.8, in terms of the single-particle states given in Table 9.7

$J^{\pi}_{k_2}(\tau_1\tau_2)$	$(\tau_1 \otimes \tau_2)_{J^{\pi}}$	x
$0^+_2(\pi\pi)$	$(\pi p_{3/2} \otimes \pi p_{3/2})_{0^+}$	0.997
$2^+_3(\pi\pi)$	$(\pi p_{3/2} \otimes \pi p_{3/2})_{2^+}$	0.877
$0^+_1(\nu\nu)$	$(\nu f_{7/2} \otimes \nu f_{7/2})_{0^+}$	0.985
$2^+_1(\nu\nu)$	$(\nu f_{7/2} \otimes \nu f_{7/2})_{2^+}$	0.989
$2^+_2(\pi\nu)$	$(\pi p_{3/2} \otimes \nu f_{7/2})_{2^+}$	0.999
$3^+_2(\pi\nu)$	$(\pi p_{3/2} \otimes \nu f_{7/2})_{3^+}$	0.987
$4^+_2(\pi\nu)$	$(\pi p_{3/2} \otimes \nu f_{7/2})_{4^+}$	0.998
$5^+_2(\pi\nu)$	$(\pi p_{3/2} \otimes \nu f_{7/2})_{5^+}$	0.981

The corresponding amplitudes are given in the third column

position of the main maxima remains practically unchanged. We also mention that the amount of quartet eigenstates resulting from the diagonalization of the system (9.21) is much larger, but only few states plotted in Fig. 9.4 have large HF's (in fact, 'enhancement factor' would be a more appropriate name for the HF, which is the generally used term, in the present case of very strong α resonances).

From Fig. 9.4, one can see that our calculations predict a much more complex resonance structure for the $\alpha + {}^{40}$Ca reaction than a simple rotational spectrum,

Table 9.10 The experimental (Refs. [39, 40], second column) and theoretical (third column) values of the energies of the quasimolecular resonances versus the spin (first column)

J^{π}	E_{exp} [MeV]	E_{th} [MeV]	E_{th} [MeV]
0^+	6.67	7.01	5.10
2^+	7.22	7.13	5.69
4^+	7.75	7.34	7.09
6^+		7.53	

In the last column are given the values of Ref. [26] for comparison

given by the first high-lying α-like states. We predict strong resonances at about 10 MeV for $J = 0, 2, 4, 6$ and 15 MeV for $J = 0$. We mention here that a similar fragmentation due to the core structure can be seen in the spectroscopic factor of the α + Ar system, given by Fig. 9.2 of the Ref. [28, 29].

For states with odd spins, we obtained HF's by two orders of magnitude smaller than the unity implying no α resonances for odd spins contrary to experiment (the strong 1^- resonance obtained in [40]). This result is consistent with the conclusion of the experimental paper [41]. Here it has been shown that this resonance is difficult to be assigned to an α-cluster state. Our belief is that these states are actually connected with the $1 \otimes 3$ partitions of the four-particle basis, which we did not consider in our analysis. This is also suggested by Ref. [28, 29].

Concerning the low lying states, their spectrum is more compressed than the experimental one, at variance with the ^{208}Pb case where we obtained a very good agreement. Here, the ^{40}Ca core is "softer" than ^{208}Pb (less distance between the major shells) and this may be the explanation for the compression. This feature, however, seems to affect much less the high-lying part of the spectrum describing quasimolecular states, as can be seen from Table 9.10. As suggested in Ref. [30], it could be that the ^{36}Ar core would give a better description of the low-lying states.

9.6 Superfluid α-Emitters

9.6.1 Deformed Nuclei

Most of α-particle emitters have a superfluid character. A superfluid deformed nucleus can be described by using the following schematic Hamiltonian

$$H_{\tau} = \sum_{m>0} (\varepsilon_{\tau m} + \varepsilon_{\tau \overline{m}} - \lambda_{\tau}) a^{\dagger}_{\tau j m} a_{\tau j \overline{m}} - G_{\tau} \sum_{m} a^{\dagger}_{\tau m} a^{\dagger}_{\tau \overline{m}} \sum_{m'} a_{\tau \overline{m}'} a_{\tau m'}, \qquad (9.44)$$

where we introduced the short-hand notation $m \equiv (m^{\pi})$ and by $a^{\dagger}_{\tau \overline{m}}$ we denoted the time reversed state. We will diagonalize this Hamiltonian within the Bardeen–Cooper–Schriffer (BCS) approach, by using quasiparticle operators, defined in terms of particle operators as follows

$$\begin{pmatrix} \alpha^{\dagger}_{\tau m} \\ \alpha_{\tau \overline{m}} \end{pmatrix} = \begin{pmatrix} u_{\tau m} & -v_{\tau m} \\ v_{\tau m} & u_{\tau m} \end{pmatrix} \begin{pmatrix} a^{\dagger}_{\tau m} \\ a_{\tau \overline{m}} \end{pmatrix}. \tag{9.45}$$

By using the inverse transformation, one obtains the Hamiltonian in the quasiparticle representation. The BCS equations are obtained from the linearized equation of motion

$$[H, a^{\dagger}_{\tau m}] = E_{\tau m} a^{\dagger}_{\tau m}, \tag{9.46}$$

and the Hamiltonian in this approximation becomes

$$H_{\tau} \approx \sum_{m} (2\varepsilon_{\tau m} - \lambda_{\tau}) v^2_{\tau m} - \frac{\Delta^2_{\tau}}{G_{\tau}} + \sum_{m} E_{\tau m} \alpha^{\dagger}_{\tau m} \alpha_{\tau m}. \tag{9.47}$$

The quasiparticle energy is given in terms of the gap parameter as follows

$$\begin{aligned} E_{\tau m} &= \sqrt{(\varepsilon_{\tau m} - \lambda_{\tau})^2 + \Delta^2_{\tau}} \\ \Delta_{\tau} &= G_{\tau} \sum_{m} u_{\tau m} v_{\tau m}. \end{aligned} \tag{9.48}$$

The BCS amplitudes squared have simple expressions

$$\begin{pmatrix} u^2_{\tau m} \\ v^2_{\tau m} \end{pmatrix} = \frac{1}{2} \left(1 \pm \frac{\varepsilon_{\tau m} - \lambda_{\tau}}{E_{\tau m}} \right). \tag{9.49}$$

The Lagrange multipliers λ_{τ} are determined from the number of particle conservation laws for protons and neutrons, i.e.

$$2 \sum_{m} v^2_{\tau m} = N_{\tau}. \tag{9.50}$$

In computing the preformation amplitude, one uses the following expansion of the parent wave function in terms of the daughter state

$$|P\rangle = \sum_{m>0} X_{\pi m} a^{\dagger}_{\pi m} a^{\dagger}_{\pi \overline{m}} \sum_{m'>0} X_{\nu m'} a^{\dagger}_{\nu m'} a^{\dagger}_{\nu \overline{m}'} |D\rangle. \tag{9.51}$$

For the ground state to ground state transitions, by using the inverse to (9.45) transformation, one obtains $X_{\tau m} = u_{\tau m} v_{\tau m}$.

By expanding the deformed state in partial spherical waves

$$a^{\dagger}_{\tau m} = \sum_{lj} a^{\dagger}_{\tau l j m}, \tag{9.52}$$

where we introduced the spherical components (9.25)

$$a^{\dagger}_{\tau ljm} \rightarrow \psi_{\tau ljm}(\mathbf{x}) \equiv f_{\tau ljm}(r)[Y_l(\hat{r}) \otimes \chi_{\frac{1}{2}}(\mathbf{s})]_{jm}, \tag{9.53}$$

one obtains

$$\begin{aligned} |P\rangle = &\sum_{m>0} X_{\pi m} \sum_{12} (-)^{j_2-m} a^{\dagger}_{\pi l_1 j_1 m} a^{\dagger}_{\pi l_2 j_2,-m} \\ &\times \sum_{m'>0} X_{\nu m'} \sum_{34} (-)^{j_4-m'} a^{\dagger}_{\nu l_3 j_3 m'} a^{\dagger}_{\nu l_4 j_4,-m'} |D\rangle. \end{aligned} \tag{9.54}$$

By using standard angular momentum recoupling, one obtains in the configuration space

$$\begin{aligned} |P\rangle \rightarrow &\sum_{L_\alpha} \sum_{L_\pi L_\nu} \langle L_\pi, 0; L_\nu, 0 | L_\alpha, 0\rangle \\ &\times \left\{ \sum_{12} \left[\psi_{\pi l_1 j_1} \otimes \psi_{\pi l_2 j_2}\right]_{L_\pi} \otimes \sum_{34} \left[\psi_{\nu l_3 j_3} \otimes \psi_{\nu l_4 j_4}\right]_{L_\nu} \right\}_{L_\alpha,0} \\ &\times \sqrt{2} \sum_{m>0} X_{\pi m} (-)^{j_2-m} \langle j_1, m; j_2, -m | L_\pi, 0\rangle \\ &\times \sqrt{2} \sum_{m'>0} X_{\nu m'} (-)^{j_4-m'} \langle j_3, m'; j_4, -m' | L_\nu, 0\rangle |D\rangle. \end{aligned} \tag{9.55}$$

Now we expand each l, j, m component in the ho basis (9.26). One finally obtains the following relation

$$\begin{aligned} |P\rangle = &\sum_{L_\alpha} \sum_{L_\pi L_\nu} \langle L_\pi, 0; L_\nu, 0 | L_\alpha, 0\rangle \\ &\times \left\{ \sum_{12} \left[\varphi^{(\beta)}_{n_1 l_1 j_1} \otimes \varphi^{(\beta)}_{n_2 l_2 j_2}\right]_{L_\pi} \otimes \sum_{34} \left[\varphi^{(\beta)}_{n_3 l_3 j_3} \otimes \varphi^{(\beta)}_{n_4 l_4 j_4}\right]_{L_\nu} \right\}_{L_\alpha,0} \\ &\times B_\pi(n_1 l_1 j_1 n_2 l_2 j_2; L_\pi) B_\nu(n_3 l_3 j_3 n_4 l_4 j_4; L_\nu) |D\rangle, \end{aligned} \tag{9.56}$$

where the B_π-coefficient is

$$\begin{aligned} B_\pi(n_1 l_1 j_1 n_2 l_2 j_2; L_\pi) \equiv &\sqrt{2} \sum_{m>0} X_{\pi m} (-)^{j_2-m} \langle j_1, m; j_2, -m | L_\pi, 0\rangle \\ &\times c_{n_1}(\pi l_1 j_1 m) c_{n_2}(\pi l_2 j_2 m). \end{aligned} \tag{9.57}$$

By using a similar expression for neutrons, after the standard recoupling procedure to the relative and cm coordinates, the preformation amplitude becomes a superposition of different multipoles, i.e.

$$\mathscr{F}_\alpha(\mathbf{R}) = \sum_{L_\alpha} \mathscr{F}^{(\alpha)}_{L_\alpha}(\mathbf{R}) = \sum_{L_\alpha} \sum_{N_\alpha} W(N_\alpha L_\alpha) \phi^{(4\beta)}_{N_\alpha L_\alpha}(\mathbf{R}), \tag{9.58}$$

where W coefficient will contain the corresponding Clebsch–Gordan coefficient instead of the amplitude $X_{\pi\pi\nu\nu}(L_\pi L_\nu; L_\alpha)$ in (9.36), i.e.

$$
\begin{aligned}
W(N_\alpha L_\alpha) \equiv 8 \sum_{N_\pi N_\nu} \sum_{L_\pi L_\nu} G_\pi(N_\pi L_\pi) G_\nu(N_\nu L_\nu) \langle L_\pi, 0; L_\nu, 0 | L_\alpha, 0 \rangle \\
\times \sum_{n_\alpha} \langle n_\alpha 0 N_\alpha L_\alpha; L_\alpha | N_\pi L_\pi N_\nu L_\nu; L_\alpha \rangle \mathscr{I}_{n_\alpha 0}^{(\beta\beta_\alpha)}.
\end{aligned}
\tag{9.59}
$$

Here the G-coefficients are given by Eq. 9.31.

We define the microscopic spectroscopic factor as follows

$$
\begin{aligned}
S_{gs} &= \int_0^\infty R^2 dR \int d\hat{R} |\mathscr{F}_\alpha(\mathbf{R})|^2 = \sum_{L_\alpha} \int_0^\infty R^2 dR \left[\sum_{N_\alpha} W(N_\alpha L_\alpha) \mathscr{R}_{N_\alpha L_\alpha}^{(4\beta)}(R) \right]^2 \\
&= \sum_{L_\alpha N_\alpha} W^2(N_\alpha L_\alpha).
\end{aligned}
\tag{9.60}
$$

This quantity is an estimate of the α-particle content inside the parent pairing wave function.

9.6.2 Superdeformed Nuclei

Superdeformed nuclei have been intensively investigated, both experimentally [42] and theoretically [43]. One of the important questions raised by these studies is the possibility of an α-decay branch connecting members of superdeformed bands in the mother nucleus with the ground band in the daughter nucleus.

An estimation of the probability of observing α-decay from superderformed nuclei requires a comparison of α-decay with other competing modes. This can be done by using the same techniques as in the analysis of the structure of superdeformed bands, i.e. by means of a cranked Hartree–Fock–Bogoliubov (HFB) approximation [44]. Within this approach, one would be able to describe, for instance, the expected influence of large quadrupole deformations and pairing correlations on the rotational motion and, therefore, on the α-particle formation process.

As a rule, the structure of mother and daughter nuclei concerning deformation and spin, are very different. The dynamics of the rotating system of protons ($\tau = \pi$) and neutrons ($\tau = \nu$) in the daughter nucleus is, therefore, determined by the following constrained Hamiltonian

$$
\begin{aligned}
H_{D\tau}^{(\omega)} &= \sum_{kl} \varepsilon_{D\tau;kl} a_{\tau k}^\dagger a_{\tau l} - G_{D\tau} \sum_{kl>0} a_{\tau k}^\dagger a_{\tau \bar{k}}^\dagger a_{\tau \bar{l}} a_{\tau l}. \\
\varepsilon_{D\tau;kl} &= (E_{D\tau m} - \lambda_{D\tau}) \delta_{kl} - \hbar\omega_D (j_x)_{D;kl},
\end{aligned}
\tag{9.61}
$$

where the index k labels the available quantum numbers, which in this case are the energy E and the projection m of the angular momentum. The operator $a_{\tau k}^\dagger$ creates the daughter sp state $|\psi_{D\tau k}\rangle$, $E_{D\tau m}$ is the sp energy in the deformed daughter nucleus and $\lambda_{D\tau}$ the Lagrange multiplier that takes into account the conservation in average of the number of particles. In the same way, the last term takes into account the average conservation of the angular momentum along the x-axis, with the angular frequency ω as a Lagrange multiplier, i.e

$$\sum_{\tau kl}(j_x)_{D\tau;kl}\rho_{D\tau;kl} = \sqrt{J_D(J_D+1)}; \quad \rho_{D\tau;kl} = \sum_i v_{D\tau;li}v^*_{D\tau;ki}, \tag{9.62}$$

where $v_{D\tau;ki}$ are the HFB amplitudes defined below by Eq. 9.63. A similar Hamiltonian is taken for the mother nucleus.

The HFB transformation can be written in a general form as (where we dropped all indices)

$$\begin{pmatrix}\alpha^\dagger\\ \alpha\end{pmatrix} = \begin{pmatrix}U_D^T & V_D^T\\ V_D^+ & U_D^+\end{pmatrix}\begin{pmatrix}a^\dagger\\ a\end{pmatrix}; \quad \begin{pmatrix}\beta^\dagger\\ \beta\end{pmatrix} = \begin{pmatrix}U_P^T & V_P^T\\ V_P^+ & U_P^+\end{pmatrix}\begin{pmatrix}b^\dagger\\ b\end{pmatrix}, \tag{9.63}$$

where $a^\dagger(b^\dagger)$ creates a normal particle and $\alpha^\dagger$ $(\beta^\dagger)$ creates a quasiparticle in the daughter (mother) nucleus. The HFB amplitudes defined above are found by the standard procedure described for instance in Ref. [44].

We assume that the two nuclei have different deformations. Yet, their wave functions are connected by the Hermitian transformation

$$\begin{pmatrix}b^\dagger\\ b\end{pmatrix} = \begin{pmatrix}C^* & 0\\ 0 & C\end{pmatrix}\begin{pmatrix}a^\dagger\\ a\end{pmatrix}; \quad C^*_{kk'} = \langle 0|a_k b^\dagger_{k'}|0\rangle \equiv \langle \psi_{Dk}|\psi_{Pk'}\rangle. \tag{9.64}$$

From this equation and Eq. 9.63, one can readily obtain a relation between the vectors $(\beta^\dagger, \beta)$ and $(\alpha^\dagger, \alpha)$.

The two HFB vacua $|\Psi_X\rangle, X = D, P$ are connected according to the Thouless theorem, i.e. [45]

$$|\Psi_{\tau P}\rangle = \langle\Psi_{\tau D}|\Psi_{\tau P}\rangle exp\left(\sum_{k<k'} Z_{\tau D;kk'}\alpha^\dagger_{\tau k}\alpha^\dagger_{\tau k'}\right)|\Psi_{\tau D}\rangle; \quad Z_{\tau D} = (V_{\tau D}U^{-1}_{\tau D})^* = -Z^T_{\tau D}. \tag{9.65}$$

In order to compute the formation amplitude we expand, as usually, the mother wave function in terms of the product of the daughter wave function times the two-neutron and two-proton wave functions

$$\Psi_{\tau D+2} = \sum_{k>0} X_{\tau k}\mathscr{A}\left[\psi_{\tau D;k}(1)\psi_{\tau D;\bar{k}}(2)\right]\Psi_{\tau D} * . \tag{9.66}$$

If one neglects the antisymmetrization between the core and the cluster [46] the coefficient X_k is given by the pairing density

$$X_{\tau k} = \langle \Psi_{\tau D+2} | a^{\dagger}_{\tau k} a^{\dagger}_{\tau \bar{k}} | \Psi_{\tau D} \rangle = \langle \Psi_{\tau D+2} | \Psi_{\tau D} \rangle (V_{\tau D} U^{\dagger}_{\tau D} + U^{*}_{\tau D} Z^{*}_{D\tau} U^{\dagger}_{\tau D})_{k\bar{k}}, \quad (9.67)$$

For gs–gs transitions, one obtains the previous expression $X_{\tau k} = u_{\tau k} v_{\tau k}$.

Using standard recoupling technique, one gets the formation amplitude (9.58) in terms of the B-coefficients (9.57), depending on the above quantities $X_{\tau k}$.

In Ref. [47], we assumed that the mother nucleus decays by electromagnetic transitions to the head of a super deformed band. From here, we considered that α-decay proceeds following two different possibilities:

1. direct α-decay to the ground state of the daughter nucleus and
2. α-decay to an excited state belonging to the superdeformed band of the daughter nucleus followed by an electromagnetic transitions to the ground state. The first process involves large transferred angular momenta and Q-values, while for the second one those quantities are both rather small.

We found that the α-decay probability from the head of a superdeformed band in ^{192}Pb to the corresponding state in ^{188}Hg is about 14 orders of magnitude larger than the corresponding probability for the ground state to ground state transition. This is therefore a likely candidate to observe α-decay transitions from superdeformed bands.

9.6.3 Spherical Nuclei

For spherical superfluid nuclei the parent wave function can be expanded in terms of four-body creation operators acting on the daughter wave function in a similar way to (9.51), i.e.

$$|P\rangle = \sum_{j_1} X_{\pi j_1} \left[a^{\dagger}_{\pi j_1} \otimes a^{\dagger}_{\pi j_1} \right]_0 \sum_{j_2} X_{\nu j_2} \left[a^{\dagger}_{\nu j_2} \otimes a^{\dagger}_{\nu j_2} \right]_0 |D\rangle, \quad (9.68)$$

where we defined sp index $j \equiv (\varepsilon l j)$. The coefficients $X_{\tau} j$ can be written in terms of the BCS occupation amplitudes, as follows

$$X_{\tau j} = \frac{1}{2} \langle D | \left[a_{\tau j} a_{\tau j} \right]_0 | P \rangle \approx \frac{\hat{j}}{2} u^{(D)}_{\tau j} v^{(P)}_{\tau j}, \quad (9.69)$$

where $\hat{j} = \sqrt{2j+1}$. In the general relation (9.36) one has $L_{\pi} = L_{\nu} = L_{\alpha} = 0$ with $X_{\pi\pi\nu\nu}(L_{\pi} L_{\nu}; L_{\alpha}) = 1$. As a result, the preformation amplitude contains only the monopole component, i.e.

$$\mathscr{F}^{(\alpha)}_{0}(R_{\alpha}) = \sum_{N_{\alpha}} W(N_{\alpha} 0) \phi^{(4\beta)}_{N_{\alpha} 0}(R). \quad (9.70)$$

where $R = R_{\alpha}/2$ and

$$W(N_\alpha 0) = 8 \sum_{n_\alpha N_\pi N_\nu} G_\pi(N_\pi 0) G_\nu(N_\nu 0) \langle n_\alpha 0 N_\alpha 0; 0 | N_\pi 0 N_\nu 0; 0 \rangle \mathscr{I}^{(\beta,\beta_\alpha)}_{n_\alpha 0}, \quad (9.71)$$

and the proton G-coefficient is the particular case of (9.31)

$$G_\pi(N_\pi 0) = \sum_{n_1 n_2 lj} B_\pi(n_1 ljn_2 lj;0) \langle (ll)0 \left(\frac{11}{22}\right) 0;0 | \left(l\frac{1}{2}\right) j \left(l\frac{1}{2}\right) j;0 \rangle$$
$$\times \sum_{n_\pi} \langle n_\pi 0 N_\pi 0; 0 | n_1 l n_2 l; 0 \rangle \mathscr{I}^{(\beta\beta_\alpha)}_{n_\pi 0} . \quad (9.72)$$

The B-coefficient is given by

$$B_\pi(n_1 ljn_2 lj;0) \equiv \sqrt{2} \sum_\varepsilon X_{\pi j} c_{n_1}(\pi\varepsilon lj) c_{n_2}(\pi\varepsilon lj). \quad (9.73)$$

9.7 α-Decay in Superheavy Nuclei

In the last years, the investigation of superheavy nuclei by using α-decay chains became a very active field of the nuclear physics. The synthesis of elements with $Z > 104$ was suggested by Flerov [48]. The real predictions were performed by using the cold valleys in the potential energy surface between different combinations, giving the same compound nucleus, within the so-called fragmentation theory [49]. Thirty years ago [50–52], it was shown that the most favorable combinations with $Z \geq 104$ are connected with the so-called Pb potential valley, i.e. the same valley of the heavy cluster emission [53, 54]. Due to the double magicity of ^{48}Ca, similar with ^{208}Pb, at the same time in Ref. [51] it was proposed ^{48}Ca as a projectile on various transuranium targets. Indeed, the production of many superheavy elements with $Z \leq 118$ during last three decades was based on this idea [55–59].

The production of superheavy nuclei is a difficult experimental undertaking. It is even more difficult to measure, once the superheavy nucleus has been detected, spectroscopic properties associated with the short lived nucleus. Usually, these nuclei are deformed and are created at large values of angular momenta, i.e. large values of K. In their decay down to the ground state, they may follow a complicated path that ends at a still relatively large value of K, i.e. much larger than the K-values corresponding to the few states lying below.

The nuclei with $Z > 104$ are unstable within the liquid drop model due to the strong Coulomb repulsion between protons. Their existence is strongly connected with the shell effects, which are estimated as corrections of the liquid drop picture by using the Strutinsky procedure [60, 61]. The closure property of the mean field in this region was evidenced in Ref. [62]. The formation of superheavy compound systems by fusion was intensively explored [63–66]. The investigation of experimental data concerning fusion and fission of superheavy nuclei with $Z = 112, 114, 116$, together with data on survival probability of these nuclei in evaporation channels with 3–4 neutrons, revealed the fact that the fission barriers are quite high, leading to a relative high stability of such systems [67].

Several papers were devoted to the investigation of α-decay half lives in this region. They used different phenomenological methods based on the Gamow penetration picture [68], like the Viola and Seaborg formula [62], the generalized liquid drop model [69–71], the preformed cluster decay model [72], the empirical A. Brown formula [73], or the double folding approach to estimate the barrier [74–76]. The fission theories like the superasymmetric fission model were also used to estimate half lives of superheavy α-emitters [77].

In the last decade, a considerable computational effort was devoted to the microscopic estimate of α and cluster-decay Q-values within the Non-Relativistic (Skyrme–Hartree–Fock) and Relativistic Mean Field Theory, in the region of superheavy nuclei. The half lives were mainly obtained by using the Viola–Seaborg formula (2.95) with a special set of parameters for the superheavy region, like for instance in Ref. [78]. The main achievements in this field are analyzed in Refs. [79, 80].

In this section, we analyze a special issue, connected with isomeric states in superheavy nuclei. Recently the investigation of high-spin two quasiparticle (2qp) isomeric states has shown a possible enhanced stability of superheavy nuclei [81]. There were predicted low-lying deformed 2qp configurations in the vicinity of the Fermi surface, by using potential energy surface (PES) calculations. They are formed by the broken pairs of quasiparticles. Aligned configurations can achieve a high angular momentum intrinsic projection. It turns out that the high-spin isomerism increases the fission barrier and decreases the probability of α-particle emission. This phenomenon is somehow similar to the experimentally observed enhanced stability of odd-mass superheavy nuclei in comparison with the even–even ones, due to the blocking effect.

In Ref. [81], the HF's for α-decays from 2qp states in superheavy nuclei were analyzed. Preformation amplitudes for transitions between ground states and from 2qp configurations were estimated by using a microscopic approach, given below. Let us first consider proton 2qp states, i.e. two protons in the parent nucleus move in quasiparticle states, while the neutrons remain in the BCS condensate. For neutrons in the 2qp states and protons in the vacuum, the formalism is exactly the same. The proton 2qp state is then

$$|K_\pi\rangle = \alpha^\dagger_{\pi K_a}\alpha^\dagger_{\pi K_b}|\mathrm{BCS}_\pi\rangle, \tag{9.74}$$

where $\alpha^\dagger_{\pi K}$ is the creation quasiparticle operator. Here, $K_\pi = K_a + K_b$ and the parity is the product of the two parities. In the proton BCS state the Nilsson states K_a and K_b are blocked. Therefore the parent state can be expanded as

$$|P\rangle = \alpha^\dagger_{\pi K_a}\alpha^\dagger_{\pi K_b}\sum_{K'>0} X^\nu_{K'} a^\dagger_{\nu K'} a^\dagger_{\nu \overline{K'}}|D\rangle. \tag{9.75}$$

Proceeding as before, one can write the 2qp wave function in a spherical basis, according to the transformation

$$\alpha^{\dagger}_{\pi K_a}\alpha^{\dagger}_{\pi K_b} \rightarrow u_{\pi K_a}u_{\pi K_b}a^{\dagger}_{\pi K_a}a^{\dagger}_{\pi K_b}$$
$$\rightarrow \sqrt{2}u_{\pi K_a}u_{\pi K_b}\sum_{L_\pi}\sum_{l_1 j_1}\sum_{l_2 j_2}\langle j_1,K_a;j_2,K_b|L_\pi,K_\pi\rangle\left[\psi_{\pi l_1 j_1}(\xi_1)\psi_{\pi l_2 j_2}(\xi_2)\right]_{L_\pi,K_\pi}, \tag{9.76}$$

where $K_\pi = K_a + K_b$. Following the same steps that led to Eq. 9.58, the formation amplitude corresponding to transitions from 2qp excited states becomes

$$\mathscr{F}_{K_\pi}(\mathbf{R}) = \sum_{L_\alpha} F_{L_\alpha K_\pi}(\mathbf{R}) = \sum_{L_\alpha}\sum_{N_\alpha} W(N_\alpha L_\alpha K_\pi)\phi^{(4\beta)}_{N_\alpha L_\alpha K_\pi}(\mathbf{R}). \tag{9.77}$$

The amplitude W is given in the Appendix (14.11) [82]. Notice that now the ho components depend upon the intrinsic projection K_π.

With the formation amplitude thus obtained one can evaluate the spectroscopic factor, which is defined as

$$S_{K_\pi} = \int_0^\infty R^2 dR \int d\hat{R}|\mathscr{F}_{K_\pi}(\mathbf{R})|^2 = \sum_{L_\alpha N_\alpha} W^2(N_\alpha L_\alpha K_\pi). \tag{9.78}$$

A similarly expression for the ground state was already given by Eq. 9.60. We can now define the hindrance factor (HF) which, as already mentioned, is the ratio between the mean values of the ground state to ground state formation amplitudes squared and the one corresponding to the transition from the state under study. As seen from Eq. 9.78, an equivalent definition is given by the ratio between the corresponding spectroscopic factors, i.e.

$$\mathrm{HF} = \frac{S_{gs}}{S_{K_\pi}}, \tag{9.79}$$

where S_{gs} is given by (9.60). Each Nilsson state is expanded in spherical components, by using seven partial waves, i.e. for a given K the spherical components carry the j-values $j = K, j = K + 1,\ldots, j = K + 6$. For convenience of presentation we denote the largest of those components by $K^\pi(j)$. For instance, for $\tau = \nu$ or π the label $\frac{7}{2}^-(9/2)$ means that $K^\pi = 7/2^-$ and that the largest spherical component is the one corresponding to the partial wave $j = 9/2$.

In Fig. 9.5, we show the components of the formation amplitudes corresponding to the decay of ^{254}Fm from the gs–gs transition (Fig. 9.5a) and from the 2qp state $8^+_\pi = \frac{7}{2}^-(7/2)\otimes\frac{9}{2}^-(9/2)$ to gs (Fig. 9.5b). In the first case, the component $F_{L_\alpha}(\mathbf{R})$ (see Eq. 9.58) as a function of R for different values of L_α is plotted, while in Fig. 9.5b we did the same for the function $F_{L_\alpha K_\pi=8^+}(\mathbf{R})$ (Eq. 9.77). The important feature of this figure is that in both cases, the monopole component, i.e. the component $L_\alpha = 0$, is overwhelmingly dominant.

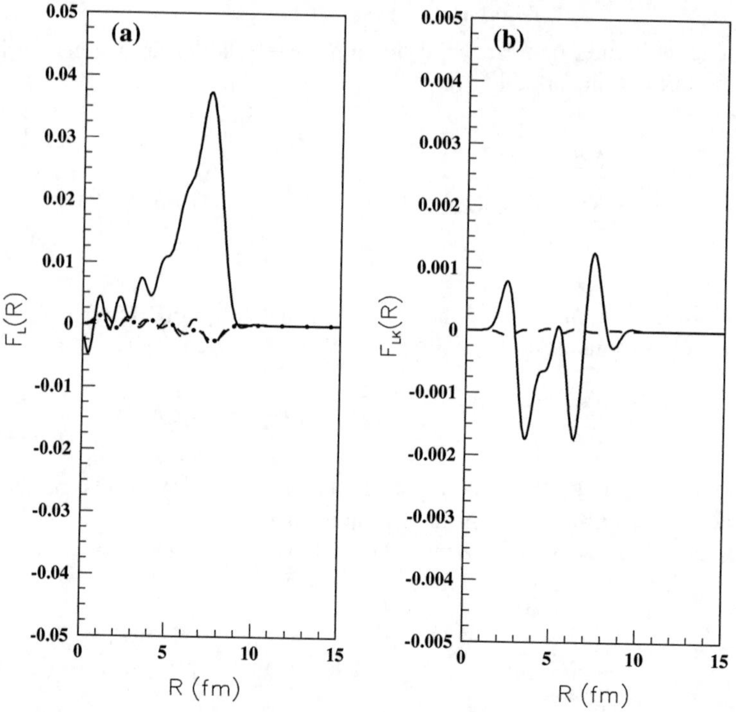

Fig. 9.5 **a** Formation amplitude components $F_{L_\alpha}(\mathbf{R})$ of the gs–gs transition (Eq. 9.58) for the α-decay of ^{254}Fm. The component $L_\alpha = 0$ corresponds to the solid line, $L_\alpha = 2$ to the dashed line and $L_\alpha = 4$ to the dot-dashed line. **b** The same as in **a**, but for the transition from the 2qp excited state $8^+_\pi = \frac{7}{2}^-(7/2) \otimes \frac{9}{2}^-(9/2)$ (Eq. 9.77)

Therefore, in Eqs (9.58) and (9.77) one can neglect all components $L_\alpha \neq 0$ and the results would not be affected substantially. However, in our calculations we do not use this approximation and all possible values of L_α are included.

The HF's (9.79) are very sensitive to the deformation, and thus can be a powerful tool to determine the structure of superheavy nuclei. One striking feature in Table 9.11 is that when the HF departs from its mean value by a large factor then the deformation also departs from its mean value. Thus, in ^{272}Rf the deformation is $\beta_2 = 0.18$ and HF $= 149$ for the proton 2qp state. For the neighboring nucleus ^{270}Rf it is $\beta_2 = 0.20$ and HF $= 11$ while for ^{258}Sg is $\beta_2 = 0.24$ and HF $= 17$. The same feature is found for neutron excitations for the state 10^+_ν in ^{272}Rf, where for $\beta_2 = 0.18$ it is HF $= 1210$, to be compared with HF $= 307$ for ^{270}Rf and HF $= 19$ for ^{258}Sg. Even in other cases where $\beta_2 = 0.20$ the values of HF depart significantly from their mean values, but the differences are not so striking. It is worthwhile to point out that these features are indeed induced by the changes in the structure of the 2qp state and not to the gs that is inherent in the HF, since this would significantly affect the value of S_0, which in Table 9.11 is seen to be rather constant.

Table 9.11 Quadrupole deformations and hindrance factors corresponding to the low-lying and high-spin 2qp state 8^+_π (proton excitations) and 10^+_ν (neutrons) in superheavy nuclei with $100 \leq Z \leq 110$

Nucleus	β_2	$HF[8^+_\pi = \frac{7}{2}^- (7/2) \otimes \frac{9}{2}^- (9/2)]$	$HF[10^+_\nu = \frac{9}{2}^+ (9/2) \otimes \frac{11}{2}^+ (11/2)]$	S_0
^{250}Fm	0.24	1.43×10^1	2.19×10^1	2.03×10^{-3}
^{252}Fm	0.24	1.57×10^1	3.11×10^1	2.95×10^{-3}
^{254}Fm	0.24	1.43×10^1	2.33×10^1	2.33×10^{-3}
^{256}Fm	0.23	1.72×10^1	3.05×10^1	3.20×10^{-3}
^{258}Fm	0.22	1.33×10^1	3.75×10^1	2.97×10^{-3}
^{252}No	0.24	1.52×10^1	2.88×10^1	2.97×10^{-3}
^{254}No	0.24	1.42×10^1	2.58×10^1	2.53×10^{-3}
^{256}No	0.23	1.58×10^1	2.72×10^1	2.82×10^{-3}
^{258}No	0.23	1.27×10^1	2.05×10^1	1.98×10^{-3}
^{260}No	0.23	1.31×10^1	3.29×10^1	2.72×10^{-3}
^{258}Rf	0.23	1.48×10^1	2.50×10^1	2.66×10^{-3}
^{260}Rf	0.23	1.58×10^1	2.07×10^1	2.32×10^{-3}
^{262}Rf	0.23	1.36×10^1	2.72×10^1	2.48×10^{-3}
^{264}Rf	0.23	1.27×10^1	3.06×10^1	2.42×10^{-3}
^{266}Rf	0.23	1.11×10^1	4.20×10^1	2.40×10^{-3}
^{268}Rf	0.21	1.06×10^1	9.89×10^1	2.43×10^{-3}
^{270}Rf	0.20	1.14×10^1	3.07×10^2	2.51×10^{-3}
^{272}Rf	0.18	1.49×10^2	1.21×10^3	2.62×10^{-3}
^{258}Sg	0.24	1.66×10^1	1.89×10^1	2.41×10^{-3}
^{260}Sg	0.24	1.62×10^1	1.88×10^1	2.39×10^{-3}
^{262}Sg	0.23	1.67×10^1	1.99×10^1	2.35×10^{-3}
^{264}Sg	0.23	1.62×10^1	2.17×10^1	2.26×10^{-3}
^{266}Sg	0.23	1.51×10^1	2.46×10^1	2.17×10^{-3}
^{268}Sg	0.23	1.35×10^1	3.36×10^1	2.18×10^{-3}
^{270}Sg	0.22	1.33×10^1	7.07×10^1	2.24×10^{-3}
^{272}Sg	0.20	1.75×10^1	2.71×10^2	2.39×10^{-3}
^{262}Hs	0.23	2.64×10^1	1.61×10^1	2.23×10^{-3}
^{264}Hs	0.23	2.61×10^1	1.79×10^1	2.18×10^{-3}
^{266}Hs	0.23	2.54×10^1	1.89×10^1	2.10×10^{-3}
^{268}Hs	0.23	2.39×10^1	2.22×10^1	2.02×10^{-3}
^{270}Hs	0.22	2.37×10^1	3.77×10^1	2.00×10^{-3}
^{272}Hs	0.21	2.61×10^1	9.37×10^1	2.10×10^{-3}
^{264}Ds	0.23	6.40×10^1	1.64×10^1	2.15×10^{-3}
^{266}Ds	0.23	6.34×10^1	1.75×10^1	2.13×10^{-3}
^{268}Ds	0.22	7.03×10^1	2.06×10^1	2.04×10^{-3}
^{270}Ds	0.22	6.53×10^1	2.52×10^1	1.95×10^{-3}
^{272}Ds	0.21	6.61×10^1	4.61×10^1	1.93×10^{-3}
^{274}Ds	0.20	7.17×10^1	1.14×10^2	1.98×10^{-3}

In the last column, the α-decay spectroscopic factors corresponding to ground state to ground state transition are given

This case shows that hindrance factors are outstanding tools to extract information of the structure of superheavy nuclei. Even small changes in the parameters that determine the nuclear structure induce very large differences in the HF. This is to be compared with the conclusions drawn from the analysis of experimental HF measured in light lead isotopes. There, it was shown that an abrupt change in the HF may be a signal that a nuclear phase transition has occurred.

In Table 9.11, HF's are so large that, if these 2qp high-spin states are reached, then it is likely that the nucleus will decay neither by γ emission nor by α-decay to the daughter gs. Instead the decay would proceed from the parent 2qp state to a similar 2qp state in the daughter nucleus. If such state does not exist in the daughter nucleus then the parent nucleus will live a long time before decaying.

The 2qp states in Table 9.11 are induced by aligned configurations. We have also studied non-aligned states in the same superheavy nuclear region, i.e. with $100 \leq Z \leq 110$. We show these in Table 9.12. As before, one notices that if the deformation parameter β_2 does not change very much from nucleus to nucleus, then the HF remains rather constant and does not depart abruptly from its mean value, as can be seen for the states 9^-_ν, 10^-_ν and 7^-_π. Even for the 12^-_ν states the deformation is rather constant and small, with the mean value $\beta_2 = 0.198$, and the HF is also rather constant. In this case it is interesting to notice that the largest value of the HF just corresponds to the smallest value of the deformation, which is $\beta_2 = 0.18$ in ^{272}Rf. Similar features can be seen in the $11^-{}_\nu$ state, although here the deformation is a bit larger. As expected according to our analysis above, a rather large change in the deformation, which occurs in going from ^{272}Rf with $\beta_2 = 0.18$ to ^{270}Sg with $\beta_2 = 0.22$, is accompanied by a corresponding change of the HF, in this case by a factor of about seven.

It is important to point out that the high HF's for ^{270}Ds corresponding to the states 9^-_ν and 10^-_ν in this Table, indicating that these may be isomeric states, is confirmed by one of the few experimental data available in superheavy nuclei. In Ref. [83], the measured half life corresponding to α-decay from the excited state is given to be about 6 ms, which is significantly larger than the value 0.1 ms, corresponding to the gs–gs decay. This is consistent with the predicted large value of HF, although perhaps, it is even more important the fact that it confirmed our conclusion that if high-spin states with large HF are reached, then they will likely be isomeric states.

Bearing this in mind, it is appropriate to remark that the values of the HF are very large in all cases of Table 9.12. They become as large as HF $= 8.3 \times 10^4$ for the state 12^-_ν in the nucleus ^{272}Rf. Therefore, the analysis above performed on the consequences of large HF, as well as the possibility of reaching an isomeric state or that the α chain of decays may not proceed through ground states, is even more compelling here.

The reason why the HF in Table 9.12 are so much larger than those in Table 9.11 is because in Table 9.11 the Nilsson states are aligned while they are not in Table 9.12.

Finally, in Table 9.13 we showed hindrance factors corresponding to nuclei with $Z > 110$. Some of these nuclei have been detected in experiments carried out

Table 9.12 Hindrance factors for high-spin 2qp low-lying states together with ground state spectroscopic factors for α-emitters with $100 \leq Z \leq 110$

Nucleus	β_2	HF	S_0
$9^-_\nu = \frac{11}{2}^-(15/2) \otimes \frac{7}{2}^+(9/2)$			
^{254}Fm	0.23	2.58×10^3	2.33×10^{-3}
^{256}Fm	0.23	5.53×10^3	3.20×10^{-3}
^{258}Fm	0.22	1.10×10^4	2.97×10^{-3}
^{256}No	0.23	2.49×10^3	2.82×10^{-3}
^{258}No	0.23	2.91×10^3	1.98×10^{-3}
^{260}No	0.23	8.84×10^3	2.72×10^{-3}
^{258}Rf	0.23	2.06×10^3	2.66×10^{-3}
^{260}Rf	0.23	3.16×10^3	2.32×10^{-3}
^{262}Rf	0.23	6.82×10^3	2.48×10^{-3}
^{260}Sg	0.23	1.63×10^3	2.39×10^{-3}
^{262}Sg	0.23	2.69×10^3	2.35×10^{-3}
^{264}Sg	0.24	5.71×10^3	2.26×10^{-3}
^{262}Hs	0.23	1.43×10^3	2.23×10^{-3}
^{264}Hs	0.24	2.12 n	2.18×10^{-3}
^{266}Hs	0.23	4.73×10^3	2.10×10^{-3}
^{264}Ds	0.23	1.41×10^3	2.15×10^{-3}
^{266}Ds	0.23	2.39×10^3	2.13×10^{-3}
^{268}Ds	0.22	4.66×10^3	2.04×10^{-3}
^{270}Ds	0.22	1.15×10^4	1.95×10^{-3}
$10^-_\nu = \frac{11}{2}^-(15/2) \otimes \frac{9}{2}^+(11/2)$			
^{256}Fm	0.23	2.08×10^4	3.20×10^{-3}
^{258}Fm	0.22	4.18×10^4	2.97×10^{-3}
^{258}No	0.23	8.82×10^3	1.98×10^{-3}
^{260}No	0.23	2.93×10^4	2.72×10^{-3}
^{260}Rf	0.23	9.83×10^3	2.32×10^{-3}
^{262}Rf	0.23	2.15×10^4	2.48×10^{-3}
^{262}Sg	0.23	8.08×10^3	2.35×10^{-3}
^{264}Sg	0.23	1.75×10^4	2.26×10^{-3}
^{264}Hs	0.23	6.25×10^3	2.18×10^{-3}
^{266}Hs	0.23	1.61×10^4	2.10×10^{-3}
^{266}Ds	0.23	1.06×10^4	2.13×10^{-3}
^{268}Ds	0.22	2.07×10^4	2.04×10^{-3}
^{270}Ds	0.22	5.45×10^4	1.95×10^{-3}
$7^-_\pi = \frac{7}{2}^+(13/2) \otimes \frac{7}{2}^-(9/2)$			
^{250}Fm	0.24	7.64×10^3	2.03×10^{-3}
^{252}Fm	0.24	7.14×10^3	2.95×10^{-3}
^{254}Fm	0.24	5.07×10^3	2.33×10^{-3}
^{256}Fm	0.23	6.92×10^3	3.20×10^{-3}
^{252}No	0.24	3.48×10^3	2.97×10^{-3}
^{254}No	0.24	2.52×10^3	2.53×10^{-3}

(continued)

Table 9.12 (continued)

Nucleus	β_2	HF	S_0
$7^-_\pi = \frac{9}{2}^+(13/2) \otimes \frac{5}{2}^-(7/2)$			
^{260}Rf	0.23	5.14×10^2	2.32×10^{-3}
^{262}Rf	0.23	4.78×10^2	2.48×10^{-3}
^{264}Rf	0.23	4.75×10^2	2.42×10^{-3}
^{258}Sg	0.24	1.04×10^3	2.41×10^{-3}
^{260}Sg	0.24	9.04×10^2	2.39×10^{-3}
^{262}Sg	0.23	8.53×10^2	2.35×10^{-3}
^{264}Sg	0.23	8.42×10^2	2.26×10^{-3}
^{266}Sg	0.23	8.19×10^2	2.17×10^{-3}
$12^-_\nu = \frac{13}{2}^-(15/2) \otimes \frac{11}{2}^+(11/2)$			
^{270}Rf	0.20	4.19×10^4	2.51×10^{-3}
^{272}Rf	0.18	8.34×10^4	2.62×10^{-3}
^{272}Sg	0.20	4.06×10^4	2.39×10^{-3}
^{272}Hs	0.21	1.92×10^4	2.10×10^{-3}
^{274}Ds	0.20	3.36×10^4	1.99×10^{-3}
$11^-_\nu = \frac{13}{2}^-(15/2) \otimes \frac{9}{2}^+(9/2)$			
^{268}Rf	0.21	9.53×10^3	2.43×10^{-3}
^{270}Rf	0.20	1.84×10^4	2.51×10^{-3}
^{272}Rf	0.18	3.75×10^4	2.62×10^{-3}
^{270}Sg	0.22	5.47×10^3	2.24×10^{-3}
^{272}Sg	0.20	1.46×10^4	2.39×10^{-3}
^{272}Hs	0.21	5.74×10^3	2.10×10^{-3}
^{274}Ds	0.20	6.59×10^3	1.99×10^{-3}
$10^-_\pi = \frac{9}{2}^-(9/2) \otimes \frac{11}{2}^+(13/2)$			
^{270}Ds	0.22	1.22×10^4	1.95×10^{-3}

in Dubna [84], but no spectroscopic measurements have been extracted so far. Therefore, it may be interesting to present the predictions of our calculations. We have chosen to give the HF's corresponding to 2qp states induced by aligned Nilsson states. As expected, due to this alignment, the order of magnitude of the HF's here is the same as those in Table 9.12. The deformations in these nuclei turn out to be all the same and small, i.e. $\beta_2 = -0.1$, and therefore the HF's are all the same within a small factor. This is specially remarkable for the state 9^+_ν, while for the state 10^+_π the differences are somehow larger, running from HF = 34 for $Z = 112$, $A = 280$ to HF = 166 for $Z = 120$, $A = 296$. But the HF's are still large and the discussions about this feature carried out in relation to Table 9.11 are also valid in this case.

As a conclusion, one can say that HF's analyzed in this section are in all cases very large, but we found that they can be divided in two sets. For 2qp states, which form an aligned configuration, the HF is between 10 and 100. For non-aligned configurations, the HF's are larger than 100 and as large as 10^5. This indicates that the decay of the nucleus by α-decay may be strongly hindered. If in the decay path mentioned above, the nucleus ends in such 2qp state then it would not decay by α particle emission to the ground state of the daughter nucleus, but rather to

Table 9.13 Hindrance factors for 10^+_π and 12^+_ν high-spin 2qp isomers together with ground state spectroscopic factors for α-emitters with $Z > 110$

Nucleus	β_2	$HF[10^+_\pi = \frac{9}{2}^+(9/2) \otimes \frac{11}{2}^+(11/2)]$	$HF[12^+_\nu = \frac{11}{2}^-(11/2) \otimes \frac{13}{2}^-(13/2)]$	S_0
$^{280}112$	−0.10	3.43×10^1	2.47×10^1	5.55×10^{-3}
$^{282}112$	−0.10	3.99×10^1	2.47×10^1	5.47×10^{-3}
$^{284}112$	−0.10	4.92×10^1	2.48×10^1	5.32×10^{-3}
$^{286}112$	−0.10	6.18×10^1	2.30×10^1	4.81×10^{-3}
$^{288}112$	−0.10	8.35×10^1	2.15×10^1	4.31×10^{-3}
$^{282}114$	−0.10	3.92×10^1	2.57×10^1	6.05×10^{-3}
$^{284}114$	−0.10	4.67×10^1	2.51×10^1	6.00×10^{-3}
$^{286}114$	−0.10	5.53×10^1	2.40×10^1	5.62×10^{-3}
$^{288}114$	−0.10	7.16×10^1	2.35×10^1	5.25×10^{-3}
$^{290}114$	−0.10	9.66×10^1	2.21×10^1	4.67×10^{-3}
$^{284}116$	−0.10	4.86×10^1	2.71×10^1	6.75×10^{-3}
$^{286}116$	−0.10	5.72×10^1	2.82×10^1	6.71×10^{-3}
$^{287}116$	−0.10	6.80×10^1	2.66×10^1	6.40×10^{-3}
$^{290}116$	−0.10	7.50×10^1	2.58×10^1	5.40×10^{-3}
$^{292}116$	−0.10	1.15×10^2	2.33×10^1	5.11×10^{-3}
$^{286}118$	−0.10	5.80×10^1	2.91×10^1	7.30×10^{-3}
$^{287}118$	−0.10	6.71×10^1	3.15×10^1	7.26×10^{-3}
$^{290}118$	−0.10	8.14×10^1	3.13×10^1	6.97×10^{-3}
$^{292}118$	−0.10	1.02×10^2	2.93×10^1	6.30×10^{-3}
$^{294}118$	−0.10	1.39×10^2	2.70×10^1	5.59×10^{-3}
$^{287}120$	−0.10	6.36×10^1	3.00×10^1	6.82×10^{-3}
$^{290}120$	−0.10	7.82×10^1	3.32×10^1	7.22×10^{-3}
$^{292}120$	−0.10	9.34×10^1	3.34×10^1	6.82×10^{-3}
$^{294}120$	−0.10	1.21×10^2	3.33×10^1	6.33×10^{-3}
$^{296}120$	−0.10	1.66×10^2	3.16×10^1	5.63×10^{-3}

an excited state having a structure similar to the parent 2qp state. In this case, the α-decay chain would not proceed through gs to gs channels but rather from excited state to excited state. However, if such daughter state does not exist then the parent nucleus would remain in the 2qp state a long time, becoming an isomer.

9.8 Two Proton Superfluid Emitters

In this section we built the two proton correlated wave function, which plays the role of the initial distribution in the dynamics of the two proton emission, described in Sect. 7.1. We start by pointing out that two proton emitters are medium nuclei and the mass of the daughter nucleus is much larger than the mass carried by the decaying protons. Therefore, we neglect recoil effects. The relative coordinates are $\mathbf{r} = \mathbf{r}_1 - \mathbf{r}_2$, $\mathbf{R} = (\mathbf{r}_1 + \mathbf{r}_2)/2$, forming a set which in few-body theory is called the Jacobi "T" system. In our non-recoil approximation it coincides with the Jacobi "Y" system [85]. We consider that the parent as well as the daughter nuclei are spherical. Thus, the parent ground state (gs) $|P\rangle$ can be written in terms of the daughter gs $|D\rangle$ as

$$|P\rangle = \sum_{\varepsilon lj} X_{\pi\varepsilon lj} \left[a^{\dagger}_{\varepsilon lj} a^{\dagger}_{\varepsilon lj}\right]_0 |D\rangle *, \tag{9.80}$$

where $a^{\dagger}_{\varepsilon lj}$ denotes, as usually, a single particle (sp) state expanded the in an ho basis defined by Eq. 9.25. By repeating the same operations as in the previous Section, i.e. transforming from *jj* to *LS* coupling and using the transformation from absolute to center of mass (cm) and relative coordinates, one obtains the two-particle wavefunction as

$$\begin{aligned}\Psi(\mathbf{r},\mathbf{R},\mathbf{S}) &= \sum_{\varepsilon lj} X_{\pi\varepsilon lj} \mathscr{A}\left\{\psi_{\varepsilon lj}(\mathbf{r}_1,\mathbf{s}_1)\psi_{\varepsilon lj}(\mathbf{r}_2,\mathbf{s}_2)\right\}_0 \\ &= \sum_{I=0,1}\sum_{\lambda\Lambda} F_{I\lambda\Lambda}(r,R)\left\{\left[Y_\lambda(\hat{r})Y_\Lambda(\hat{R})\right]_I \left[\chi_{\frac{1}{2}}(\mathbf{s}_1)\chi_{\frac{1}{2}}(\mathbf{s}_2)\right]_I\right\}_0.\end{aligned} \tag{9.81}$$

The radial multipoles are given by

$$\begin{aligned}F_{I\lambda\Lambda}(r,R) &= \sum_{nN} G_{In\lambda N\Lambda}\mathscr{R}^{(\beta/2)}_{n\lambda}(r)\mathscr{R}^{(2\beta)}_{N\Lambda}(R) \\ G_{In\lambda N\Lambda} &\equiv \sqrt{2}\sum_{n_1n_2lj}\sum_{\varepsilon} X_{\pi\varepsilon lj}c_{n_1}(\varepsilon lj)c_{n_2}(\varepsilon lj) \\ &\times \langle (ll)I(\frac{1}{2}\frac{1}{2})I;0|(l\frac{1}{2})j(l\frac{1}{2})j;0\rangle\langle n\lambda N\Lambda;I|n_1ln_2l;I\rangle.\end{aligned} \tag{9.82}$$

where standard notation has been used. We stress on the fact that we did not perform the integration over the proton relative coordinates, as it was the case in deriving the expression of the preformation amplitude.

The multipole components of the corresponding proton density are given by

$$\rho(\mathbf{r},\mathbf{R}) = \Psi(\mathbf{r},\mathbf{R},\mathbf{S})\Psi^*(\mathbf{r},\mathbf{R},\mathbf{S}) = \sum_L \rho_L(r,R)[Y_L(\hat{r})Y_L(\hat{R})]_0, \tag{9.83}$$

where the L-multipole density is

$$\begin{aligned}\rho_L(r,R) &\equiv \sum_{I=0,1}\sum_{\lambda\Lambda}\sum_{\lambda'\Lambda'}\frac{(-)^{\lambda+\Lambda}}{\sqrt{2I+1}}\langle(\lambda\Lambda)I(\lambda'\Lambda')I;0|(\lambda\lambda')L(\Lambda\Lambda')L;0\rangle \\ &\times F_{I\lambda\Lambda}(r,R)F_{I\lambda'\Lambda'}(r,R).\end{aligned} \tag{9.84}$$

The nuclear structure properties are described by the wave function amplitudes $X_{\pi\varepsilon lj}$ in Eq. 9.82. They depend upon the BCS amplitudes as in Eq. 9.69. For the central field we use a standard Woods–Saxon potential with the universal parametrization [86]. The strength of the potential is adjusted to reproduce the available experimental proton sp states. The u and v occupation amplitudes are evaluated as usually, i.e. by solving the BCS equations and adjusting the proton

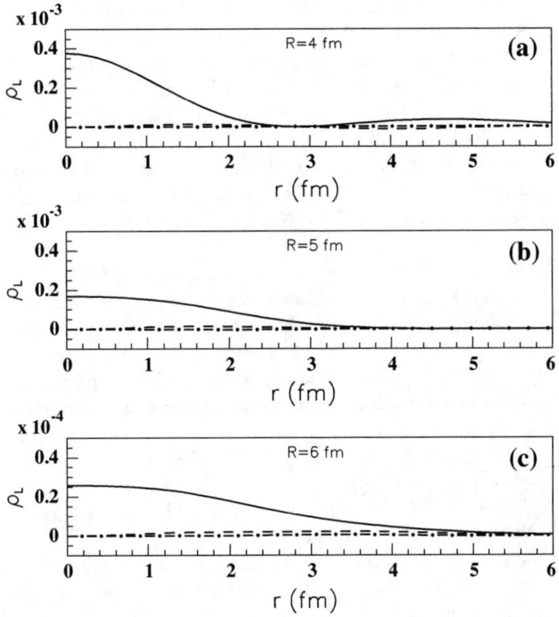

Fig. 9.6 The multipole components ρ_L of the two-proton density (9.84) versus the relative radius for the cm radius $R = 4$ fm **a**, 5 fm **b** and 6 fm **c**. The *solid line* corresponds to $L = 0$, the *dashed line* to $L = 2$ and the *dot-dashed line* to $L = 4$

pairing gap to its experimental value. In our calculations we include radial quantum numbers up to $n_{\max} = 8$.

In Fig. 9.6 we plot the proton density (Eq. 9.84) as function of the relative distance r at the cm distances $R = 4$ fm (a), $R = 5$ fm (b) and $R = 6$ fm (c), for the parent nucleus ^{45}Fe. One sees that inside the nucleus the monopole component $L = 0$ (solid line) is much larger than the ones corresponding to $L = 2$ (dashed line) and to $L = 4$ (dot-dashed line), as perhaps expected due to the collective character of the pairing states. Our calculations also showed that the singlet component $I = 0$ is by far dominant. Thus, the boundary condition for the three-body problem described in Sect. 7.1 is given by monopole component $F_{000}(r, R)$, defined by Eq. 9.82.

From Fig. 9.6 it becomes clear that the two protons are emitted following the same clustering features as those induced by the pairing interaction on the parent two-proton density and the decay proceeds through a rather narrow window around that center of mass. However, it would be incorrect to make an analogy with the α-decay and consider the decaying system as a simple di-proton. The decay proceeds neither through a di-proton particle nor as an uncorrelated two proton channel but rather as a configuration in between these two extrema. Perhaps, it is even more important that the formalism predicts that the angular distribution is a powerful tool to directly visualize the two-nucleon clustering induced by collective pairing modes.

References

1. Mang, H.J.: Calculation of α-transition probabilities. Phys. Rev. **119**, 1069–1075 (1960)
2. Săndulescu, A.: Reduced widths for favoured alpha transitions. Nucl. Phys. A **37**, 332–343 (1962)

3. Mang, H.J.: Alpha decay. Ann. Rev. Nucl. Sci. **14**, 1–28 (1964)
4. Liotta, R.J., Pomar, C.: Multi-step shell-model treatment of Sux-Particle system. Nucl. Phys. A **362**, 137–162 (1981)
5. Liotta, R.J., Pomar, C.: A graphical procedure to evaluate the many-body shell-model equations. Nucl. Phys. A **382**, 1–19 (1982)
6. Delion, D.S., Suhonen, J.: Microscopic Description of α-like Resonances. Phys. Rev. C **61**, 024304/1–12 (2000)
7. Heenen, P.-H.: Microscopic R-Matrix theory in a generator coordinate basis: (II) Study of resonance widths and application to α-^{16}O scattering. Nucl. Phys. A **272**, 399–412 (1976)
8. Takigawa, N., Lee, S.Y.: Nuclear Gory effect and α-^{40}Ca Scattering. Nucl. Phys. A **292**, 173–189 (1977)
9. Manngård P., PhD Thesis, Åbo Akademi, Åbo (1996)
10. Cwiok, S., Dudek, J., Nazarewicz, W., Skalski, J., Werner, T.: Single-particle energies, wave functions, quadrupole moments and g-factors in an axially deformed Woods-Saxon potential with applications to the two-centre-type nuclear problem. Comput. Phys. Commun. **46**, 379–399 (1987)
11. Vertse, T., Pál, K.F., Balogh, A.: GAMOW, a program for calculating the resonant state solution of the Radial Schrödinger Equation in an arbitrary optical potential. Comput. Phys. Commun. **27**, 309–322 (1982)
12. Curuchet, P., Vertse, T., Liotta, R.J.: Resonant random phase approximation. Phys. Rev. C **39**, 1020–1031 (1989)
13. Brussaard, P.J., Glaudemans, P.W.M.: Shell-Model Applications in Nuclear Spectroscopy. North-Holland, Amsterdam (1977)
14. Schuck, P., Wittman, R., Ring, P.: Lett. Nuovo Cim. **17**, 107 (1976)
15. Strottman, D.: Weak coupling calculation of ^{212}Po using realistic matrix elements. Phys. Rev. C **20**, 1150–1154 (1979)
16. Firestone, R.B., Shirley, V.S., Chu, S.Y.F., Baglin, C.M., Zipkin, J.: Table of Isotopes CD-ROM, 8th edn, Version 1.0. Wiley, New York (1996)
17. Dodig-Crnkovic, G., Janouch, F.A., Liotta, R.J., Sibanda, L.J.: Absolute α-decay rates in ^{212}Po. Nucl. Phys. A **444**, 419–435 (1985)
18. Dodig-Crnkovic, G., Janouch, F.A., Liotta, R.J.: An exact shell-model treatment of α-clustering and absolute α-decay. Nucl. Phys. A **501**, 533–545 (1989)
19. Blendowske, R., Fliessbach, T., Walliser, H.: Microscopic calculation of the ^{14}C decay of Ra nuclei. Nucl. Phys. A **464**, 75–89 (1987)
20. Blendowske, R., Fliessbach, T., Walliser, H.: Systematics of cluster-radioactivity-decay constants as suggested by microscopic calculations. Phys. Rev. Lett. **61**, 1930–1933 (1988)
21. Barker, F.C.: ^{12}O Ground-state decay by ^{2}He emission. Phys. Rev. C **63**, 047303/1–2 (2001)
22. Barmore, B., Kruppa, A.T., Nazarewicz, W., Vertse, T.: A new approach to deformed proton emitters: non-adiabadic coupled-channels. Nucl. Phys. A **682**, 256c–263c (2001)
23. Ohkubo, S. (ed.): Alpha-clustering and molecular structure of medium-weight and heavy nuclei. Prog. Theor. Phys. Suppl. **132**, 1–228 (1998)
24. Fioravanti, R., Viano, G.A.: Rotational bands and surface waves in α-^{40}Ca elastic scattering. Phys. Rev. C **55**, 2593–2603 (1997)
25. Ohkubo, S.: Alpha-cluster model theory of ^{44}Ti and effective two-body interaction. Phys. Rev. C **38**, 2377–2385 (1988)
26. Langanke, K.: Explanation of the backangle anomaly and isotope effect within a microscopic study of elastic α-scattering on even Ca isotopes. Nucl. Phys. A **377**, 53–83 (1982)
27. Wada, T., Horiuchi, H.: Resonating-group-method study of α+^{40}Ca elastic scattering and ^{44}Ti structure. Phys. Rev. C **38**, 2063–2077 (1988)
28. Sakuda, T., Ohkubo, S.: Coexistence of α-clustering and shell structure in ^{40}Ca. Z. Phys. A **349**, 361–362 (1994)
29. Sakuda, T., Ohkubo, S.: Structure study of ^{40}Ca by α+^{36}Ar cluster model. Phys. Rev. C **49**, 149–155 (1994)
30. Yamada, T., Ohkubo, S.: Core-excited α-cluster structure in ^{44}Ti. Z. Phys. A **349**, 363–365 (1994)

31. Ohkubo, S., Hirabayashi, Y., Sakuda, T.: α-Cluster structure of ^{44}Ti in core-excited α+^{40}Ca model. Phys. Rev. C **57**, 2760–2762 (1998)
32. Delion, D.S., Suhonen, J.: Microscopic description of α + Ca Quasimolecular resonances. Phys. Rev. C **63**, 061306(R)/1–5 (2001)
33. Vertse T., Curuchet P., Civitarese O., Ferreira, L.S., Liotta, R.J.: Application of Gamow resonances to continuum nuclear spectra. Phys. Rev. C **37**, 876–879 (1988)
34. Vertse, T., Liotta, R.J., Maglione, E.: Exact and approximate calculation of giant resonances. Nucl. Phys. A **584**, 13–34 (1995)
35. Langanke, K.: Explanation of the backangle anomaly and isotope effect within a microscopic study of elastic α-scattering on even Ca isotopes. Nucl. Phys. A **377**, 53–83 (1982)
36. Dudek, J., Szymanski, Z., Werner, T.: Woods-Saxon potential parameters optimized to the high spin spectra in the Lead region. Phys. Rev. C **23**, 920–925 (1981)
37. Allatt, R.G., et. al.: Fine structure in ^{192}Po α decay and shape coexistence in ^{188}Pb. Phys. Lett. B **437**, 29–34 (1998)
38. Barker, F.C.: Width of the ^{12}O ground state. Phys. Rev. C **59**, 535–538 (1999)
39. Frekers, D., et al.: Resonances in low energy ^{40}Ca(α, α)-scattering and quasimolecular band in ^{44}Ti. Z. Phys. A **276**, 317–324 (1976)
40. Frekers, D., Santo, R., Langanke, K.: Identification of quasimolecular resonances in low energy α-^{40}Ca scattering and effects of compound nucleus excitation. Nucl. Phys. A **394**, 189–220 (1983)
41. Yamada, T., et al.: Experimental examination of the lowest alpha cluster states in ^{44}Ti. Phys. Rev. C **41**, 2421–2424 (1990)
42. Richard B.F., Singh B.: Table of Superdeformed Nuclear Bands and Fission Isomers, LBL-35916 (1994)
43. Werner, T.R., Dudek, J.: Shape coexistence effects of super- and hyperdeformed configurations in rotating nuclei with $58 \leq Z \leq 74$. Atomic Data and Nucl. Data Tabl. **50**, 179–267 (1992)
44. de Voight, M.L.A., Dudek, J., Szymanski, Z.: High-spin phenomena in atomic nuclei. Rev. Mod. Phys. **55**, 949–1046 (1983)
45. Ring, P., Schuck, P.: The Nuclear Many-Body Problem. Springer-Verlag, New York (1980)
46. Varga, K., Lovas, R.G., Liotta, R.J.: Cluster-configuration shell model for alpha decay. Nucl. Phys. A **550**, 421–452 (1992)
47. Delion, D.S., Liotta, R.J.: Microscopic description of α decay from superdeformed nuclei. Phys. Rev. C **58**, 2073–2080 (1998)
48. Flerov, G.N.: Atom. Ener. **26**, 138 (1969)
49. Sandulescu, A., Gupta, R.K., Scheid, W., Greiner, W.: Synthesis of new elements within the fragmentation theory: application to $Z = 104$ and 106 elements. Phys. Lett. B **60**, 225–228 (1976)
50. Gupta, R.K., Parvulescu, C., Sandulescu, A., Greiner, W.: Further possibilities with Pb-targets for synthesizing super-heavy elements. Z. Phys. A **283**, 217–218 (1977)
51. Gupta, R.K., Sandulescu, A., Greiner, W.: Interaction barriers, Nuclear shapes and the optimum choice of a compound nucleus reaction for producing super-heavy elements. Phys. Lett. **67**B, 257–261 (1977)
52. Gupta, R.K., Sandulescu, A., Greiner, W.: Synthesis of Fermium and Transfermium elements using Calcium-48 beam. Z. Naturforsch **32**a, 704 (1977)
53. Săndulescu, A., Poenaru, D.N., Greiner, W.: Fiz. Elem. Chastits At Yadra **11**, 1334 (1980)
54. Săndulescu, A., Poenaru, D.N., Greiner, W.: New type of decay of heavy nuclei intermediate between fission and α decay. Sov. J. Part. Nucl. **11**, 528 (1980)
55. Oganessian, Yu.Ts.: The synthesis and decay properties of the heaviest elements. Nucl. Phys. A **685**, 17c–36c (2001)
56. Oganessian, Yu.Ts., et. al.: Heavy element research at Dubna. Nucl. Phys. A **734**, 109–123 (2004)
57. Hofmann, S., Münzenberg, G.: The discovery of heaviest elements. Rev. Mod. Phys. **72**, 733–767 (2000)
58. Hofmann, S., Münzenberg, G., Schädel, M.: On the discovery of superheavy elements. Nucl. Phys. News **14**(4), 5–13 (2004)
59. Nazarewicz, W., et al.: Theoretical description of superheavy nuclei. Nucl. Phys. A **701**, 165c–171c (2002)

60. Strutinsky, V.M.: Shell effects in nuclear masses and deformation energies. Nucl. Phys. A **95**, 420–442 (1967)
61. Strutinsky, V.M.: "Shells" in deformed nuclei. Nucl. Phys. A **122**, 1–33 (1968)
62. Smolanczuk, R.: Properties of the hypothetical spherical superheavy nuclei. Phys. Rev. C **56**, 812–824 (1997)
63. Denisov, V.Yu., Hofmann, S.: Formation of superheavy elements in cold fusion reactions. Phys. Rev. C **61**, 034606/1–15 (2000)
64. Smolanczuk, R.: Formation of superheavy elements in cold fusion reactions. Phys. Rev. C **63**, 044607/1–8 (2001)
65. Zagrebaev, V.I., Arimoto, Y., Itkis, M.G., Oganessian, Yu.Ts.: Synthesis of superheavy nuclei: How accurately can we describe it and calculate the cross sections? Phys. Rev. C **65**, 014607/1–13 (2001)
66. Adamian, G.G., Antonenko, N.V., Scheid, W.: Isotopic trends of the production of superheavy nuclei in cold fusion reactions. Phys. Rev. C **69**, 044601(R)/1–5 (2004)
67. Itkis, M.G., Oganessioan, Yu.Ts., Zagrebaev, V.I.: Fission barriers of superheavy nuclei. Phys. Rev. C **65**, 044602/1–7 (2002)
68. Gamow, G.: Zur Quantentheorie des Atomkernes. Z. Phys. **51**, 204–212 (1928)
69. Royer, G.: Alpha emission and spontaneous fission through quasimolecular shapes. J. Phys. G: Nucl. Part. Phys. **26**, 1149–1170 (2000)
70. Royer, G., Gherghescu, R.: On the formation and alpha decay of superheavy elements. Nucl. Phys. A **699**, 479–492 (2002)
71. Zhang, H., Zuo, W., Li, J., Royer, G.: α Decay half lives of new superheavy nuclei within a generalized liquid drop model. Phys. Rev. C **74**, 017304/1–4 (2006)
72. Kumar, S., Balasubramaniam, M., Gupta, R.K., Münzenberg, G., Scheid, W.: The formation and decay of superheavy nuclei produced in ^{48}Ca-induced reactions. J. Phys. G: Nucl. Part. Phys. **29**, 625–639 (2003)
73. Typel, S., Brown, B.A.: Skyrme Hartree-Fock calculations for the α decay Q-values of superheavy nuclei. Phys. Rev. C **67**, 034313/1–6 (2003)
74. Roy Chowdhury, P., Samanta, C., Basu, D.N.: α Decay half-lives of new superheavy elements. Phys. Rev. C **73**, 014612/1–6 (2006)
75. Mohr, P.: α-Nucleus potentials, α-decay half-lives, and shell closures for superheavy nuclei. Phys. Rev. C **73**, 031301(R)/1–5 (2006)
76. Buck, B., Merchand, A.C.: A consistent cluster model treatment of exotic decays and alpha decays from heavy nuclei. J. Phys. G **15**, 615–635 (1989)
77. Poenaru, D.N., Plonski, I.-H., Greiner, W.: α-Decay half lives of superheavy nuclei. Phys. Rev. C **74**, 014312/1–5 (2006)
78. Sobiczewski, A., Patyk, Z., Cwiok, S.: Deformed superheavy nuclei. Phys. Lett. B **224**, 1–4 (1989)
79. Gambhir, Y.K., Bhagwat, A., Gupta, M.: Microscopic description of superheavy nuclei. Ann. Phys. (NY) **320**, 429–452 (2005)
80. Dupré, Z.A., Bürvenich, T.J.: Predictions of α-decay half lives based on potential from self-consistent mean-field models. Nucl. Phys. A **767**, 81–91 (2006)
81. Xu, F.R., Zhao, E.G., Wyss, R., Walker, P.M.: Enhanced stability of superheavy nuclei due to high-spin isomerism. Phys. Rev. Lett. **92**, 252501/1–4 (2004)
82. Delion, D.S., Liotta, R.J., Wyss, R.: α Decay of high-spin isomers in superheavy nuclei. Phys. Rev. C **76**, 044301/1–8 (2007)
83. Hofmann, S., et al.: The new isotope ^{270}110 and its decay products ^{266}Hs and ^{262}Sg. Eur. Phys. J. A **10**, 5–10 (2001)
84. Itkis, M.G. et. al.: FUSION06: Reaction Mechanism and Nuclear Structure at the Coulomb Barrier, AIP Conference Proceedings, vol. **853**, p. 231 (2006).
85. Grigorenko, L.V., Johnson, R.C., Mukha, I.G., Thomson, I.J., Zhukov, M.V.: Two-proton radioactivity and three-body decay: general problems and theoretical approach. Phys. Rev. C **64**, 054002/1–12 (2001)
86. Dudek, J., Nazarewicz, W., Werner, T.: Discussion of the improved parametrisation of the Woods-Saxon potential for deformed nuclei. Nucl. Phys. A **341**, 253–268 (1980)

Chapter 10
Selfconsistent Emission Theory

10.1 General Framework

It was shown that the usual shell model space using $N = 6–8$ major shells underestimates the experimental decay width by several orders of magnitude [1, 2], due to the exponential decrease of bound single particle wave functions [3]. An answer to the problem would be the inclusion of the sp narrow resonances lying in continuum [4–6], i.e. Gamow states. In spite of the fact that the true asymptotic behaviour of the wave functions is achieved, the value of the half life is still not reproduced [7]. Only the background components in continuum can describe the right order of magnitude of experimental decay widths [8–12]. The inclusion of the background contribution becomes important because an important part of the α-clustering process proceeds through such states.

The problem of considering the continuum part of the spectrum in microscopic calculations is rather involved, but very important especially for drip line nuclei [13]. The idea to replace the integration over the real spectrum in continuum by sp Gamow resonances plus an integration along a contour in the complex plane including these resonances was considered by Berggren [14]. The calculation is very much simplified, if one considers that in some physical processes only the narrow resonances are relevant and the integration, giving the background, can be neglected [15, 16]. This was shown to be an adequate approach, for instance in giant resonances [17] and in the nucleon decay processes [18]. To estimate the decay width, the states in continuum can be taken into account effectively by including a cluster component [19], or by considering a sp basis with a larger harmonic oscillator (ho) parameter for states in continuum [12].

The α-decaying state can be described as a sp resonance, namely using the matching between logarithmic derivatives of the preformation amplitude and external Coulomb function. The derivative of the α-particle preformation factor, estimated within the shell model, is almost constant along any neutron chain and therefore is not consistent with the decreasing behaviour of Q values along such chains [20, 21]. We will show that the slope of the preformation amplitude can be

D. S. Delion, *Theory of Particle and Cluster Emission*, Lecture Notes in Physics, 819, DOI: 10.1007/978-3-642-14406-6_10, © Springer-Verlag Berlin Heidelberg 2010

corrected by changing the ho parameter of sp components. These components are connected with an α-cluster term, not predicted by the standard shell model [19]. Indeed, the even–odd pair staggering of binding energies found along the α-lines with $N - Z =$ const. can be explained in terms of a "pairing" in the isospin space between proton and neutron pairs, considered as bosons [22, 23]. The generalization of this approach in terms many-body Greens's functions was performed in Refs. [24, 25]. This fact suggests that α-particles are already preformed at least in the low density region of the nuclear surface and they can, indeed, explain the above inconsistency. We will discuss all these points within the so-called selfconsistent emission theory.

In a selfconsistent emission theory, the product between the reduced width and penetrability in the decay width (8.96) should not depend upon the matching radius R. This condition is not a trivial one in the microscopic theory and we will discus this point below.

Let us consider, for simplicity, the case of the α-decay between ground states of even–even nuclei. In the decay width (2.83) derived in Sect. 2.5 in Chap. 2

$$\Gamma = \sum_L \hbar v |N_L|^2, \tag{10.1}$$

the main ingredients are the scattering amplitudes N_L, which are the coefficients of the outgoing Coulomb-Hankel waves $H_L^{(+)}$. We have shown that, by introducing the so-called external fundamental system of solutions $\mathscr{H}^{(+)}(R)$, the scattering amplitudes are fully determined in terms of te internal wave function components (4.11), i.e.

$$N_L = \sum_{L'} \left[\mathscr{H}^{(+)}(R)\right]^{-1}_{LL'} f_{L'}^{(\mathrm{int})}(R). \tag{10.2}$$

Partial decay widths in (10.1) should not depend upon the matching radius R. In other words, each scattering amplitude in (10.2) should be a constant. In a phenomenological theory, this condition is automatically fulfilled due to the fact that internal and external components satisfy the same equations. This becomes more clear for spherical emitters, where the scattering amplitude is the ratio between internal and external solutions, i.e.

$$N_L = \frac{f_L^{(\mathrm{int})}(R)}{H_L^{(+)}(\chi, \kappa R)}. \tag{10.3}$$

In a microscopic theory, this statement is a priori not true and the check of the condition, called also *plateau condition*

$$\frac{\partial N_L}{\partial R} = 0, \tag{10.4}$$

is a test of selfconsistency. In next sections, we will analyze in detail how this goal can be achieved.

10.2 A Simple Cluster Model

The analysis performed within the MSM has shown that the microscopic preformation factor is not able to describe the absolute value of the decay width. We used a two-body interaction in order to build the α-particle preformation amplitude. This function can be interpreted as a wave function of the α-particle on the nuclear surface $\Psi^{(\alpha)}$, generated by some effective Hamiltonian H_α, given by (9.39). It is clear that a modified interaction is necessary in order to reproduce the absolute value of the decay width, i.e.

$$H_\alpha \Psi^{(\alpha)} \to H_{\text{int}} \Psi^{(\text{int})} = E_\alpha \Psi^{(\text{int})}, \tag{10.5}$$

where in general the Hamiltonian operators are non-local. Thus, the internal wave function becomes

$$\Psi^{(\text{int})}(\mathbf{R}) = \Psi^{(\alpha)}(\mathbf{R}) + \psi^{(\alpha)}(\mathbf{R}). \tag{10.6}$$

A value $\psi^{(\alpha)} = 0$ would correspond to shell model preformation amplitude, fully predicting the total decay width, independently upon the radius R. In general, this statement is by far not true.

First of all, we mention that the monopole component of the microscopic preformation amplitude computed within the pairing approach (9.35) represents more than 90% of the total value. This is shown in Fig. 10.1 for the decay process $^{226}\text{Ra} \to {}^{222}\text{Ra} + \alpha$. One sees that $f_0(R)$ has a Gaussian-like structure and the wave function $f_0(R)/R$ can be also very well approximated by a Gaussian beyond the largest maximal value, which is the most important region for the decay process

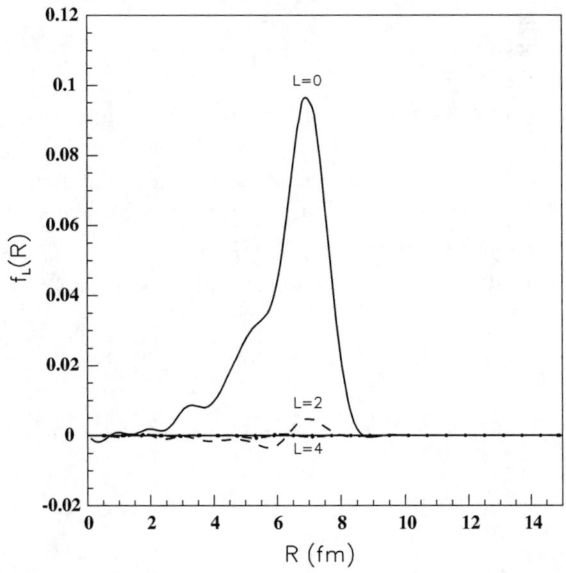

Fig. 10.1 Microscopic preformation amplitude versus radius for $L = 0$ (*solid curve*), $L = 2$ (*dashed curve*) and $L = 4$ (*dot-dashed curve*). The quadrupole deformation is $\beta = 0.2$. The decay process is $^{226}\text{Ra} \to {}^{222}\text{Rn} + \alpha$

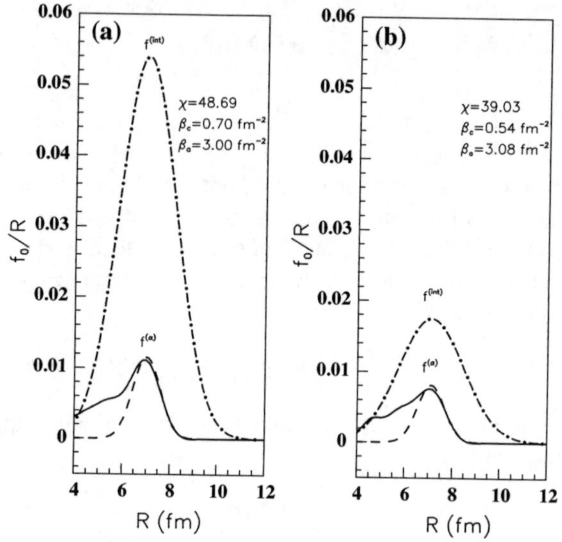

Fig. 10.2 Monopole microscopic preformation amplitude versus radius (*solid curve*), its Gaussian fit (*dashed curve*) and the internal (microscopic + cluster) amplitude (*dot-dashed curve*) for **a** the decay process $^{226}\text{Ra} \to {}^{222}\text{Rn} + \alpha$ and **b** the decay process $^{220}\text{Ra} \to {}^{216}\text{Rn} + \alpha$

$$\mathscr{F}_0^{(\alpha)} = \frac{f_0^{(\alpha)}(R)}{R} = \sum_N W(N0)\mathscr{R}_{N0}^{(4\beta)}(R) \approx A_0 e^{-\beta_a(R-R_m)^2/2}. \tag{10.7}$$

Here, β_a is the width of the distribution and R_m is the nuclear surface radius $R_m \approx R_0 = 1.2A_D^{1/3}$. This is shown in Fig. 10.2a for the decay process $^{226}\text{Ra} \to {}^{222}\text{Ra} + \alpha$ and in Fig. 10.2b for the decay process $^{220}\text{Ra} \to {}^{216}\text{Ra} + \alpha$. The microscopic amplitude is plotted by the solid line while its Gaussian fit is represented by a dashed line.

Thus, for the internal total wave function a reasonable assumption is to consider a similar ansatz

$$\frac{f_0^{(\text{int})}(R)}{R} = C_0 e^{-\beta_c(R-R_m)^2/2}. \tag{10.8}$$

This function is nothing else than the cm cluster wave function $\psi_{cm}^{(\beta_c)}(R - R_m)$ in (9.3).

In order to estimate the total internal wave function, one considers that the channel radius is located inside the Coulomb barrier, where the nuclear interaction has vanishing values, i.e. $\mathscr{H}_{LL'}^{(+)}(R) \approx \delta_{LL'} H_L^{(+)}(\chi, kR) \approx \delta_{LL'} G_L(\chi, kR)$. Thus, one obtains the following matching conditions for function and its derivative

$$\begin{aligned} C_0 e^{-\beta_c(R-R_m)^2/2} &= N_0 \frac{G_0(\chi, kR)}{R} \\ -\beta_c(R - R_m) C_0 e^{-\beta_c(R-R_m)^2/2} &= N_0 \left[\frac{G_0(\chi, kR)}{R}\right]' \\ N_0 &= \sqrt{\frac{\Gamma_{\text{exp}}}{\hbar v}}. \end{aligned} \tag{10.9}$$

The cluster frequency can be expressed in terms of the Coulomb parameter as follows

$$\begin{aligned}\beta_c(R-R_m) &= -\frac{[G_0(\chi,kR)/kR]'}{G_0(\chi,kR)/kR} \\ &= -\frac{d}{d\rho}\left[\chi(\alpha-\sin\alpha\,\cos\alpha)+\frac{1}{2}\ln ctg\,\alpha-\ln\rho\right].\end{aligned} \tag{10.10}$$

Here we used the WKB representation of the irregular Coulomb function given in Appendix (14.2), i.e.

$$\begin{aligned}G_0(\chi,\rho) &= (ctg\,\alpha)^{\frac{1}{2}}e^{\chi(\alpha-\sin\alpha\,\cos\alpha)} \\ \cos^2\alpha &= \frac{\rho}{\chi}, \quad \rho = kR.\end{aligned} \tag{10.11}$$

It turns out that the product $\beta_c(R-R_m)$ has a week dependence upon the radius, in the region around the touching configuration $R_C \approx 9.1$ fm, as it is seen in Fig. 10.3. Anyway the dependence of the cluster frequency multiplied by $R - R_m$ upon the Coulomb parameter is an universal function. This function is plotted in Fig. 10.4 for a typical value of the reduced radius $\rho = 10$. This dependence will play a central role in the selfconsistent theory of the α-decay, given in Sect. 10.4. We obtained the dot-dashed curves, representing $f_0^{(int)}$ in Fig. 10.2a, b, using a matching radius $R = 10$ fm, because at this value the α-daughter interaction is practically given only by the Coulomb part. As a consequence of this choice, the obtained values of the ho parameter β_c are close to the values of the free alpha-particle, i.e. $\beta_\alpha \approx 0.5$ fm^{-2}.

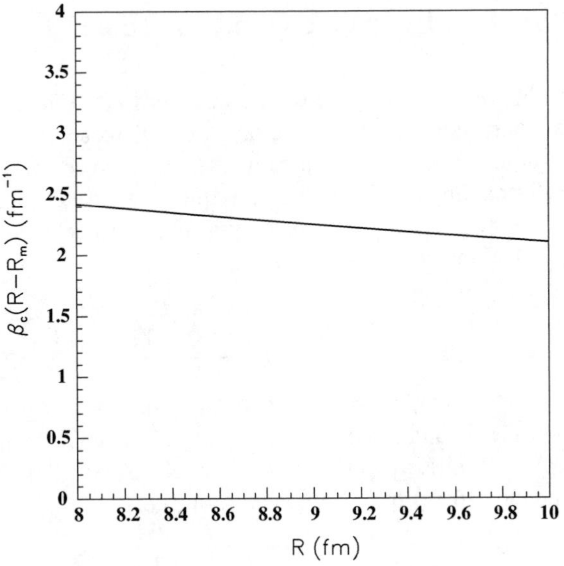

Fig. 10.3 Dependence of the quantity $\beta_c(R-R_m)$ versus the radius. The decay process is $^{226}Ra \rightarrow {}^{222}Rn + \alpha$

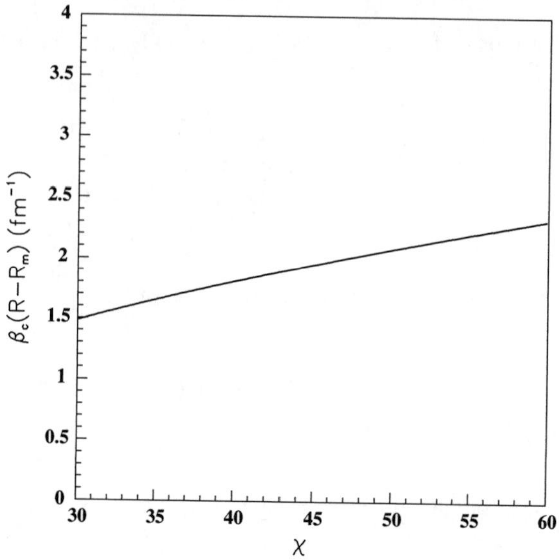

Fig. 10.4 Universal dependence of the quantity $\beta_c(R - R_m)$ versus the Coulomb parameter χ for $\rho = 10$

By comparing the Fig. 10.2a with b one sees that, due to the matching conditions (10.9), the total preformation factor has an important variation, proportional to the Coulomb parameter χ, while the microscopic part is almost unchanged concerning both the amplitude A_0 and width β_a. Thus, the selfconsistency condition (10.4) is not satisfied by the microscopic part. This feature suggests that an important cluster component, different from the microscopic part estimated in terms of single particle constituents, should exist on the nuclear surface.

10.3 Two Harmonic Oscillator Model

In this section, we propose a simple procedure to achieve the above described selfconsistency between the internal microscopic amplitude and the external Coulomb function. We already used sp representation in order to compute the preformation amplitude. For this purpose, we used the stationary Schrödinger equation describing the sp motion of a particle with the reduced mass μ_0 in a spherical mean field $V_\tau(r, s)$, i.e.

$$\mathbf{H}_\tau\psi(\mathbf{x}) \equiv \left[-\frac{\hbar\omega}{2\beta_0}\Delta + V_\tau(r,s)\right]\psi(\mathbf{x}) = \varepsilon\psi(\mathbf{x}), \tag{10.12}$$

where $\tau = \pi, \nu$ is the isospin projection, $\mathbf{x} = (\mathbf{r}, \mathbf{s})$ the set of spatial and spin coordinates, and the ho parameter is $\beta_0 = \mu_0\omega/\hbar$, with $\hbar\omega = 41\,A^{-1/3}$.

As we already pointed out, the wave function that solves the eigenvalue problem (10.12) has a separable form, i.e.

$$\psi_{\tau\varepsilon ljm}(\mathbf{x}) = u_{\tau\varepsilon lj}(r)\mathcal{Y}_{jm}^{(l\frac{1}{2})}(\hat{r}, s),$$
$$\mathcal{Y}_{jm}^{(l\frac{1}{2})}(\hat{r}, s) = \left\{Y_l(\hat{r})\chi_{\frac{1}{2}}(s)\right\}_{jm}, \tag{10.13}$$

where l is the angular momentum of the particle, j total spin and m spin projection on the z-axis. Let us introduce the same short-hand notation as in the previous paragraph, i.e. $(\tau\varepsilon lj) \to (\tau j)$. The radial part of the wave function satisfies (4.27) with the spherical potential, i.e.

$$-\frac{\hbar\omega}{2\beta_0}\left[\frac{1}{r}\frac{d^2}{r^2}r - \frac{l(l+1)}{r^2}\right]u_{\tau j}(r) \equiv H_l^{(\beta_0)}u_{\tau j}(r) = [\varepsilon - V_\tau(r,s)]u_{\tau j}(r). \tag{10.14}$$

The radial wave functions $u_{\tau j}$ is usually expanded in an ho basis, as in (4.29), i.e.

$$u_{\tau j}(r) = \sum_{n=0}^{\infty} c_n(\tau j)\mathcal{R}_{nl}^{(\beta)}(r), \tag{10.15}$$

where we used the new radial function $u_{\tau j} = f_{\tau j}/r$, with respect to Sect. 4.3 in Chap. 4. Here $\mathcal{R}_{nl}^{(\beta)}(r) = \langle r|\beta nl\rangle$ is the standard ho radial wave function with radial quantum number n and angular momentum l (see Appendix (14.8)). The ho parameter β determines the representation. It may be convenient to choose it according to the number of shells that one includes in the basis and, therefore, it might differ from the value of β_0. We define β as

$$\beta = f\beta_0 = f\frac{\mu_0\omega}{\hbar}, \tag{10.16}$$

and the constant f is chosen so that one obtains an ho spectrum that fits best the discrete part of $V_\tau(r)$ (this usually is a realistic Woods–Saxon potential).

As discussed above, this approach is not enough for a microscopic description of the α or cluster decay processes, for which it is very important to reproduce the wave function at distances larger than the nuclear radius. From the previous section it became clear that, in order to have a proper asymptotic behaviour of the wave functions at large distances, it is necessary to use a cluster component. A semi-phenomenological analysis of the α-decay process from ^{212}Po was given in Ref. [19]. Here, the sp wave function was written as a sum of two components: a standard shell model plus a phenomenological cluster component

$$\Psi = a\Psi_{SM} + b\Psi_\alpha. \tag{10.17}$$

Here, we proceed in a similar, but fully microscopic way. That is, instead of the expansion (10.15) for the sp radial wave function, we use the following representation

$$u_{\tau j}(r) = \sum_{2n_1+l=N_1\le N_0} c_{n_1}^{(1)}(\tau j)\mathcal{R}_{n_1 l}^{(\beta_1)}(r) + \sum_{2n_2+l=N_2>N_0} c_{n_2}^{(2)}(\tau j)\mathcal{R}_{n_2 l}^{(\beta_2)}(r), \tag{10.18}$$

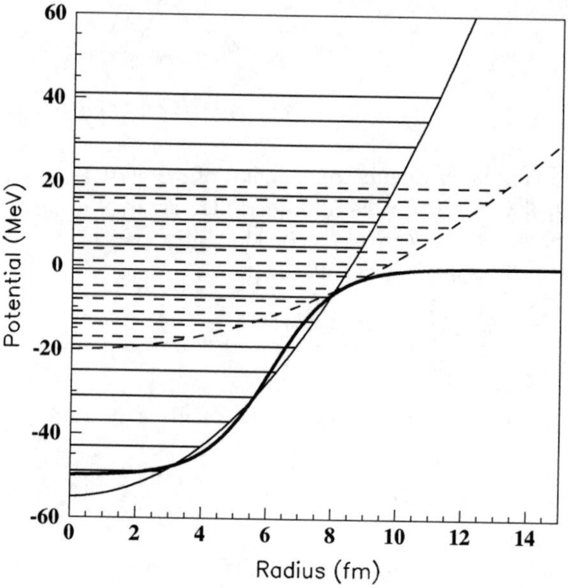

Fig. 10.5 Two harmonic oscillator potential

where β_1 is the ho parameter corresponding to the ho potential which fits the Woods–Saxon interaction in the region of the discrete spectrum, while β_2 corresponds to an ho potential that describes the clustering properties. Notice that these two different ho potentials are both centered at the origin of coordinates, as it is shown in Fig. 10.5. Let us also mention that both representations (10.17), as well as (10.18), are the generalizations of the Feshbach decomposition (8.53) using non-orthogonal components.

Below we describe how to evaluate the amplitudes c in (10.18). But it is worthwhile to point out here that, although the representation used in (10.18) is discrete, the density of states is larger at high energies. In fact, it can be made as large as one wishes by properly decreasing the parameter β_2. This is just what is needed to describe processes occurring at distances larger than the nuclear surface.

For each of the ho potentials the radial wave function $\mathscr{R}$ satisfies

$$H_l^{(\beta_i)}\mathscr{R}_{nl}^{(\beta_i)}(r) = \hbar\omega\left(2n + l + \frac{3}{2} - \frac{\beta_i r^2}{2}\right)\mathscr{R}_{nl}^{(\beta_i)}(r), \tag{10.19}$$

where

$$H_l^{(\beta_i)} = -\frac{\hbar\omega}{2\beta_i}\left[\frac{1}{r}\frac{d^2}{r^2}r - \frac{l(l+1)}{r^2}\right], \tag{10.20}$$

and β_i is the ho parameter of the potential i.

The diagonalization of the potential V in terms of the two ho representations can be performed by inserting the expanded wave function (10.18) in Schrödinger equation. Then, one obtains the following set of equations

$$\begin{pmatrix} \mathscr{H}^{(11)}_{n_1n'_1} & \mathscr{H}^{(12)}_{n_1n'_2} \\ \mathscr{H}^{(21)}_{n_2n'_1} & \mathscr{H}^{(22)}_{n_2n'_2} \end{pmatrix} \begin{pmatrix} c^{(1)}_{n'_1}(\tau j) \\ c^{(2)}_{n'_2}(\tau j) \end{pmatrix} = \varepsilon \begin{pmatrix} \mathscr{I}^{(11)}_{n_1n'_1} & \mathscr{I}^{(12)}_{n_1n'_2} \\ \mathscr{I}^{(21)}_{n_2n'_1} & \mathscr{I}^{(22)}_{n_2n'_2} \end{pmatrix} \begin{pmatrix} c^{(1)}_{n'_1}(\tau j) \\ c^{(2)}_{n'_2}(\tau j) \end{pmatrix}, \tag{10.21}$$

where the amplitudes c are as in (10.18) and the overlap integrals are defined by

$$\mathscr{I}^{(ik)}_{n_in'_k} \equiv \langle \beta_i n_i l | \beta_k n'_k l \rangle = \langle \mathscr{R}^{(\beta_i)}_{n_i l} | \mathscr{R}^{(\beta_k)}_{n'_k l} \rangle, \tag{10.22}$$

while the Hamiltonian kernels are given by

$$\begin{aligned} \mathscr{H}^{(ik)}_{n_in'_k} = \hbar\omega f_k \left[\left(2n'_k + l + \frac{3}{2} \right) \mathscr{I}^{(ik)}_{n_in'_k} - \frac{1}{2} \langle \beta_i n_i l | \beta_k r^2 | \beta_k n'_k l \rangle \right] \\ + \langle \beta_i n_i l | V_\tau | \beta_k n'_k l \rangle. \end{aligned} \tag{10.23}$$

The parameters β_k are connected with the standard ho parameter by the multiplier f_k

$$f_k = \frac{\beta_k}{\beta_0}; \quad k = 1, 2. \tag{10.24}$$

The overlap integrals (10.22) can be written in terms of Gamma functions [26].

A method to diagonalize the Hermitean system (10.21), corresponding to the representation of the Schrödinger operator in a non-orthogonal basis, is described in Appendix 14.9.

In the section devoted to the description of the α-decay from ^{212}Po, it was shown that the first $N_0 = 6$ shells in (10.18) contribute with less than 1% to the total decay width, while most of the contribution is given by the cluster part of the spectrum. In this way the usual shell model space should be enlarged to be able to describe four-body correlations during the α-decay process.

Our analysis showed that the bound states and narrow resonances have relative stable positions, while the density of the other states in continuum becomes higher by decreasing $f = f_2/f_1$. In this way, one obtains a quasicontinuum description of the background. The optimal value of the parameter f is taken in such a way that one obtains the right value of the total decay width (10.1), using a minimal amount of major shells. The expression of the decay width within the two ho basis is rather involved and the necessary details can be found in Ref. [12]. In Fig. 10.6, one can see that, if one considers only $N_0 = 6$ major shells, the total decay width is not reproduced (dot-dashed line). The same happens if one takes $N = 9$ shells with $f = 1$ (dashed line), while in the case of $f = 0.2$ (solid line) for the last three shells one practically obtains a plateau beyond the geometrical touching point, reproducing the experimental value.

10.4 Selfconsistent Description of the α-Decay

In spite of the fact that the two ho representations are able to describe the absolute values of the decay width, the shell model estimate of the α-particle preformation

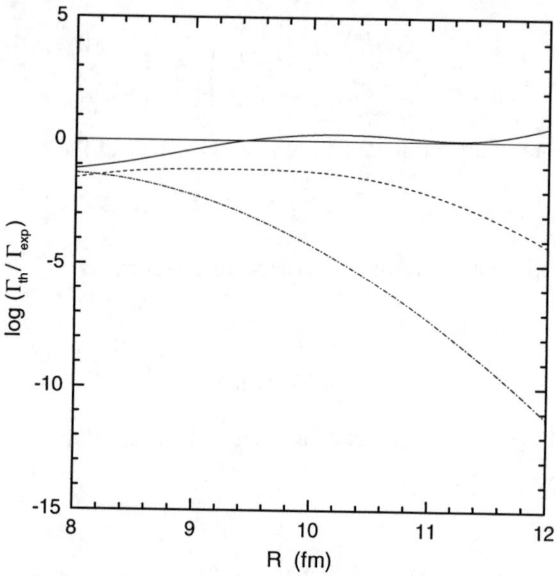

Fig. 10.6 The total decay width if one considers $N_0 = 6$ major shells (*dot-dashed curve*), $N = 9$ shells with $f = f_2/f_1 = 1$ (*dashed curve*) and $f = 0.2$ (*solid curve*) for the last three major shells

factor continues to remain not consistent with the decreasing behaviour of Q values along any neutron chain, or α-line, where $I = N - Z =$ const. [20, 21]. In our approach, we consider α-decay by treating the decaying state as a resonance, namely using the matching between logarithmic derivatives of the internal preformation amplitude and Coulomb function. It turns out that this condition is not satisfied along any neutron chain if one uses the standard hell model estimate for the preformation factor, due to the fact that its slope is almost constant with respect to the neutron or proton number. Yet, it is possible to correct the slope of the preformation amplitude by changing the ho parameter of single particle components corresponding to the cluster part of the expansion (10.18) [27]. Recently, a similar idea was used in Ref. [28]. We will show that, in order to fulfil the so-called plateau condition (10.4), it is necessary to consider an ho parameter as being a linear function of the Coulomb parameter, as predicted in Fig. 10.4.

As we already pointed out, the decay width is directly connected with the Q value, computed as

$$E_\alpha = B(Z-2, N-2, \beta) + B(2,2,0) - B(Z,N,\beta), \tag{10.25}$$

where $B(Z, N, \beta)$ is the binding energy, depending upon the charge, neutron numbers and quadrupole deformation parameter. This quantity is given by the Weizsäker type relation, for instance like in Ref. [29]

$$\begin{aligned} B(Z,N,\beta) = {} & a_{\text{vol}}A - a_{\text{surf}}A^{2/3} - a_{\text{Coul}}Z^2A^{-1/3} - E_{\text{sym}}(A,I) - a_{\text{pair}}A^{-1/2} \\ & + E_{\text{def}}(Z,N,\beta) + E_{\text{shell}}(\beta). \end{aligned} \tag{10.26}$$

Along any α-line with $I = N - Z =$ const. the Coulomb term has a much stronger variation versus Z (quadratic) than the other ones. Therefore, the Q value depends linearly upon the charge number and the shell model dependence practically disappears.

The standard procedure to estimate the decay width within the microscopic approach was described in the previous chapter. We have seen that most of the contribution comes from the second part of the expansion (10.18). Thus, in order to simplify calculations we will consider only one ho parameter in the preformation amplitude, corresponding to the cluster term.

It turns out that the expression of the decay width (10.1) for deformed emitters can be simplified. First of all let us point out that the major effect is given by the quadrupole deformation of the barrier [9, 10, 30]. The monopole component of the internal wave function has the dominant role, and the decay width can be approximated by

$$\Gamma(R) = \Gamma_0(R)D(R), \tag{10.27}$$

where $\Gamma_0(R)$ is the standard spherical decay width

$$\Gamma_0(R) = \hbar v \left[\frac{f_0^{(int)}(R)}{G_0(\chi, \kappa R)}\right]^2, \tag{10.28}$$

and the deformation function $D(R)$ is given in terms of the so-called Fröman matrix [31] by (5.10). As usually, by $G_0(\chi, kR)$ we denoted the monopole irregular Coulomb function, depending upon the reduced matching radius κR and the Coulomb parameter

$$\chi = 2\frac{Z_1 Z_2 e^2}{\hbar v}. \tag{10.29}$$

Thus, the decay width contains a ratio between the internal and external solutions. As we already pointed out, it does not depend upon the matching radius R within the local potential approach, because the internal and external wave functions satisfy the same equation and, therefore, are proportional and the plateau condition is automatically satisfied.

In the previous section, we have shown that the situation becomes different when the value of the internal wave function $f_0^{(int)}(R)$ is given by an independent *microscopic approach*, namely by the preformation amplitude (9.70). In order to compute this amplitude we expand sp wave functions within the ho basis, i.e.

$$\psi_{j_\tau m}(\mathbf{r}, s) = \sum_{n=0}^{n_{max}} c_{nj_\tau} \mathcal{R}_{nl}(\beta_0 r^2) \left[Y_l(\hat{r}) \otimes \chi_{\frac{1}{2}}(s)\right]_{j_\tau m}, \quad \tau = \pi, \nu. \tag{10.30}$$

The radial ho wave function is defined in terms of the Laguerre polynomial. We use one ho parameter β_0, which is connected with the standard ho parameter using a scaling factor f_0 as follows

$$\beta_0 = f_0 \beta_N = f_0 \frac{M_N \omega}{\hbar} \approx \frac{f_0}{A^{1/3}}, \tag{10.31}$$

where A is the mass number. The preformation amplitude (9.70) can be written in terms of Laguerre polynomials as follows

$$\mathscr{F}_0(\beta_0, n_{\max}, P_{\min}; R) = e^{-4\beta_0 R^2/2} \sum_N W_N(\beta_0, n_{\max}, P_{\min}) \mathscr{N}_{N0}(4\beta_0) L_N^{1/2}(4\beta_0 R^2), \tag{10.32}$$

We stress on the fact that the exponential term is similar to the cm α-particle wave function, but it depends upon the single particle ho parameter β_0. The expansion coefficients are given in terms of recoupling Talmi-Moshinsky brackets by (9.71). In our sp basis we consider only those states with pairing density products $P_\tau = (j+1/2)^{1/2} u_{\tau j} v_{\tau j}$ larger than the minimal value $P_{\min}$, taken as a parameter.

The Coulomb parameter χ is the most important ingredient governing the penetrability of the α-particle through the barrier. The irregular Coulomb function $G_0(\chi, kR)$ depends exponentially on it, as it is shown in (10.11). The decay width has also an exponential dependence upon the quadrupole deformation. As it is clear from Fig. 5.1, the function $D(R)$ in (10.27) practically does not depend upon the radius. The largest correction, due to the quadrupole deformation, gives a factor of three for heavy nuclei and a factor of five in superheavy ones.

The preformation amplitude, given by (10.32), is a coherent superposition of many terms and, therefore, the transitions between ground states are less sensitive to the mean field parameters. Thus, in our analysis we use the universal parametrisation of the Woods–Saxon potential [32] and we considered the gap parameter estimated by $\Delta_\tau = 12/\sqrt{A}$ [33], where A is the mass number of the mother nucleus. The quadrupole deformation parameters in the Fröman matrix are taken from Ref. [34].

The preformation factor is very sensitive with respect to the maximal sp radial quantum number $n_{\max}$, the sp ho parameter β_0 and the amount of spherical configurations taken in the BCS calculation, given by $P_{\min} = \min\{P_\tau\}$. Beyond $n_{\max} = 9$ the results saturate, if one considers in the BCS basis sp states with $P \geq P_{\min} = 0.02$. We improved the description of the continuum by choosing a sp scale parameter $f_0 < 1$ in (10.31). This parameter is not independent from $P_{\min}$. It turns out that the common choice of f_0 and $P_{\min}$ ensures not only the right order of magnitude for the decay width, but also the above mentioned continuity of the derivative.

The logarithm of the decay width can be approximated by the following linear ansatz

$$\log_{10}\left[\frac{\Gamma(R)}{\Gamma_{\exp}}\right] = \gamma_0 + \gamma_1 R. \tag{10.33}$$

In the ideal case the coefficients should vanish, i.e. $\gamma_0 = \gamma_1 = 0$, in order to have a proper description of the decay width. The sensitivity of these parameters versus

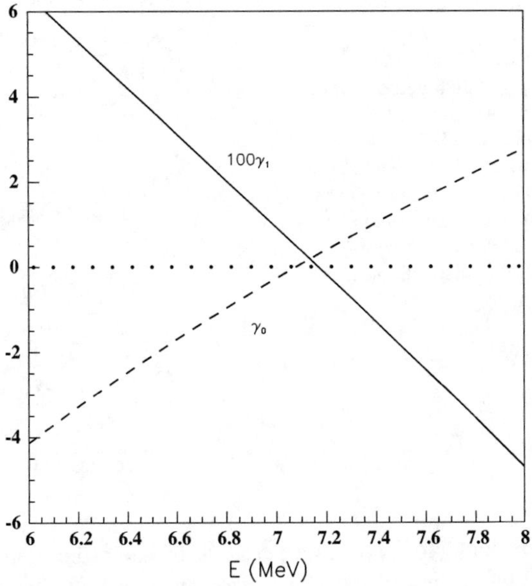

Fig. 10.7 The slope parameter $100\gamma_1$ (*solid line*) and the ratio parameter γ_0 (*dashed line*) in (10.33) versus the α-particle Q-value for $^{200}Rn \rightarrow {}^{196}Po + \alpha$

the Q value is shown in Fig. 10.7. The plateau condition (10.4) can also be written as follows

$$\gamma_1(\chi) = 0, \tag{10.34}$$

for given parameters n_{max}, β_0, P_{min}.

We analyzed α-decay chains from even–even nuclei with $N > 126$ given in Table 10.1.

It turns out that the values $n_{max} = 9$, $f_0 = 0.8$ and $P_{min} = 0.025$ give the best fit concerning the parameters γ_0 and γ_1. From Fig. 10.8a, we see that the quantity

Table 10.1 Even–even α-decay chains in the region $Z > 82$, $N > 126$

I	N_1	Z_1	No	Ref.
38	130	92	1	[35]
40	130	90	2	[35]
42	130	88	3	[35]
44	130	86	6	[35]
46	132	86	8	[35]
48	134	86	12	[35]
50	136	86	9	[35]
52	142	90	7	[35]
54	146	92	5	[35]
56	150	94	4	[35]
58	154	94	2	[35]
60	172	112	3	[36]

In the first column of each table the isospin projection $I = N - Z$ is given. In the next columns the initial neutron and proton numbers, the number of states/chain and the reference are given

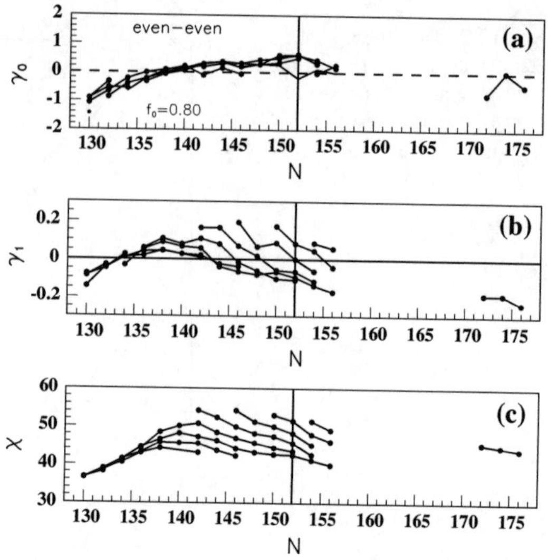

Fig. 10.8 **a** The ratio parameter γ_0, defined by (10.33), versus the neutron number for $f_0 = 0.8$, $P_{min} = 0.025$ and different even–even α-chains in Table 10.1. **b** The slope parameter γ_1, defined by (10.33), versus the neutron number. **c** The Coulomb parameter χ, defined by (10.29), versus the neutron number

$\gamma_0 \approx \log_{10}(\Gamma/\Gamma_{exp})$ has a variation of one order of magnitude around $\gamma_0 = 0$, but the description of the slope γ_1 given in Fig. 10.8b is by far not satisfactory. The reason for the variation of the slope parameter γ_1 is the relative strong dependence of the Coulomb parameter χ upon the neutron number along α-chains. In Fig. 10.8c, we give the values of this parameter for the even–even chains, which is in an obvious correlation with the slope parameter γ_1. Therefore, the derivative of the microscopic preformation amplitude changes along α-chains much slower in comparison with that of the Coulomb function. As we pointed out, the term given by the shell correction disappears in the Q value (except the magic numbers) and it remains a linear in Z dependence. Thus, the most important effect is given by the Coulomb repulsion. In order to stress on this dependence we performed the same analysis in the region $Z > 82,\ 82 < N < 126$, given in Table 10.2.

In Fig. 10.9a, b we plotted the parameters γ_0, γ_1 depending upon the neutron number. We used the same parameters, i.e. $n_{max} = 9, f_0 = 0.8, P_{min} = 0.025$. One can see that indeed their values are very close to zero. The decay widths are reproduced within a factor of two. We point out the small decrease of parameters

Table 10.2 Even–even α-decay chains in the region $Z > 82,\ 82 < N < 126$

I	N_1	Z_1	No	Ref.
28	114	86	1	[35]
30	116	86	2	[35]
32	118	86	3	[35]
34	120	86	3	[35]
36	122	86	2	[35]
38	124	86	1	[35]

The quantities are the same as in Table 10.1

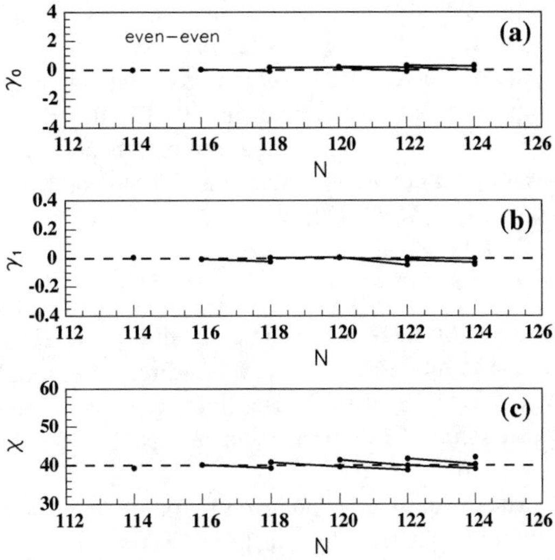

Fig. 10.9 **a** The ratio parameter γ_0, defined by (10.33), versus the neutron number for $f_0 = 0.8$, $P_{\min} = 0.025$ and different even–even α-chains in Table 10.2. **b** The slope parameter γ_1, defined by (10.33), versus the neutron number. **c** The Coulomb parameter χ, defined by (10.29), versus the neutron number

along considered α-chains is correlated with a similar behaviour of the Coulomb parameter χ in Fig. 10.9c.

Our estimate shows that the correlation coefficient between γ_1 and χ is larger than 0.7. This allows us to introduce a supplementary, but universal, correcting procedure for the preformation factor. Thus, let us define a variable size parameter f by a relation similar to (10.31), namely

$$\beta = f\beta_N. \tag{10.35}$$

The parameter χ enters in the exponent of the Coulomb function (10.11). This fact suggests a similar correction of the preformation factor, i.e.

$$\begin{aligned}\overline{\mathscr{F}}_0(\beta, \beta_m, n_{\max}, P_{\min}; R) &= e^{-4\beta R^2/2} \sum_N W_N(\beta_m, n_{\max}, P_{\min}) \\ &\times \mathscr{N}_{N0}(4\beta_m) L_{N0}^{(1/2)}(4\beta_m R^2).\end{aligned} \tag{10.36}$$

We suppose a linear dependence of the size parameter f upon the Coulomb parameter

$$\beta - \beta_m = (f - f_m)\beta_N = f_1(\chi - \chi_m)\beta_N. \tag{10.37}$$

This ansatz is confirmed by Fig. 10.4, connecting almost linearly the ho parameter of a Gaussian α-particle wave function with the Coulomb parameter. The above relation (10.36) can be rewritten as follows

$$\begin{aligned}\overline{\mathscr{F}}_0(\beta, \beta_m, n_{\max}, P_{\min}; R) &= e^{-4(\beta-\beta_m)R^2/2} \mathscr{F}_0(\beta_m, n_{\max}, P_{\min}; R) \\ &= \mathscr{F}_0(\beta - \beta_m, 0, 0; R) \mathscr{F}_0(\beta_m, n_{\max}, P_{\min}; R)\end{aligned}, \tag{10.38}$$

i.e. the usual preformation amplitude is multiplied by a cluster preformation amplitude with $n_{max} = 0$. Thus, one has to multiply the right hand side of the expansion (10.32) by this factor.

In our calculations, we used the parameters $f_m = 0.83$, $\chi_m = 55$. For the proportionality coefficient in (10.37) the regression analysis gives the value $f_1 = 8.0 \times 10^{-4}$. The situation in the superheavy chain can be described by assuming a quadratic dependence of the coefficient f_1 upon the number of clusters $N_\alpha = (N - N_0)/2$ with $N_0 = 126$, namely

$$f_1 \rightarrow f_1 + f_2 N_\alpha^2. \tag{10.39}$$

A quadratic in N_α dependence of the Q value was also empirically found in Ref. [22]. The final results are given in Fig. 10.10a, b. We considered a correcting term with $f_2 = 1.28 \times 10^{-6}$. The improvement of the slope parameter is obvious. The mean value of this parameter and its standard deviation for even–even chains is $\gamma_1 = -0.001 \pm 0.034$.

The quadratic dependence in (10.39) can be also interpreted in terms of the total number of interacting clustering pairs, namely $N_\alpha^2 \approx 2N_\alpha(N_\alpha - 1)/2$ [22]. We mention here that the effect of proton–neutron correlations on the α-decay rate was also evidenced using the experimental double difference of binding energies [37]. The even–odd pair staggering of binding energies along α-lines was interpreted in Ref. [23] as an α-condensate of proton and neutron boson pairs, similar to the usual pairing condensate among protons or neutrons.

It is interesting to compare these results with those given by spectroscopic factors. The microscopic spectroscopic factor S_{gs}, given by (9.60), is plotted in Fig. 10.11 versus the neutron number by open circles. In the region with $N < 126$, the values are comparable with those given by the experimental phenomenological

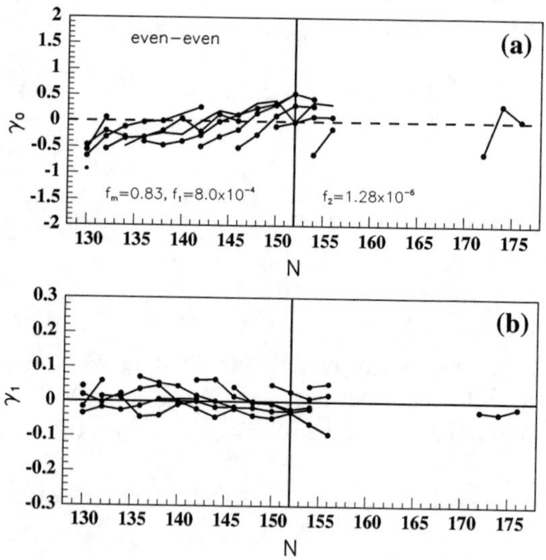

Fig. 10.10 **a** The parameter γ_0 versus the neutron number for different even–even α-chains in Table 10.1. The preformation parameters are $f_m = 0.83$, $f_1 = 8.0 \times 10^{-4}$, $f_2 = 1.28 \times 10^{-6}$, $P_{min} = 0.025$. **b** Same as in **a**, but for the slope parameter γ_1

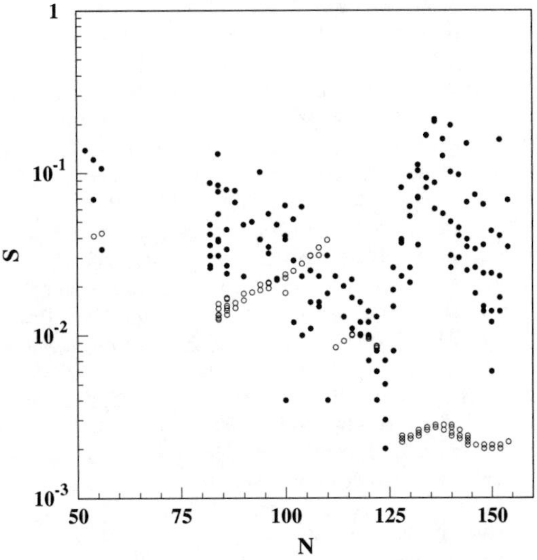

Fig. 10.11 Phenomenological spectroscopic factor (2.142), plotted by *dark circles* and microscopic spectroscopic factor (9.60), plotted by *open circles*, versus the neutron number. Data for $Z \geq 82$ are taken from Tables 6.1 and 6.2

spectroscopic factor S, given by (2.142), plotted by solid circles. This means that the α-clustering in this region can be qualitatively described within the pairing model. The jump of the microscopic spectroscopic factor around $N = 115$ is connected with $Z = 82$ proton magic number [the border between the regions 2 and 3 in (2.109)]. On the other hand in the region above ^{208}Pb, i.e. $N > 126$ and $Z > 82$ the values of S_{gs} are much smaller than the phenomenological ones. From Fig. 10.11 one cleary see that here the ratio $S/S_{gs} \gg 1$ and therefore an additional α-clustering component is necessary here.

References

1. Fliessbach, T., Mang, H.J., Rasmussen, J.O.: The reduced Width amplitude in the reaction theory for composite particles. Phys. Rev. C **13**, 1318–1323 (1976)
2. Tonozuka, I., Arima, A.: Surface α-clustering and α-decays of ^{211}Po. Nucl. Phys. A **323**, 45–60 (1979)
3. Fliessbach, T., Okabe, S.: Surface alpha-clustering in the lead region. Z. Phys. A **320**, 289–294 (1985)
4. Janouch, F.A., Liotta, R.: Influence of α-cluster formation on α decay. Phys. Rev. C **27**, 896–898 (1983)
5. Dodig-Crnkovic, G., Janouch, F.A., Liotta, R.J.: The continuum and the alpha-particle formation. Phys. Scr. **37**, 523–525 (1988)
6. Lenzi, S.M., Dragun, O., Maqueda, E.E., Liotta, R.J., Vertse, T.: Description of alpha clustering including continuum configurations. Phys. Rev. C **48**, 1463–1465 (1993)
7. Delion, D.S., Suhonen, J.: Microscopic description of α-like resonances. Phys. Rev. C **61**, 024304/1–12 (2000)
8. Insolia, A., Curutchet, P., Liotta, R.J., Delion, D.S.: Microscopic description of alpha decay of deformed nuclei. Phys. Rev. C **44**, 545–547 (1991)
9. Delion, D.S., Insolia, A., Liotta, R.J.: Anisotropy in alpha decay of odd-mass deformed nuclei. Phys. Rev C **46**, 884–888 (1992)

10. Delion, D.S., Insolia A., Liotta, R.J.: Alpha widths in deformed nuclei: microscopic approach. Phys. Rev C **46**, 1346–1354 (1992)
11. Delion, D.S., Insolia, A., Liotta, R.J.: Microscopic description of the anisotropy in alpha decay. Phys. Rev. C **49**, 3024–3028 (1994)
12. Delion, D.S., Insolia, A., Liotta, R.J.: New single particle basis for microscopic description of decay processes. Phys. Rev. C **54**, 292–301 (1996)
13. Dobaczewski, J., Nazarewicz, W., Werner, T.R., Berger, J.F., Chinn, C.R., Decharge, J.: Mean-field description of ground-state properties of drip-line nuclei: pairing and continuum effects. Phys. Rev. C **53**, 2809–2840 (1996)
14. Berggren, T.: The use of resonant states in eigenfunction expansions of scattering and reaction amplitudes. Nucl. Phys. A **109**, 265–287 (1968)
15. Vertse, T., Curuchet, P., Civitarese, O., Ferreira, L.S., Liotta, R.J.: Application of gamow resonances to continuum nuclear spectra. Phys. Rev. C **37**, 876–879 (1988)
16. Curuchet, P., Vertse, T., Liotta, R.J.: Resonant random phase approximation. Phys. Rev. C **39**, 1020–1031 (1989)
17. Vertse, T., Liotta, R.J., Maglione, E.: Exact and approximate calculation of giant resonances. Nucl. Phys. A **584**, 13–34 (1995)
18. Maglione, E., Ferreira, L.S., Liotta, R.J.: Nucleon decay from deformed nuclei. Phys. Rev. Lett. **81**, 538–541 (1998)
19. Varga, K., Lovas, R.G., Liotta, R.J.: Cluster-configuration shell model for alpha decay. Nucl. Phys. A **550**, 421–452 (1992)
20. Rasmussen, J.O.: Alpha-decay barrier penetrabilities with an exponential nuclear potential: even-even Nuclei. Phys. Rev. **113**, 1593–1598 (1959)
21. Akovali, Y.A.: Review of alpha-decay data from double-even nuclei. Nucl. Data Sheets **84**, 1–114 (1998)
22. Dussel, G., Caurier, E., Zuker, A.P.: Mass predictions based on α-line systematics. At. Data Nucl. Data Tables **39**, 205–211 (1988)
23. Gambhir, Y.K., Ring, P., Schuck, P.: Nuclei: A superfluid condensate of α particles? A study within the interacting-boson model. Phys. Rev. Lett. **51**, 1235–1238 (1983)
24. Dussel, G.G., Fendrik, A.J., Pomar, C.: Microscopic description of four-body excitations in heavy nuclei. Phys. Rev. C **34**, 1969–1973 (1986)
25. Dussel, G.G., Fendrik: Microscopic description of four-body excitations in heavy nuclei. Phys. Rev. C **34**, 1097–1109 (1986)
26. Eisenberg, J.M., Greiner, W.: Nuclear Theory I, Nuclear Models. North-Holland, Amsterdam (1970)
27. Delion, D.S., Săndulescu, A.: Toward a self-consistent microscopis α-decay theory. J. Phys. G **28**, 617–625 (2002)
28. Schuck, P., Tohsaki, A., Horiuchi, H., Röpke, G.: In: Nazarewicz, W., Vretenar, D. (eds.) The Nuclear Many Body Problem 2001, p. 271. Kluwer Academic Publishers, Dordrecht (2002)
29. Spanier, L., Johansson, S.A.E.: A modified Bethe-Weizsäcker mass formula with deformation and shell corrections and few free parameters. At. Data Nucl. Data Tables **39**, 259–264 (1988)
30. Delion, D.S., Liotta, R.J.: Microscopic description of α decay from superdeformed nuclei. Phys. Rev. C **58**, 2073–2080 (1998)
31. Fröman, P.P.: Alpha decay from deformed nuclei. Mat. Fys. Skr. Dan. Vid. Selsk. **1**,no. 3 (1957)
32. Dudek, J., Szymanski, Z., Werner, T.: Woods-Saxon potential parameters optimized to the high spin spectra in the lead region. Phys. Rev. C **23**, 920–925 (1981)
33. Bohr A., Mottelson, B.: Nuclear Structure. Benjamin, New York (1975)
34. Möller, P., Nix, R.J., Myers, W.D., Swiatecki, W.: Nuclear ground-state masses and deformations. At. Data Nucl. Data Tables **59**, 185–381 (1995)
35. Buck, B., Merchand, A.C., Perez, S.M.: Half-lives of favoured alpha decays from nuclear ground states. At. Data Nucl. Data Tables **54**, 53–73 (1993)
36. Oganessian, Y.T., et al.: Synthesis of superheavy nuclei in the ^{48}Ca+^{244}Pu reaction: ^{288}114. Phys. Rev. C **62**, 041604(R)/1–4 (2000)
37. Kaneko, K., Hasegawa, M.: Decay and proton–neutron correlations. Phys. Rev. C **67**, 041306(R)/1–5 (2003).

Chapter 11
QRPA Description of the α-Decay to Excited States

11.1 Description of the Excited States

For transitions between ground states the preformation amplitude is a coherent superposition of many single-particle configurations, including states in continuum. Therefore, the decay width is not very sensitive to the nuclear mean field parameters. The situation changes for transitions to excited states. It turns out that the decay width is very sensitive to the structure of the wave function in the daughter nucleus. The fine structure in spectra of emitted α-particles cannot be explained only by different Q-values, due to the excitation energy of the daughter nucleus. It is known for a long time that the experimental HF's are not constant [1].

As we already pointed out, the first attempts to calculate HF's in vibrational nuclei within the Quasiparticle Random Phase Approximation (QRPA) were performed in Refs. [2–4]. Recently, the description of α-decay transitions to 2^+ states in vibrational nuclei within QRPA was performed in a series of papers [5–7]. This chapter is based on results of these works.

Let us consider an α-decay process

$$P \rightarrow D(n) + \alpha, \tag{11.1}$$

where $n = 0$ labels the ground state and $n > 0$ some excited state in the daughter nucleus. We will consider low-lying collective particle–hole excitations within the spherical QRPA. Indeed, in Refs. [8, 9] it was shown that the influence of the deformation on α-decay is mainly given by the Coulomb barrier. The exited states are described by a phonon operator acting on the daughter gs, taken as a vacuum

$$|D(n)\rangle = \Gamma^{\dagger}_{\lambda\mu}(n)|D\rangle, \tag{11.2}$$

where n is the eigenstate index. It is defined in terms of the basis pair operators

D. S. Delion, *Theory of Particle and Cluster Emission*, Lecture Notes in Physics, 819, DOI: 10.1007/978-3-642-14406-6_11, © Springer-Verlag Berlin Heidelberg 2010

$$\begin{pmatrix} \Gamma^{\dagger}_{\lambda\mu}(n) \\ (-)^{\lambda-\mu}\Gamma_{\lambda-\mu}(n) \end{pmatrix} = \begin{pmatrix} \mathscr{X}^{(n)}_{\lambda}(\tau_1 j_1 \tau_2 j_2) & -\mathscr{Y}^{(n)}_{\lambda}(\tau_1 j_1 \tau_2 j_2) \\ -\mathscr{Y}^{(n)}_{\lambda}(\tau_1 j_1 \tau_2 j_2) & \mathscr{X}^{(n)}_{\lambda}(\tau_1 j_1 \tau_2 j_2) \end{pmatrix} \times \begin{pmatrix} \overline{\mathscr{A}}^{\dagger}_{\lambda\mu}(\tau_1 j_1 \tau_2 j_2) \\ (-)^{\lambda-\mu}\overline{\mathscr{A}}_{\lambda-\mu}(\tau_1 j_1 \tau_2 j_2) \end{pmatrix}. \tag{11.3}$$

Here we have used the same short-hand notation $j \equiv (\varepsilon l j)$ as in the previous chapters, where ε is the sp energy, l angular momentum and j total spin. The summation on the pair index is taken only over $j_1 \leq j_2$ if $\tau_1 = \tau_2$ and over all combinations $j_1\, j_2$ if $\tau_1 \neq \tau_2$. The normalized basis pair operators are defined as follows

$$\overline{\mathscr{A}}^{\dagger}_{\lambda\mu}(\tau_1 j_1 \tau_2 j_2) = \frac{1}{\Delta_{\tau_1 j_1;\tau_2 j_2}}\left[\alpha^{\dagger}_{\tau_1 j_1}\alpha^{\dagger}_{\tau_2 j_2}\right]_{\lambda\mu}, \tag{11.4}$$

with the normalization

$$\Delta_{\tau_1 j_1;\tau_2 j_2} = \sqrt{1+\delta_{\tau_1 j_1;\tau_2 j_2}}, \tag{11.5}$$

where we have denoted the angular momentum coupling by the square brackets.

The excited λ-pole states are obtained by diagonalizing the QRPA matrix based on the phonon operators (11.3). This matrix equation has the form

$$\begin{pmatrix} \mathbf{A}^{\lambda} & \mathbf{B}^{\lambda} \\ -\mathbf{B}^{\lambda} & -\mathbf{A}^{\lambda} \end{pmatrix}\begin{pmatrix} \mathscr{X}^{(n)}_{\lambda} \\ \mathscr{Y}^{(n)}_{\lambda} \end{pmatrix} = E^{(n)}_{\lambda}\begin{pmatrix} \mathscr{X}^{(n)}_{\lambda} \\ \mathscr{Y}^{(n)}_{\lambda} \end{pmatrix}, \tag{11.6}$$

where

$$\begin{aligned} \mathbf{A}^{\lambda}(a_1a_2, a_3a_4) &= \langle \mathrm{BCS}|\overline{\mathscr{A}}_{\lambda\mu}(a_1a_2)\hat{H}\overline{\mathscr{A}}^{\dagger}_{\lambda\mu}(a_3a_4)|\mathrm{BCS}\rangle \\ &= (E^{\mathrm{qp}}_{a_1}+E^{\mathrm{qp}}_{a_2})\delta_{a_1a_3}\delta_{a_2a_4} \\ &+ (v_{a_1}v_{a_2}v_{a_3}v_{a_4}+u_{a_1}u_{a_2}u_{a_3}u_{a_4})\langle a_1a_2;\lambda|\bar{v}|a_3a_4;\lambda\rangle \\ &+ (u_{a_1}v_{a_2}u_{a_3}v_{a_4}+v_{a_1}u_{a_2}v_{a_3}u_{a_4})\langle a_1a_2^{-1};\lambda|\hat{V}_{\mathrm{res}}|a_3a_4^{-1};\lambda\rangle \\ &+ (u_{a_1}v_{a_2}v_{a_3}u_{a_4}+v_{a_1}u_{a_2}u_{a_3}v_{a_4})\langle a_1a_2^{-1};\lambda|\hat{V}_{\mathrm{res}}|a_4a_3^{-1};\lambda\rangle \\ &\times (-1)^{j_{a_3}+j_{a_4}+\lambda+1}, \end{aligned} \tag{11.7}$$

$$\begin{aligned} \mathbf{B}^{\lambda}(a_1a_2, a_3a_4) &= \langle \mathrm{BCS}|\overline{\mathscr{A}}_{\lambda\mu}(a_1a_2)(-1)^{\lambda-\mu}\overline{\mathscr{A}}_{\lambda-\mu}(a_3a_4)\hat{H}|\mathrm{BCS}\rangle \\ &= -(u_{a_1}u_{a_2}v_{a_3}v_{a_4}+v_{a_1}v_{a_2}u_{a_3}u_{a_4})\langle a_1a_2;\lambda|\bar{v}|a_3a_4;\lambda\rangle \\ &+ (u_{a_1}v_{a_2}v_{a_3}u_{a_4}+v_{a_1}u_{a_2}u_{a_3}v_{a_4})\langle a_1a_2^{-1};\lambda|\hat{V}_{\mathrm{res}}|a_3a_4^{-1};\lambda\rangle \\ &+ (u_{a_1}v_{a_2}u_{a_3}v_{a_4}+v_{a_1}u_{a_2}v_{a_3}u_{a_4})\langle a_1a_2^{-1};\lambda|\hat{V}_{\mathrm{res}}|a_4a_3^{-1};\lambda\rangle \\ &\times (-1)^{j_{a_3}+j_{a_4}+\lambda+1}, \end{aligned} \tag{11.8}$$

where $\hat{H} = \hat{T} + \hat{V}_{\mathrm{res}}$ is the residual nuclear Hamiltonian, $a_1 \equiv \tau_1 j_{a\_1}$, etc., E^{qp} are the quasiparticle energies, and $\langle a_1a_2;\lambda|\bar{v}|a_3a_4;\lambda\rangle$ is the antisymmetrized and normalized shell model two-body interaction matrix element. The particle–hole matrix element is defined here as

$$\langle a_1 a_2^{-1}; \lambda | \hat{V}_{\rm res} | a_3 a_4^{-1}; \lambda \rangle = -\frac{1}{\Delta_{a_1 a_2} \Delta_{a_3 a_4}} \sum_{\lambda'} (2\lambda' + 1) \Delta_{a_1 a_4} \Delta_{a_2 a_3} \times \langle a_1 a_4; \lambda' | \bar{v} | a_3 a_2; \lambda' \rangle \begin{Bmatrix} j_{a_1} & j_{a_2} & \lambda \\ j_{a_3} & j_{a_4} & \lambda' \end{Bmatrix}, \tag{11.9}$$

where the matrix in curly brackets denotes a $6j$ symbol. Naturally, the $\mathbf{A}^\lambda$ and $\mathbf{B}^\lambda$ matrices contain the $\pi\pi\pi\pi$, $\pi\pi\nu\nu$ and $\nu\nu\nu\nu$ parts yielding to eigenvectors containing both the proton–proton and neutron–neutron two-quasiparticle amplitudes.

11.2 α-Particle Preformation Amplitude

The amplitude of the decay process (11.1), called the α-particle preformation amplitude, is given as an overlap integral over the internal coordinates of the daughter nucleus and the emitted cluster, i.e.

$$\mathscr{F}_{\lambda\mu}^{(n)}(\mathbf{R}_\alpha) \equiv \langle D | (-)^{\lambda-\mu} \Gamma_{\lambda-\mu}(n) \Psi_\alpha^* | P \rangle = \int d\mathbf{x}_A \, d\mathbf{x}_\alpha [\Psi_D^{(n)}(\mathbf{x}_D)]^* \Psi_\alpha^*(\mathbf{x}_\alpha) \Psi_P(\mathbf{x}_P), \tag{11.10}$$

where $\mathbf{x}$ denotes the internal coordinates. For $n = 0$ one uses the ground state in the daughter nucleus instead of a QRPA excitation. We have described the method to compute the α-particle preformation amplitude for the $n = 0$ case in the previous chapter. Here we generalize this procedure by considering excited states of the daughter nucleus.

For the core-cluster cluster we neglected antisymmetrization because we estimated the overlap for distances beyond the geometrical nuclear radius, where the Pauli principle becomes less important [10].

The α-particle internal wave function is a product of the lowest ho and singlet-spin wave functions (9.6) [11] in terms of Moshinsky coordinates (9.5). In order to estimate the overlap integral (11.10), let us expand the mother wave function in terms of four-particle states times the collective excitation acting on the daughter wave function. We consider the following symmetrized form

$$\begin{aligned} |P\rangle = \frac{1}{2} \Big\{ & \sum_{j_1=j_2} \sum_{j_3 \le j_4} X(\pi j_1 \pi j_2; \nu j_3 \nu j_4) \\ & \times \left[a_{\pi j_1}^\dagger \otimes a_{\pi j_2}^\dagger \right]_0 \hat{\lambda} \left[\left[a_{\nu j_3}^\dagger \otimes a_{\nu j_4}^\dagger \right]_\lambda \otimes \Gamma_\lambda^\dagger(n) \right]_0 |D\rangle \\ & + \sum_{j_1=j_2} \sum_{j_3 \le j_4} X(\nu j_1 \nu j_2; \pi j_3 \pi j_4) \\ & \times \left[a_{\nu j_1}^\dagger \otimes a_{\nu j_2}^\dagger \right]_0 \hat{\lambda} \left[\left[a_{\pi j_3}^\dagger \otimes a_{\pi j_4}^\dagger \right]_\lambda \otimes \Gamma_\lambda^\dagger(n) \right]_0 |D\rangle \Big\}, \end{aligned} \tag{11.11}$$

where $\hat{\lambda} = \sqrt{2\lambda + 1}$. We find each coefficients by multiplying from the left with the conjugate operators and acting on the daughter state $\langle A|$. In this way, one obtains for the first coefficient

$$\begin{aligned} X(\pi j_1 \pi j_2; \nu j_3 \nu j_4) &= \frac{\delta_{j_1 j_2}}{2\hat{\lambda}(1+\delta_{j_3 j_4})} \\ &\times \langle D| \left[\Gamma_\lambda(n) \otimes \left[a_{\nu j_4} \otimes a_{\nu j_3} \right]_\lambda \right]_0 \left[a_{\pi j_1} \otimes a_{\pi j_1} \right]_0 |P\rangle, \end{aligned} \tag{11.12}$$

and a similar expression for the second term. By expressing the particle operators in terms of quasiparticle operators and inverting (11.3), one obtains

$$\begin{aligned} X(\pi j_1 \pi j_2; \nu j_3 \nu j_4) &= X_\pi^{(0)}(j_1 j_2) X_\nu^{(n)}(j_3 j_4), \\ X(\nu j_1 \nu j_2; \pi j_3 \pi j_4) &= X_\nu^{(0)}(j_1 j_2) X_\pi^{(n)}(j_3 j_4), \end{aligned} \tag{11.13}$$

where the two terms are given, respectively, by

$$\begin{aligned} X_\tau^{(0)}(j_1 j_2) &= \delta_{j_1 j_2} \frac{\hat{j}_1}{2} u_{\tau j_1} v_{\tau j_1}, \\ X_\tau^{(n)}(j_3 j_4) &= \frac{1}{\sqrt{1+\delta_{j_3 j_4}}} \left[u_{\tau j_3} u_{\tau j_4} \mathscr{Y}_\lambda^{(n)}(\tau j_3 \tau j_4) - v_{\tau j_3} v_{\tau j_4} \mathscr{X}_\lambda^{(n)}(\tau j_3 \tau j_4) \right]. \end{aligned} \tag{11.14}$$

In particular for a gs to gs transition one obtains the well-known expansion

$$|P\rangle = \sum_{j_1 j_2} X_\pi^{(0)}(j_1 j_1) X_\nu^{(0)}(j_2 j_2) \left[a_{\pi j_1}^\dagger \otimes a_{\pi j_1}^\dagger \right]_0 \left[a_{\nu j_2}^\dagger \otimes a_{\nu j_2}^\dagger \right]_0 |D\rangle. \tag{11.15}$$

We stress on the fact that this procedure to expand the mother wave function in terms of excitation operators acting on daughter nucleus can be extended in a straightforward way, for instance, to treat two-particle–two-hole excitations.

The coordinate representation of the pair-creation operators connecting the same kind of particles has the following form

$$\begin{aligned} \left[a_{\tau j_1}^\dagger \otimes a_{\tau j_2}^\dagger \right]_{\lambda\mu} &\to \mathscr{A}\{\psi_{\tau j_1}(\mathbf{x}_1) \otimes \psi_{\tau j_2}(\mathbf{x}_2)\}_{\lambda\mu} \\ &\equiv \frac{1}{\sqrt{2}} \left\{ \left[\psi_{\tau j_1}(\mathbf{x}_1) \otimes \psi_{\tau j_2}(\mathbf{x}_2) \right]_{\lambda\mu} - \left[\psi_{\tau j_1}(\mathbf{x}_2) \otimes \psi_{\tau j_2}(\mathbf{x}_1) \right]_{\lambda\mu} \right\}, \end{aligned} \tag{11.16}$$

where $\mathbf{x} \equiv (\mathbf{r}, s)$. Taking into account that the $\tau\tau$ components of the α-particle wave function (9.6) are antisymmetric, one obtains the following relation for the overlap integral

$$\begin{aligned} \mathscr{F}_{\lambda\mu}^{(n)}(R_\alpha) &= \frac{1}{2} \sum_{\tau_1 \neq \tau_2} \sum_{j_1} \sum_{j_3 \leq j_4} X_{\tau_1}^{(0)}(j_1 j_1) X_{\tau_2}^{(n)}(j_3 j_4) \\ &\times \int d\mathbf{x}_\alpha \Psi_\alpha^*(\mathbf{x}_\alpha) \left[\psi_{\tau_1 j_1}(\mathbf{x}_1) \otimes \psi_{\tau_1 j_1}(\mathbf{x}_2) \right]_0 \left[\psi_{\tau_2 j_3}(\mathbf{x}_3) \otimes \psi_{\tau_2 j_4}(\mathbf{x}_4) \right]_{\lambda\mu}, \end{aligned} \tag{11.17}$$

which depends on the cm Moshinsky coordinate.

We solve the Schrödinger equation to determine sp basis (9.25) by using the ho representation (9.26).

Let us first consider the pair $\pi\pi$ building block, which is given by the product of sp wave functions entering the ho basis. By changing the j–j to the L–S coupling scheme, only the singlet component gives contribution in the overlap with the α-particle wave function. Then, the configuration part can be written in the relative+cm system using the Talmi–Moshinsky transformation. Finally, the overlap of the $\pi\pi$ part of the wave function with the two-proton part of the α-particle wave function (9.6) is given by the following expansion, in terms of the ho wave functions [12]

$$\int d\mathbf{x}_\pi [\phi_{00}^{(\beta_\alpha)}(\mathbf{r}_\pi)\chi_{00}^{(\pi)}(12)]^* \sum_{j_1 \leq j_2} X_\pi^{(k)}(j_1 j_2)\mathscr{A}\{\psi_{\pi j_1}(\mathbf{x}_1) \otimes \psi_{\pi j_2}(\mathbf{x}_2)\}_{\lambda\mu}$$
$$= \sum_{N_\pi} G_\pi^{(k)}(N_\pi \lambda)\phi_{N_\pi \lambda\mu}^{(\beta)}(\mathbf{R}_\pi), \quad k = 0, n, \tag{11.18}$$

where we have introduced the G_π-coefficient by

$$G_\pi^{(k)}(N_\pi\lambda) \equiv \sum_{12} B_\pi^{(k)}(n_1 l_1 j_1 n_2 l_2 j_2; \lambda)\left\langle (l_1 l_2)\lambda\left(\frac{1}{2}\frac{1}{2}\right)0;\lambda \middle| \left(l_1\frac{1}{2}\right)j_1\left(l_2\frac{1}{2}\right)j_2;\lambda\right\rangle$$
$$\times \sum_{n_\pi} \langle n_\pi 0 N_\pi \lambda; \lambda | n_1 l_1 n_2 l_2; \lambda\rangle \mathscr{I}_{n_\pi 0}^{(\beta\beta_\alpha)} \tag{11.19}.$$

Here, as usually, the first angular brackets denote the recoupling jj–LS, while the second ones denote the Talmi–Moshinsky coefficients, and the coefficients $B_\pi^{(k)}$ are expressed in terms of the coefficients given by Eqs 11.14 and 9.25 as

$$B_\pi^{(k)}(n_1 l_1 j_1 n_2 l_2 j_2; \lambda) \equiv \sum_{\varepsilon_1 \leq \varepsilon_2} X_\pi^{(k)}(\varepsilon_1 l_1 j_1 \varepsilon_2 l_2 j_2) c_{n_1}(\pi\varepsilon_1 l_1 j_1) c_{n_2}(\pi\varepsilon_2 l_2 j_2). \tag{11.20}$$

Furthermore, the overlap integral between the spherical ho radial functions is defined as

$$\mathscr{I}_{n0}^{(\beta\beta_\alpha)} \equiv \langle \mathscr{R}_{n0}^{(\beta)} | \mathscr{R}_{00}^{(\beta_\alpha)} \rangle. \tag{11.21}$$

One also obtains a similar expansion for G_ν.

By performing the transformation of the $\pi\pi$–$\nu\nu$ product to the relative+cm coordinates of the α-particle, one finally obtains the overlap integral, defining the preformation amplitude. It is written as a superposition of the ho functions in the cm coordinate $R = R_\alpha/2$ of the α-cluster system:

$$\mathscr{F}_{\lambda\mu}^{(n)}(R) Y_{\lambda\mu}(\hat{R}) = \sum_{N_\alpha} W^{(n)}(N_\alpha \lambda)\phi_{N_\alpha \lambda}^{(4\beta)}(\mathbf{R}), \tag{11.22}$$

where the $W^{(n)}$-coefficient is given by

$$W^{(n)}(N_\alpha\lambda) \equiv 4\sum_{N_\pi N_\nu} G_\pi^{(0)}(N_\pi 0)G_\nu^{(n)}(N_\nu\lambda)\sum_{n_\alpha}\langle n_\alpha 0 N_\alpha\lambda;\lambda|N_\pi 0 N_\nu\lambda;\lambda\rangle\mathscr{I}_{n_\alpha 0}^{(\beta\beta_\alpha)}$$
$$+4\sum_{N_\pi N_\nu} G_\pi^{(n)}(N_\pi\lambda)G_\nu^{(0)}(N_\nu 0)\sum_{n_\alpha}\langle n_\alpha 0 N_\alpha\lambda;\lambda|N_\pi\lambda N_\nu 0;\lambda\rangle\mathscr{I}_{n_\alpha 0}^{(\beta\beta_\alpha)}. \quad (11.23)$$

The preformation amplitude connecting the ground states of the two nuclei is obtained in a similar way

$$\mathscr{F}_0^{(0)}(R) = \sum_{N_\alpha} W^{(0)}(N_\alpha 0)\mathscr{R}_{N_\alpha 0}^{(4\beta)}(\mathbf{R}), \quad (11.24)$$

where the $W^{(0)}$-coefficient is also given below and it is already known from the previous paragraph

$$W^{(0)}(N_\alpha 0) \equiv 8\sum_{N_\pi N_\nu} G_\pi^{(0)}(N_\pi 0)G_\nu^{(0)}(N_\nu 0)\sum_{n_\alpha}\langle n_\alpha 0 N_\alpha 0;0|N_\pi 0 N_\nu 0;0\rangle\mathscr{I}_{n_\alpha 0}^{(\beta\beta_\alpha)}. \quad (11.25)$$

The α-decay width to a final state with angular momentum λ is given by the standard quantum mechanical rule (10.1)

$$\Gamma_\lambda^{(n)} = \hbar v|N_\lambda|^2 = 2\left\{\frac{\kappa_\lambda R}{\left|H_\lambda^{(+)}(\chi,\kappa_\lambda R)\right|^2}\right\}\left\{\frac{\hbar^2}{2\mu R}\left[\mathscr{F}_\lambda^{(n)}(R)\right]^2\right\}$$
$$\equiv 2P_\lambda(E,R)\left[\gamma_\lambda^{(n)}(R)\right]^2, \quad (11.26)$$

where v is the cm velocity of the emitted α-particle. This quantity is a product of two functions, the penetrability (P_λ) and preformation probability (γ_λ), strongly depending on the radius R. However, $\Gamma_\lambda^{(n)}$ should be a constant in a region around the geometrical touching point. It can also be seen as the ratio between the internal and external solutions. Therefore the continuity of the derivatives for internal and external solutions is automatically achieved if the decay width is independent of the matching radius.

We estimate the HF of the first collective 2^+ state as the ratio between the corresponding "strengths" of the preformation factors, i.e.

$$\mathrm{HF} = \frac{\langle R[F_0^{(0)}(R)]^2\rangle}{\langle R[F_2^{(1)}(R)]^2\rangle}, \quad (11.27)$$

where by brackets we denoted the mean values, considered over the interval $R \geq R_c$, beyond the touching radius

$$R_c = 1.2(A_\alpha^{1/3} + A_D^{1/3}), \quad (11.28)$$

including the last two maxima of preformation amplitudes. In Ref. [13], we have shown that the ratio of averaged quantities in (11.27), practically, does not depend upon the matching radius. This definition is consistent with the experimental estimate given by

$$\mathrm{HF}_{\mathrm{exp}} = \frac{\Gamma^{(0)}_{\mathrm{exp}}}{\Gamma^{(1)}_{\mathrm{exp}}} \frac{P_2(E - E_{2^+}, R)}{P_0(E, R)}, \tag{11.29}$$

where $\Gamma^{(k)}$ with $k = 0$ denotes the decay width between ground states, while $k = 1$ to the first 2^+ eigenstate. According to Rasmussen [1], the hindrance factor contains the penetration through a barrier with a sharp internal potential. As we will show later the comparison between the two estimates gives a difference less than a factor of two, i.e. within the experimental error. Anyway, we prefer the estimate given by (11.29), because it depends only upon the well defined irregular function $G_l(\kappa R)$ generated by the external Coulomb repulsion and does not contain any arbitrary internal potential.

The effect of the deformation, given by the Fröman matrix [14] for the penetration part was recently investigated in Ref. [15]. Here, it was shown that for $|\beta_2| \le 0.2$ the effect of the barrier deformation is less than a factor of two. Thus, the use of a deformed penetration for both transitions will give similar corrections and will not affect their ratio. We mention that this approach was used in Ref. [16] to investigate transitions to rotational states.

Let us also compute another useful quantity, namely the spectroscopic factor, defined as

$$S^{(k)}_{\alpha} = \int_0^{\infty} |RF^{(k)}_{\lambda}(R)|^2 dR. \tag{11.30}$$

It gives the order of magnitude of the α-particle probability inside the nucleus.

Concerning the electromagnetic quadrupole transition, we used the standard relation for the reduced matrix element, namely

$$\begin{aligned} \langle 0||T_2||k\rangle &= \sum_{\tau} \sum_{j_1 \le j_2} \xi_{\tau j_1 j_2} \left[\mathscr{X}^{(k)}_{\lambda}(\tau j_1 j_2) + \mathscr{Y}^{(k)}_{\lambda}(\tau j_1 j_2) \right], \\ \xi_{\tau j_1 j_2} &= \frac{e_\tau}{\sqrt{1+\delta_{j_1 j_2}}} \langle \tau j_1 || r^2 Y_2 || \tau j_2 \rangle \left(u_{\tau j_1} v_{\tau j_2} + v_{\tau j_1} u_{\tau j_2} \right), \end{aligned} \tag{11.31}$$

where e_τ denotes the effective charge, and Y_2 the quadrupole spherical function.

11.3 Analysis of the Experimental Data

We analyzed α-decays to low-lying 2^+ states in even–even nuclei with moderate quadrupole deformations, $|\beta_2| \le 0.2$, i.e. in the so-called vibrational and transitional nuclei. At the same time, we will connect this strong interacting process with electromagnetic transitions in the daughter nucleus.

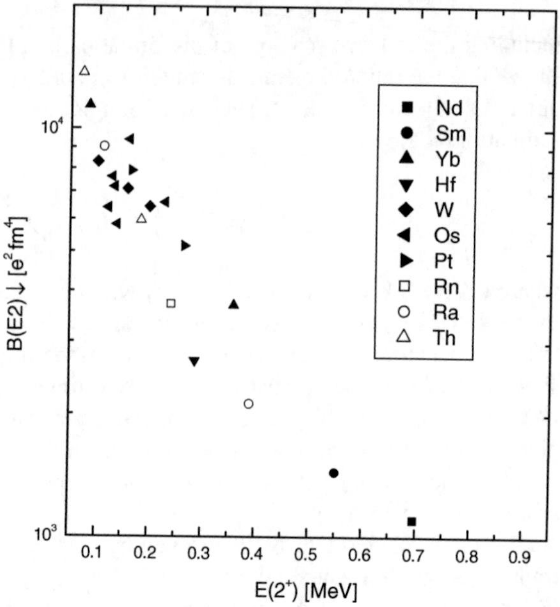

Fig. 11.1 Experimental $B(E2)$ values versus E_{2^+}

Concerning $B(E2)$ values there are many available data, e.g. Ref. [17]. In Fig. 11.1 we plotted them as a function of the excitation energy for those even-even nuclei where α-decay half lives to the ground state are measured. One can observe an universal systematic decreasing behaviour versus the excitation energy, almost independent of the considered chain. This quantity is not very sensitive to other nuclear structure details, due to the collective character of the 2^+ state. Indeed, in Ref. [18] it was shown that one has the following empirical relation between the $B(E2)$ value and E_{2^+} energy

$$B(E2) = \frac{cZ^2}{A^{2/3}E_{2^+}}, \tag{11.32}$$

where Z is the charge and A the mass number of the nucleus.

The situation is quite different concerning the α-decay. At present, the amount of available HF data is limited, see e.g. Ref. [19]. They are given in Table 11.1.

Here, in the last two columns experimental HF's, estimated according to the Rasmussen procedure $HF_{exp}^{(Ras)}$ [1] and by using Eq. (11.29) HF_{exp}, respectively, are given. One can see that they differ by a factor less than two. Anyway, we prefer to use the simple definition (11.29), which is free of any potential parameter. We plotted the logarithm of the HF_{exp} in Fig. 11.2 by a solid line, connecting experimental points along a given neutron chain. In the same figure experimental values of the 2^+ energy are represented by dashes. The labels on the abscissa correspond to the numbers in the first column of Table 11.1. We divided the figure into three regions, namely

Table 11.1 The experimental hindrance factors [19] according to Rasmussen [1] and (11.29) (last two columns)

Region	i	Z	N	A	E_{2^+} (MeV)	β_2	$HF_{exp}^{(Ras)}$	HF_{exp}
A	1	76	96	172	0.228	0.190	38.000	21.287
	2	76	98	174	0.159	0.226	3.300	1.828
	3	78	98	176	0.264	0.171	153.000	85.688
	4	78	100	178	0.171	0.254	31.300	14.043
	5	78	102	180	0.152	0.265	47.000	26.571
B	6	84	110	194	0.319	0.026	110.000	66.379
	7	84	112	196	0.463	0.136	180.000	173.541
	8	84	120	204	0.684	0.009	1.250	0.706
	9	84	122	206	0.700	−0.018	6.900	3.947
	10	84	124	208	0.686	−0.018	1.430	0.830
C	11	84	130	214	0.609	−0.008	4.800	2.774
	12	84	132	216	0.550	0.020	3.200	1.775
	13	84	134	218	0.511	0.039	1.900	1.052
	14	86	130	216	0.462	0.008	2.600	1.565
	15	86	132	218	0.324	0.040	1.450	0.852
	16	86	134	220	0.241	0.111	1.080	0.622
	17	86	136	222	0.186	0.137	0.960	0.546
	18	88	130	218	0.389	0.020	2.000	1.219
	19	88	132	220	0.178	0.103	0.960	0.572
	20	88	134	222	0.111	0.130	1.080	0.650
	21	88	136	224	0.084	0.164	0.900	0.536
	22	90	132	222	0.183	0.111	1.400	0.830
	23	90	134	224	0.098	0.164	1.000	0.579

In the first columns the three regions and the states in the abscissa of the Fig. 11.2 are given. In the next columns we give the charge and atomic numbers, the energy of the 2^+ state [19] and the quadrupole deformation [20]

$$\begin{aligned} &(\mathbf{A}): \; Z<82, \quad N<126, \\ &(\mathbf{B}): \; Z>82, \quad N<126, \\ &(\mathbf{C}): \; Z>82, \quad N>126. \end{aligned} \tag{11.33}$$

First of all, one notices an opposite tendency compared to $B(E2)$ data, namely a general decreasing trend by decreasing the excitation energy in the daughter nucleus. This is given by the solid lines along each isotopic chain. It turns out that the HF's in all the regions, apart from nuclei close to magic numbers, satisfy the rule (6.6), i.e.

$$\log_{10}(\mathrm{HF}) = aE_{2^+} + b, \tag{11.34}$$

with a common coefficient for each of the regions. As a general rule, the slope of this dependence for the region **A** is larger than for **C**, i.e. $a_{\mathbf{A}} > a_{\mathbf{C}}$. The situation in the region **B**, describing Po isotopes as daughter nuclei, is more complex. Here, one has a phase transition along Po isotopes, from large HF's, for neutron deficient isotopes, to small ones. Unfortunately the available experimental values do not describe the full chain of Po isotopes. It seems that the left side of the region **(B)**

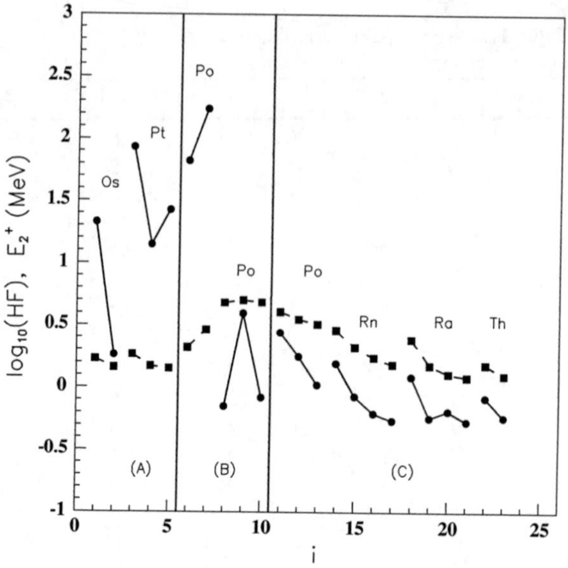

Fig. 11.2 Experimental hindrance factors (*solid lines*) and E_{2^+} (*dashed lines*) versus the state number in Table 11.1

has a similar behaviour with **(A)**, while the right one resembles the behaviour of region **(C)**.

This behaviour is less universal than the trend of electromagnetic transitions. Thus, according to the available experimental material, one concludes that the α-decay fine structure is more sensitive to concrete nuclear structure details than the corresponding electromagnetic transition. This makes the study of the α-decay to excited states an useful tool to investigate the properties of collective states.

11.4 Analysis of the QRPA Features

In our analysis we will try to explain the above discussed experimental features within our QRPA formalism for collective 2^+ excitations. We use the universal parametrisation of the Woods–Saxon potential, which is suitable for generating proton and neutron single-particle spectra, especially for nuclei around Pb region [21]. In the Woods–Saxon diagonalization procedure we consider $N = 18$ major shells, together with 20 proton and 18–19 neutron sp basis states around the corresponding Fermi surfaces. Thus, in order to simplify the calculations, we consider a smaller number of configurations than the value necessary to reproduce the absolute width [15]. Our analysis has shown that indeed the ratio of decay widths for transitions to excited and ground state, i.e. HF, is sensitive only to spherical single-particle orbitals around the Fermi level. Moreover, HF is insensitive to the used size parameter f because both preformation factors are changed in the same way. We used a standard value of this parameter, namely $f = 1$.

In Ref. [13], we used a modified surface delta residual interaction, by decoupling strength parameters for different multipolarities. In our present analysis we use realistic G-matrix elements, generated by starting from the Bonn one-boson-exchange potential [22]. The quasiparticles are generated by the monopole part of the two-body interaction. The pairing strengths have been adjusted separately for protons and neutrons to reproduce experimental gap values. In our calculations we consider only superfluid nuclei, i.e. for both mother and daughter nuclei Z and N are not magic numbers.

In order to investigate the quadrupole–quadrupole part we use three parameters, namely the proton–proton (V_π), neutron–neutron (V_ν) and proton–neutron strengths ($V_{\pi\nu}$). This allows us to schematically rewrite the residual interaction in the following form

$$\begin{aligned} &-V_\pi Q_\pi Q_\pi^\dagger - V_\nu Q_\nu Q_\nu^\dagger - V_{\pi\nu}[Q_\pi Q_\nu^\dagger + Q_\nu Q_\pi^\dagger] \\ &= -V_+ Q_+ Q_+^\dagger - V_- Q_- Q_-^\dagger - V_\pm [Q_+ Q_-^\dagger + Q_- Q_+^\dagger], \end{aligned} \tag{11.35}$$

in terms of isoscalar Q_+ and isovector quadrupole components Q_-, defined as

$$Q_+ = Q_\pi + Q_\nu, \quad Q_- = Q_\pi - Q_\nu. \tag{11.36}$$

The corresponding strengths can be written as follows

$$\begin{aligned} V_+ &= \frac{V_\pi + V_\nu + 2V_{\pi\nu}}{4}, \\ V_- &= \frac{V_\pi + V_\nu - 2V_{\pi\nu}}{4}, \\ V_\pm &= \frac{V_\pi - V_\nu}{4}. \end{aligned} \tag{11.37}$$

It turns out that our analysis depends only on two parameters, namely the following ratios

$$\begin{aligned} R_1 &\equiv V_-/V_+ = \frac{V_\pi + V_\nu - 2V_{\pi\nu}}{V_\pi + V_\nu + 2V_{\pi\nu}}, \\ R_2 &\equiv \frac{2V_\pm}{V_+ + V_-} = \frac{V_\pi - V_\nu}{V_\pi + V_\nu}, \end{aligned} \tag{11.38}$$

because, after fixing them, we adjust V_+ to obtain the experimental value of the energy E_{2^+}. In what follows, we call the second ratio R_2 the proton–neutron asymmetry.

The QRPA amplitudes $\mathscr{X}, \mathscr{Y}$ in (11.3) contain the information about the nuclear structure in the HF and $B(E2)$ values. These are the key ingredients connecting the two kinds of transition. Thus, we will characterize the "collectivity" of the first excited state, with $k = 1$, by the following proton and neutron ratios

$$Y_\tau^2/X_\tau^2 \equiv \sum_{j_1 \le j_2} \mathscr{Y}^2_{\tau j_1 j_2}(1) \Big/ \sum_{j_1 \le j_2} \mathscr{X}^2_{\tau j_1 j_2}(1), \quad \tau = \pi, \nu. \tag{11.39}$$

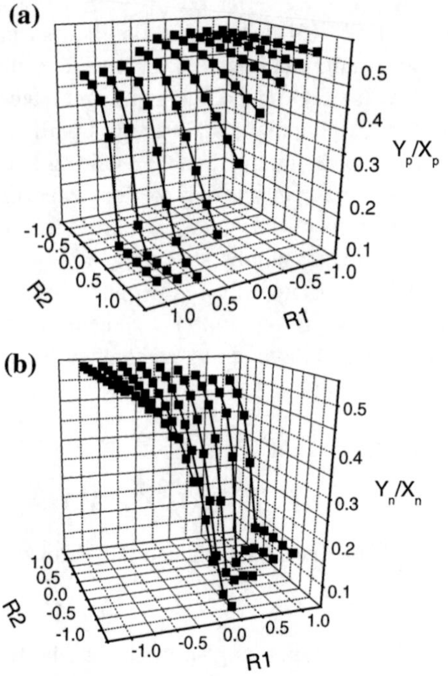

Fig. 11.3 The proton and neutron ratio of summed QRPA amplitudes squared Y^2_τ/X^2_τ, $\tau = \pi$ (**a**), $\tau = \nu$ (**b**) versus the ratios R_1 and R_2. The decay process is ^{220}Ra → ^{216}Rn

It turns out that the behaviour of these ratios is strongly dependent upon the values of R_1 and R_2. This is shown in Fig. 11.3a and b for protons and neutrons for the decay process ^{220}Ra → ^{216}Rn.

We stress on the fact that $R_1 = 0$, $R_2 = 0$ corresponds to a pure isoscalar interaction with a common nucleon–nucleon interaction strength, i.e. $V_\pi = V_\nu = V_{\pi\nu}$. In Ref. [13] we investigated this case for transitions from the Rn isotopes. Now we have extended our analysis to the interval $-1 < R_1 < 1$. In particular, when $R_1 = -1$, $R_2 = 0$ one has $V_\pi = V_\nu = 0$ and only the effective interaction between proton and neutron systems, $V_{\pi\nu}$, has a non-vanishing value. In the opposite extreme, i.e. for $R_1 = 1$, $R_2 = 0$, the effective interaction between proton and neutron systems vanishes, $V_{\pi\nu} = 0$. In our analysis we also considered a non-vanishing proton–neutron asymmetry parameter R_2.

Firstly, we have analyzed the behaviour of the $B(E2)$ value as a function of the ratio R_1 and R_2 and excitation energy E_{2^+}. In (11.31), we used the bare charges $e_\pi = 1$, $e_\nu = 0$. As a typical example in Fig. 11.4a we plot $B(E2)$ values as a function of R_1 and excitation energy, for $R_2 = 0$ by considering the decay process ^{220}Ra → ^{216}Rn. In Fig. 11.4b we give a similar plot versus R_2 and excitation energy, for $R_1 = 0$. One can see that the $B(E2)$ value decreases when R_1 and the excitation energy increase or when R_2 decreases. It is interesting that the dependence of the $B(E2)$ value on these variables is monotonic, except for the vicinity of $R_1 = 1$. This feature explains the universal behaviour of experimental data, seen in Fig. 11.1. Thus, we conclude that the QRPA is able to

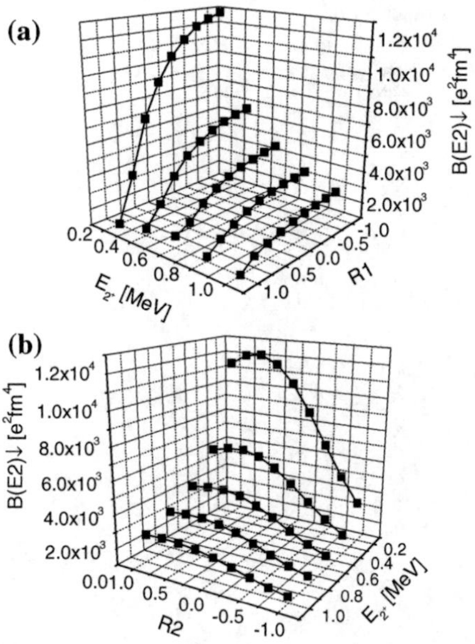

Fig. 11.4 $B(E2)$ values as a function of the E_{2^+}, R_1 for $R_2 = 0$ (**a**) and E_{2^+}, R_2 for $R_1 = 0$ (**b**). The decay process is ^{220}Ra $\rightarrow$ ^{216}Rn

reproduce the experimental behaviour of the $B(E2)$ value versus the excitation energy E_{2^+}.

Now let us consider the main quantity in our analysis, namely the HF. In the three-dimensional Fig. 11.5a we plot this quantity as a function of the ratio R_1 (with $R_2 = 0$) and E_{2^+} for the decay process ^{220}Ra $\rightarrow$ ^{216}Rn of the region **C**, while in Fig. 11.5b we plot the HF versus the ratio R_2 (with $R_1 = 0$) and E_{2^+}. One observes, first of all, that the HF has much stronger dependence upon R_1, R_2 and E_{2^+} than the $B(E2)$ value. This means that the HF is more sensitive to the Y^2_τ/X^2_τ-ratio than the electromagnetic transition. On the other hand, as a general rule, the HF increases with the increase of the excitation energy, thus reproducing the main trend of experimental data in Fig. 11.2. Moreover, our numerical analysis for several decay processes has shown that, as a general rule, the slope versus the excitation energy is larger in the region **A** than in the region **C**. Thus, once again, the QRPA is able to reproduce the main experimental trends.

11.5 Systematic Predictions

Finally, we computed the $B(E2)$ values and HF's for several vibrational and transitional nuclei, where α-decay half lives were measured. We studied the dependence of the standard mean square deviation from experimental data both for the $B(E2)$ and HF's. It turned out that this quantity has a pronounced minimum concerning the $B(E2)$ value for $R_1 = 0$, $R_2 = -0.75$ in the region (**C**) and the

Fig. 11.5 Hindrance factor as a function of the E_{2^+}, R_1 for $R_2 = 0$ (**a**) and E_{2^+}, R_2 for $R_1 = 0$ (**b**). The decay process is ^{220}Ra → ^{216}Rn

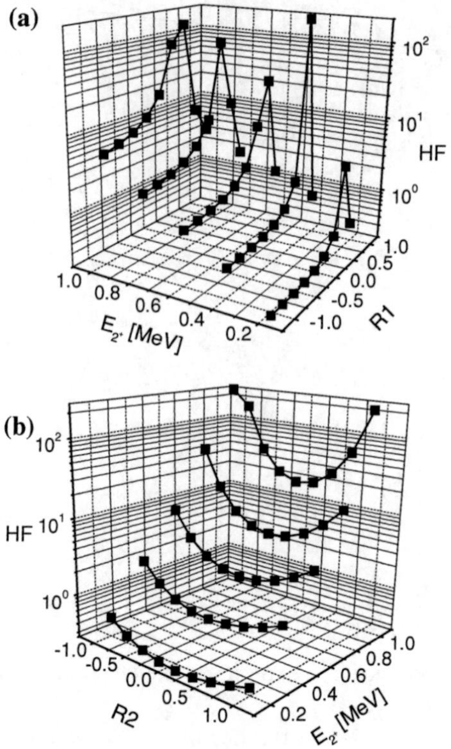

right side of the region (**B**). In the region (**A**), this quantity is not sensitive to R_1 for the interval $R_1 \in [-1, 0]$. Concerning the HF, we found out that the minimal standard square deviation is achieved close to the same values $R_1 = 0$, $R_2 = -0.75$.

Thus, we chose these values to make predictions for both the $B(E2)$ values and the HF's for the other nuclei. In the last column of Table 11.2 we give the HF's, as predictions by our model, for measured α-decays connecting ground states of even–even nuclei with moderate deformations in the region $Z < 82$. We considered only those nuclei with $E_{2^+} \geq 0.2$ MeV. Here, we also give the corresponding $B(E2)$ values. One can see that the electromagnetic transitions are rather well reproduced, together with most of the measured HF's. In Table 11.3, we give similar predictions for even-even α-emitters with $Z > 82$. We divided the Table into the above mentioned **B** (upper part) and **C** (lower part) regions. Except for some values in the region **B**, in the beginning of the Table 11.3, we obtained a good agreement with experimental data, concerning both the $B(E2)$ values and the HF's. The value $R_1 = 0$ would correspond to an isoscalar symmetry for equal strengths. Due to the fact that $R_2 = -0.75$, the neutron strength dominates over the proton one.

Table 11.2 The experimental and predicted HF's (in the seventh and eighth columns) for α-emitters below Pb

Z	A	β_2	E_{2^+}(MeV)	$B(E2)_{exp}$ (e^2 fm^4)	$B(E2)$	HF_{exp}	HF
60	144	0.149	0.696	1.10×10^3	5.16×10^2	–	1.32×10^0
62	146	0.155	0.747	–	4.66×10^2	–	2.64×10^0
62	148	0.161	0.550	1.44×10^3	9.69×10^2	–	1.11×10^0
64	148	0.156	0.784	–	5.94×10^2	–	1.73×10^0
64	150	0.161	0.638	–	8.13×10^2	–	2.00×10^0
66	150	0.153	0.804	–	6.40×10^2	–	1.48×10^0
68	152	−0.018	0.808	–	8.31×10^2	–	1.22×10^0
68	154	0.143	0.561	–	1.58×10^3	–	6.35×10^{-1}
70	154	−0.008	0.821	–	5.85×10^2	–	2.47×10^0
70	156	0.125	0.536	–	1.65×10^3	–	6.85×10^{-1}
70	158	0.161	0.358	3.70×10^3	2.31×10^3	–	5.25×10^{-1}
72	156	0.035	0.858	–	3.08×10^2	–	3.97×10^1
72	158	0.107	0.476	–	1.02×10^3	–	2.89×10^0
72	160	0.152	0.390	–	2.40×10^3	–	1.18×10^0
72	162	0.180	0.285	2.70×10^3	3.54×10^3	–	1.08×10^0
74	162	0.134	0.450	–	1.09×10^3	–	3.67×10^1
74	164	0.161	0.332	–	2.48×10^3	–	5.89×10^0
74	166	0.181	0.252	–	3.24×10^3	–	2.49×10^0
76	166	0.134	0.431	–	1.01×10^3	–	3.77×10^1
76	168	0.162	0.341	–	1.48×10^3	–	7.20×10^1
76	170	0.171	0.287	–	1.67×10^3	–	1.57×10^1
76	172	0.190	0.228	6.60×10^3	2.62×10^3	2.13×10^1	2.85×10^1
78	172	0.126	0.457	–	9.80×10^2	–	1.07×10^1
78	174	0.153	0.394	–	1.02×10^3	–	1.31×10^1
78	176	0.171	0.264	5.16×10^3	1.90×10^3	8.57×10^1	1.55×10^1

In the first four columns the charge, mass number, quadrupole deformation [20] and E_{2^+} [17] in the daughter nucleus are given. In the next two columns experimental $B(E2)\downarrow$ values [17] and theoretical predictions are given. The interaction parameters are $R_1 = 0$, $R_2 = -0.75$

The phase transition in the region **B** was reproduced by using different values of the ratio parameter R_1. Thus, the first two large HF's can be reproduced by using $R_1 = 0.5$, with the same R_2. For the moment, we do not have experimental data concerning $B(E2)$ values for these nuclei and we can only point out this interesting feature.

Finally, we should point out that the spectroscopic factors for transitions connecting ground states, computed by using (11.30), have similar values for the decays below Pb, namely $S_\alpha^{(0)} \approx 5 \times 10^{-3}$. For decays above Pb we obtained $S_\alpha^{(0)} \approx 10^{-3}$. This gives the order of magnitude of the α-decay preformation probability, in agreement with the estimate of Refs. [23, 24].

Thus, we show once again that α-decay fine structure is a more sensitive tool to probe residual interaction than the corresponding electromagnetic transition.

Table 11.3 Same as in Table 11.2, but for α-emitters above Pb

Z	A	β_2	E_{2^+} (MeV)	$B(E2)_{exp}$ (e^2 fm^4)	$B(E2)$	HF_{exp}	HF
84	194	0.026	0.319	–	$4.55 \times 10^{+2}$	$6.64 \times 10^{+1}$	$9.80 \times 10^{+1}$
84	196	0.136	0.463	–	$4.21 \times 10^{+2}$	$1.74 \times 10^{+2}$	$1.33 \times 10^{+2}$
84	200	0.009	0.666	–	$5.34 \times 10^{+2}$	–	$1.94 \times 10^{+2}$
84	202	0.009	0.677	–	$4.80 \times 10^{+2}$	–	$1.85 \times 10^{+1}$
84	204	0.009	0.684	–	$3.03 \times 10^{+2}$	7.06×10^{-1}	$3.25 \times 10^{+0}$
84	206	−0.018	0.700	–	$2.39 \times 10^{+2}$	$3.95 \times 10^{+0}$	$1.32 \times 10^{+0}$
84	212	0.045	0.727	–	$3.25 \times 10^{+2}$	–	$1.19 \times 10^{+1}$
84	214	−0.008	0.609	–	$6.41 \times 10^{+2}$	$2.77 \times 10^{+0}$	$3.92 \times 10^{+0}$
84	216	0.020	0.550	–	$8.89 \times 10^{+2}$	$1.77 \times 10^{+0}$	$4.04 \times 10^{+0}$
84	218	0.039	0.511	–	$5.81 \times 10^{+2}$	$1.05 \times 10^{+0}$	$3.37 \times 10^{+1}$
86	202	−0.104	0.504	–	$9.17 \times 10^{+2}$	–	$7.36 \times 10^{+2}$
86	204	−0.087	0.543	–	$8.96 \times 10^{+2}$	–	$4.09 \times 10^{+1}$
86	206	−0.044	0.575	–	$6.47 \times 10^{+2}$	–	$5.20 \times 10^{+0}$
86	208	−0.026	0.636	–	$2.90 \times 10^{+2}$	–	$3.84 \times 10^{+0}$
86	214	0.008	0.694	–	$6.79 \times 10^{+2}$	–	$4.63 \times 10^{+0}$
86	216	0.008	0.465	–	$1.61 \times 10^{+3}$	$1.56 \times 10^{+0}$	$1.60 \times 10^{+0}$
86	218	0.040	0.324	–	$2.57 \times 10^{+3}$	8.52×10^{-1}	9.43×10^{-1}
86	220	0.111	0.241	$3.72 \times 10^{+3}$	$3.28 \times 10^{+3}$	6.22×10^{-1}	8.05×10^{-1}
88	208	−0.104	0.520	–	$9.15 \times 10^{+2}$	–	$2.12 \times 10^{+1}$
88	210	−0.053	0.603	–	$4.85 \times 10^{+2}$	–	$5.02 \times 10^{+0}$
88	216	0.008	0.688	–	$9.81 \times 10^{+2}$	–	$5.20 \times 10^{+0}$
88	218	0.020	0.389	$2.12 \times 10^{+3}$	$2.38 \times 10^{+3}$	$1.22 \times 10^{+0}$	$1.17 \times 10^{+0}$
90	218	0.008	0.689	–	$1.49 \times 10^{+3}$	–	$9.70 \times 10^{+0}$
90	220	0.030	0.373	–	$4.29 \times 10^{+3}$	–	$2.63 \times 10^{+0}$

The interaction parameters are $R_1 = 0$, $R_2 = -0.75$, except for the first two nuclei, where $R_1 = 0.5$, $R_2 = -0.75$

References

1. Rasmussen, J.O.: Alpha-decay barrier penetrabilities with an exponential nuclear potential: even–even nuclei. Phys. Rev. **113**, 1593–1598 (1959)
2. Săndulescu, A., Dumitrescu, O.: Alpha decay to vibrational states. Phys. Lett. **19**, 404–407 (1965)
3. Săndulescu, A., Dumitrescu, O.: Alpha decay and the structure of β-vibrational states. Phys. Lett. B **24**, 212–216 (1967)
4. Cristu, M.I., Dumitrescu, O., Pyatov, N.I., Săndulescu, A.: Alpha decay and the structure of the $K^\pi = 0^+$ states in the Th-U region. Nucl. Phys. A **130**, 31–40 (1969)
5. Delion, D.S., Liotta, R.J.: Nuclear deformation in α decay. Phys. Rev. C **56**, 1782–1787 (1997)
6. Delion, D.S., Suhonen, J.: Microscopic description of the α-decay fine structure in spherical nuclei. Phys. Rev. C **64**, 064302/1–6 (2001)
7. Peltonen, S., Delion, D.S., Suhonen, J.: Systematics of the alpha decay to vibrational states. Phys. Rev. C **71**, 044315/1–9 (2005)
8. Delion, D.S., Insolia, A., Liotta, R.J.: Anisotropy in alpha decay of odd-mass deformed nuclei. Phys. Rev. C **46**, 884–888 (1992)

9. Delion, D.S., Insolia, A., Liotta, R.J.: Alpha widths in deformed nuclei: microscopic approach. Phys. Rev. C **46**, 1346–1354 (1992)
10. Fliessbach, T., Okabe, S.: Surface alpha-clustering in the lead region. Z. Phys. A **320**, 289–294 (1985)
11. Mang, H.J.: Alpha decay. Ann. Rev. Nucl. Sci. **14**, 1–28 (1964)
12. Delion, D.S., Suhonen, J.: Microscopic description of α-like resonances. Phys. Rev. C **61**, 024304/1–12 (2000)
13. Delion, D.S., Săndulescu, A., Mişicu, S., Cârstoiu, F., Greiner, W.: Quasimolecular resonances in the binary cold fission of ^{252}Cf. Phys. Rev. C **64**, 041303(R)/1–5 (2001)
14. Fröman, P.P.: Alpha decay from deformed nuclei. Mat. Fys. Skr. Dan. Vid. Selsk. **1**(3) (1957)
15. Delion, D.S., Săndulescu, A., Greiner, W.: Evidence for α clustering in heavy and superheavy nuclei. Phys. Rev. C **69**, 044318/1–19 (2004)
16. Radi, H.M.A., Shihab-Eldin, A.A., Rasmussen, J.O., Oliveira, L.F.: Relation of α-decay rotational signatures to nuclear deformation changes. Phys. Rev. Lett. **41**, 1444–1446 (1978)
17. Ramayya, A.V., et. al.: Binary and ternary fission studies with ^{252}Cf. Progr. Part. Nucl. Phys. **46**, 221–229 (2001)
18. Raman, S., Nestor, C.W., Kahane, S., Bhatt, K.H.: Low-lying collective quadrupole and octupole strengths in even–even nuclei. Phys. Rev. C **43**, 556–581 (1991)
19. Akovali, Y.A.: Review of alpha-decay data from double-even nuclei. Nucl. Data Sheets **84**, 1–114 (1998)
20. Möller, P., Nix, R.J., Myers, W.D., Swiatecki, W.: Nuclear Ground-state masses and deformations. At. Data Nucl. Data Tables **59**, 185–381 (1995)
21. Dudek, J., Szymanski, Z., Werner, T.: Woods-saxon potential parameters optimized to the high spin spectra in the lead region. Phys. Rev. C **23**, 920–925 (1981)
22. Holinde, K.: Two-nucleon forces and nuclear matter. Phys. Rep. **68**, 121–188 (1981)
23. Blendowske, R., Fliessbach, T., Walliser, H.: Microscopic Calculation of the ^{14}C Decay of Ra nuclei. Nucl. Phys. A **464**, 75–89 (1987)
24. Blendowske, R., Fliessbach, T., Walliser, H.: Systematics of cluster-radioactivity-decay constants as suggested by microscopic calculations. Phys. Rev. Lett. **61**, 1930–1933 (1988)

Chapter 12
Heavy Cluster Decays

12.1 Preformation Amplitude of Heavy Clusters

We will consider the spherical BCS approach to estimate various cluster preformation amplitudes, as in Sect. 9.6 in Chap. 9.

12.1.1 α-Particle Emission

First of all let us remind the main ingredients to estimate the α-particle preformation amplitude within the spherical approach. It is the main building block in estimating the preformation amplitude of heavy cluster emission processes. We remind that the α-particle intrinsic wave function is made up by a proton and a neutron pair, i.e.

$$\psi_\alpha(\mathbf{x}_\alpha) = \psi_{00}(\mathbf{x}_\pi)\psi_{00}(\mathbf{x}_\nu)\phi_{00}(\mathbf{r}_\alpha), \tag{12.1}$$

where

$$\begin{aligned} \psi_{00}(\mathbf{x}_\pi) &= \phi_{00}(\mathbf{r}_\pi)\chi_{00}(s_1 s_2), \\ \psi_{00}(\mathbf{x}_\nu) &= \phi_{00}(\mathbf{r}_\pi)\chi_{00}(s_1 s_2). \end{aligned} \tag{12.2}$$

The relative and cm Jacobi–Moshinsky coordinates are given by

$$\begin{aligned} \mathbf{r}_\pi &= \frac{\mathbf{r}_1 - \mathbf{r}_2}{\sqrt{2}}, \quad \mathbf{R}_\pi = \frac{\mathbf{r}_1 + \mathbf{r}_2}{\sqrt{2}} \equiv \sqrt{2}\mathbf{R}^\pi_{\mathrm{cm}}, \\ \mathbf{r}_\nu &= \frac{\mathbf{r}_3 - \mathbf{r}_4}{\sqrt{2}}, \quad \mathbf{R}_\nu = \frac{\mathbf{r}_3 + \mathbf{r}_4}{\sqrt{2}} \equiv \sqrt{2}\mathbf{R}^\nu_{\mathrm{cm}}, \\ \mathbf{r}_\alpha &= \frac{\mathbf{R}_\pi - \mathbf{R}_\nu}{\sqrt{2}}, \quad \mathbf{R}_\alpha = \frac{\mathbf{R}_\pi + \mathbf{R}_\nu}{\sqrt{2}} \equiv 2\mathbf{R}^\alpha_{\mathrm{cm}}. \end{aligned} \tag{12.3}$$

D. S. Delion, *Theory of Particle and Cluster Emission*, Lecture Notes in Physics, 819, 259
DOI: 10.1007/978-3-642-14406-6_12, © Springer-Verlag Berlin Heidelberg 2010

where $\mathbf{x}_\pi = (\mathbf{r}_1, \mathbf{s}_1, \mathbf{r}_2, \mathbf{s}_2)$ are the proton spatial-spin coordinates and $\mathbf{x}_\nu = (\mathbf{r}_3, \mathbf{s}_3, \mathbf{r}_4, \mathbf{s}_4)$ are the neutron spatial-spin coordinates. We use the ho parameter given by electron scattering experiments, i.e. $\beta = 0.5$ fm^{-2}.

The alpha-particle preformation factor is given by

$$\mathscr{F}_0^{(\alpha)}(R) = \sum_N W_{N0}\phi_{N0}^{(4\beta)}(R), \tag{12.4}$$

where the W-coefficients are given by Eq. 9.71.

12.1.2 ^{8}Be Emission

The wave functions of the heavier clusters are built from the lowest orbitals in the ho potential with appropriate parameter [1–3]. The ^{8}Be cluster is built from two α-clusters. Its wave function is given in [4]. In order to illustrate the main idea how to build the preformation amplitude, we will use a simplified version, namely the boson approximation, i.e.

$$\psi_{^8\mathrm{Be}} = \psi_{\alpha_1}(\mathbf{x}_{\alpha_1})\psi_{\alpha_2}(\mathbf{x}_{\alpha_2})\phi_{20}(\mathbf{r}_{\alpha_1\alpha_2}), \tag{12.5}$$

where the coordinates are defined as follows:

$$\begin{aligned}
\mathbf{r}_{\alpha_k} &= \mathbf{R}_{\pi_k} - \mathbf{R}_{\nu_k}, \quad \mathbf{R}_{\alpha_k} = \frac{\mathbf{R}_{\pi_k} + \mathbf{R}_{\nu_k}}{\sqrt{2}}, \qquad k = 1, 2, \\
\mathbf{r}_{\alpha_1\alpha_2} &= \frac{\mathbf{R}_{\alpha_1} - \mathbf{R}_{\alpha_2}}{\sqrt{2}}, \quad \mathbf{R}_{\alpha_1\alpha_2} = \frac{\mathbf{R}_{\alpha_1} + \mathbf{R}_{\alpha_2}}{\sqrt{2}} = \frac{\sum_{i=1}^{8}\mathbf{r}_i}{\sqrt{8}} = \sqrt{8}\mathbf{R}_{\mathrm{cm}}.
\end{aligned} \tag{12.6}$$

Note that the relative wave function of the two α-clusters in (12.5) corresponds to the four protons and four neutrons occupying the states $1p_{1/2}^{(\pi)}$ and $1p_{1/2}^{(\nu)}$, respectively. Thus, the relative wave function of the two α particles carries $n = 2$. The parent wave function can be now expanded as for the α-decay, but in terms of two proton and two neutron monopole pairs, i.e.

$$\begin{aligned}
|P\rangle &= \sum_{j_1 j_2 j_3 j_4} X_{\pi j_1} X_{\pi j_2} X_{\nu j_3} X_{\nu j_4} \\
&\times \left[a_{\pi j_1}^\dagger \otimes a_{\pi j_1}^\dagger\right]_0 \left[a_{\pi j_2}^\dagger \otimes a_{\pi j_2}^\dagger\right]_0 \left[a_{\nu j_3}^\dagger \otimes a_{\nu j_3}^\dagger\right]_0 \left[a_{\nu j_4}^\dagger \otimes a_{\nu j_4}^\dagger\right]_0 |D\rangle.
\end{aligned} \tag{12.7}$$

Here we obviously neglected the correlation between pairs, i.e. we considered a boson approximation. Then we continue the chain of calculation we performed many times, by adding the recoupling of two alpha pairs into the relative and cm parts. By integrating over relative $\alpha_1 - \alpha_2$ coordinate one obtains

$$\mathscr{F}_0^{(^8\mathrm{Be})}(R) = \sum_{N_{12}} W_{N_{12}}^{(^8\mathrm{Be})} \phi_{N_{12}0}^{(8\beta)}(R), \tag{12.8}$$

where

$$W_{N_{12}}^{(^8\mathrm{Be})} = \sum_{i_1k_1i_2k_2} \sum_{n_{12}} \sum_{N_1N_2} W_{N_10}W_{N_20}\langle n_{12}0N_{12}0;0|N_10N_20;0\rangle \mathscr{I}_{n_{12},2}^{(\beta,\beta_{^8\mathrm{Be}})}. \tag{12.9}$$

We mention that until now the spontaneous emission of ^{8}Be was observed only the ternary fission processes.

12.1.3 ^{12}C Emission

In a similar way we can construct from three α-clusters the wave function for the ^{12}C cluster as

$$\psi_{^{12}\mathrm{C}} = \psi_{\alpha_1}(\mathbf{x}_{\alpha_1})\psi_{\alpha_2}(\mathbf{x}_{\alpha_2})\psi_{\alpha_3}(\mathbf{x}_{\alpha_3})\phi_{20}(\mathbf{r}_{\alpha_1\alpha_2})\phi_{20}(\mathbf{r}_{^{12}\mathrm{C}}), \tag{12.10}$$

where in addition to the relation (12.6) with $k = 1, 2, 3$, we have also for the remaining relative and cm coordinates:

$$\begin{aligned} r_{^{12}\mathrm{C}} &= \frac{\mathbf{R}_{\alpha_1\alpha_2} - \sqrt{2}\mathbf{R}_{\alpha_3}}{\sqrt{3}}, \\ \mathbf{R}_{^{12}\mathrm{C}} &= \frac{\sqrt{2}\mathbf{R}_{\alpha_1\alpha_2} + \mathbf{R}_{\alpha_3}}{\sqrt{3}} = \frac{\sum_{i=1}^{12}\mathbf{r}_i}{\sqrt{12}} = \sqrt{12}\mathbf{R}_{\mathrm{cm}}. \end{aligned} \tag{12.11}$$

In this case the value for the ho constant is $\beta_{^{12}\mathrm{C}} = 0.35\,\mathrm{fm}^{-2}$.

Now, the parent wave function can be expanded in a similar way in terms of three proton and three neutron monopole pairs, i.e.

$$\begin{aligned} |P\rangle = \sum_{j_1j_2j_3j_4} & X_{\pi j_1}X_{\pi j_2}X_{\pi j_3}X_{\nu j_4}X_{\nu j_5}X_{\nu j_6} \\ & \times \left[a_{\pi j_1}^\dagger \otimes a_{\pi j_1}^\dagger\right]_0 \left[a_{\pi j_2}^\dagger \otimes a_{\pi j_2}^\dagger\right]_0 \left[a_{\pi j_3}^\dagger \otimes a_{\pi j_3}^\dagger\right]_0 \\ & \times \left[a_{\nu j_4}^\dagger \otimes a_{\nu j_4}^\dagger\right]_0 \left[a_{\nu j_5}^\dagger \otimes a_{\nu j_5}^\dagger\right]_0 \left[a_{\nu j_6}^\dagger \otimes a_{\nu j_6}^\dagger\right]_0 |D\rangle. \end{aligned} \tag{12.12}$$

By doing similar manipulations one obtains the following relation for the preformation amplitude

$$\mathscr{F}_0^{(^{12}\mathrm{C})}(R) = \sum_{N_{123}} W_{N_{123}}^{(^{12}\mathrm{C})} \phi_{N_{123}0}^{(12\beta)}(R), \tag{12.13}$$

where

$$W_{N_{123}}^{(^{12}\mathrm{C})} = \sum_{i_1k_1i_2k_2i_3k_3} \sum_{n_{12}n_{123}} \sum_{N_1N_2N_{12}N_3} W_{N_10}W_{N_20}W_{N_30} \times \langle n_{12}0N_{12}0;0|N_10N_20;0\rangle \mathscr{I}_{n_{12},2}^{(\beta,\beta_{^{12}\mathrm{C}})} \times \langle n_{123}0N_{123}0;0|N_{12}0N_30;0\rangle \mathscr{I}_{n_{123},2}^{(\beta,\beta_{^{12}\mathrm{C}})}. \quad (12.14)$$

12.1.4 ^{14}C Emission

We describe ^{14}C as the three α-clusters of ^{12}C and a neutron pair. Denoting by $\mathbf{r}_{13}$ and $\mathbf{r}_{14}$ the coordinates of these two neutrons, the wave function of ^{14}C is

$$\psi_{^{14}\mathrm{C}} = \psi_{\alpha_1}(\mathbf{x}_{\alpha_1})\psi_{\alpha_2}(\mathbf{x}_{\alpha_2})\psi_{\alpha_3}(\mathbf{x}_{\alpha_3})\phi_{20}(\mathbf{r}_{\alpha_1\alpha_2})\phi_{00}(\mathbf{r}_{\nu_4})\phi_{10}(\mathbf{r}_{\alpha_3\nu_4})\phi_{20}(\mathbf{r}_{^{14}\mathrm{C}}), \quad (12.15)$$

where the relative and cm coordinates of the neutron pair are

$$\mathbf{r}_{\nu_4} = \frac{\mathbf{r}_{13} - \mathbf{r}_{14}}{\sqrt{2}}, \quad \mathbf{R}_{\nu_4} = \frac{\mathbf{r}_{13} + \mathbf{r}_{14}}{\sqrt{2}}. \quad (12.16)$$

The relative and cm coordinates of the neutron pair and the third α-cluster are given, respectively, by

$$\mathbf{r}_{\alpha_3\nu_4} = \frac{\mathbf{R}_{\alpha_3} - \sqrt{2}\mathbf{R}_{\nu_4}}{\sqrt{3}}, \quad \mathbf{R}_{\alpha_3\nu_4} = \frac{\sqrt{2}\mathbf{R}_{\alpha_3} + \mathbf{R}_{\nu_4}}{\sqrt{3}},$$
$$\mathbf{r}_{^{14}\mathrm{C}} = \frac{\sqrt{3}\mathbf{R}_{\alpha_1\alpha_2} - 2\mathbf{R}_{\alpha_3\nu_4}}{\sqrt{7}}, \quad \mathbf{R}_{^{14}\mathrm{C}} = \frac{2\mathbf{R}_{\alpha_1\alpha_2} + \sqrt{3}\mathbf{R}_{\alpha_3\nu_4}}{\sqrt{7}}, \quad (12.17)$$
$$\frac{\sum_{i=1}^{14}\mathbf{r}_i}{\sqrt{14}} = \sqrt{14}\mathbf{R}_{\mathrm{cm}}.$$

For this case the value of the ho parameter is $\beta_{^{14}\mathrm{C}} = 0.33\,\mathrm{fm}^{-2}$.

Now, the parent wave function can be expanded in a similar way to ^{12}C by adding a neutron pair, i.e.

$$|P\rangle = \sum_{j_1j_2j_3j_4} X_{\pi j_1}X_{\pi j_2}X_{\pi j_3}X_{\nu j_4}X_{\nu j_5}X_{\nu j_6}X_{\nu j_7} \times \left[a^\dagger_{\pi j_1} \otimes a^\dagger_{\pi j_1}\right]_0 \left[a^\dagger_{\pi j_2} \otimes a^\dagger_{\pi j_2}\right]_0 \left[a^\dagger_{\pi j_3} \otimes a^\dagger_{\pi j_3}\right]_0 \times \left[a^\dagger_{\nu j_4} \otimes a^\dagger_{\nu j_4}\right]_0 \left[a^\dagger_{\nu j_5} \otimes a^\dagger_{\nu j_5}\right]_0 \left[a^\dagger_{\nu j_6} \otimes a^\dagger_{\nu j_6}\right]_0 \left[a^\dagger_{\nu j_7} \otimes a^\dagger_{\nu j_7}\right]_0 |D\rangle. \quad (12.18)$$

The preformation amplitude is given by

$$\mathscr{F}_0^{(^{14}\mathrm{C})}(R) = \sum_{N_{1234}} W_{N_{1234}}^{(^{14}\mathrm{C})} \phi_{N_{1234}0}^{(14\beta)}(R), \tag{12.19}$$

where

$$\begin{aligned} W_{N_{1234}}^{(^{14}\mathrm{C})} = & \sum_{n_{12}n_{123}n_4n_{1234}} \sum_{N_1N_2N_{12}N_3N_4N_{34}} W_{N_1 0} W_{N_2 0} W_{N_3 0} G_{N_4} \\ & \times \langle n_{12}0N_{12}0; 0|N_1 0N_2 0; 0\rangle \mathscr{I}_{n_{12},2}^{(\beta,\beta_{^{14}\mathrm{C}})} \\ & \times \langle n_{34}0N_{34}0; 0|N_3 0N_4 0; 0\rangle \mathscr{I}_{n_{34},1}^{(\beta,\beta_{^{14}\mathrm{C}})} \\ & \times \langle n_{1234}0N_{1234}0; 0|N_{12}0N_{34}0; 0\rangle \mathscr{I}_{n_{1234},2}^{(\beta,\beta_{^{14}\mathrm{C}})}. \end{aligned} \tag{12.20}$$

Among all analyzed cases, only the ^{14}C spontaneous binary emission from the ground state was detected. The other clusters can be emitted only in ternary emission processes. The first microscopic calculations of the ^{14}C emission were performed in Ref. [5] within the above described MSM and in Ref. [6, 7], by using an antisymmetrized version of the MSM. The experimental order of magnitude for the decay width was reproduced. In Ref. [8] it was shown that the effect of the quadrupole deformation on decay rates is about one order of magnitude. The formalism given in this Section was for the first time applied in Refs. [9–11] by considering only a pure shell model configuration and therefore it was not able to reproduce the experimental data. The Feshbach reaction theory, given in Sect. 8.3, was applied for Ra, Pa and U isotopes in Ref. [12]. Recently potential energy surfaces for heavy cluster emitters were estimated in Ref. [13]. The shell and pairing corrections obtained by the Two Center Shell Model (TCSM) were added to the Yukawa-plus-exponential deformation energy within the so-called macroscopic–microscopic model.

In order to estimate the decay width Γ we apply the above described formalism for the preformation amplitude, by using Eq. 10.27, which is factorized into the spherical width $\Gamma_0(R)$ and the deformation function $D(R)$. The sp basis to diagonalize the Woods–Saxon mean field with universal parametrization depends upon the ho parameter $\beta = f\beta_0$ (10.31), with $f = 0.7$. The decay widths, averaged in the interval $R \in [11, 13]$ fm for ^{14}C emission from Ra emitters, are given in the Table 12.1 [14].

The spectroscopic factor for these decay processes, estimated according to Eq. 11.30, is within the values given by the general scaling relation (2.112) [15].

Table 12.1 Experimental spherical and deformed decay widths

Nuleus	β_2	Γ_{exp}	Γ_0	Γ	$(\Gamma_{^{14}\mathrm{C}}/\Gamma_\alpha)_{\mathrm{exp}}$	$(\Gamma_{^{14}\mathrm{C}}/\Gamma_\alpha)_{\mathrm{th}}$	$(S_{^{14}\mathrm{C}}/S_\alpha)_{\mathrm{th}}$
^{222}Ra	0.19	4.1×10^{-33}	1.9×10^{-35}	2.1×10^{-34}	3.4×10^{-10}	2.3×10^{-11}	1.0×10^{-7}
^{224}Ra	0.18	6.3×10^{-38}	1.2×10^{-39}	1.4×10^{-38}	4.5×10^{-11}	1.7×10^{-11}	8.1×10^{-8}
^{226}Ra	0.20	2.9×10^{-43}	3.8×10^{-44}	1.1×10^{-42}	3.3×10^{-11}	2.2×10^{-10}	1.0×10^{-7}

Their ratios with respect to α-decay widths and the ratio of the corresponding spectroscopic factors

In Ref. [16] it was computed the fine structure for ^{14}C emission from ^{223}Ra by using the Landau–Zenner effect for *sp* levels provided by the TCSM. Finally we mention that the fine structure for the same process was estimated within the shell model approach in Ref. [17].

12.2 Two Center Shell Model (TCSM)

The theoretical study of the heavy cluster emission at very large mass asymmetries is limited by the difficulties encountered in the calculation of single particle levels for very deformed one-center potentials. One way to overcome this difficulty is to consider a two harmonic oscillator single particle expansion basis with one center, as it was done in previous chapters. But central potentials are not able to describe the shapes for the passage of one nucleus in two separate nuclei in a correct manner. For very large prolate deformations, the sum of single particle energies obtained from the level scheme for any smooth potential reaches an infinite value. These difficulties can also be overcomed by considering that the mean field is generated by nucleons belonging to two nuclear fragments. This dinuclear system can be treated within the TCSM [18–20]. The TCSM is based on the assumption that nucleons can be described in molecular states during the nuclear reaction. This means that the relative motion between the two centers of the fragments is slower than the rearrangement of the nucleons in the mean field. This model found a large area of uses in various fields like fission [21, 22], cluster- and α-decay [16, 23], investigation of superheavy elements [24, 25] or heavy-ion collision processes [26, 27] and can be applied to all systems behaving like nuclear molecules. Unfortunately, analytical solutions for the TCSM can be obtained only for semi-symmetric shapes as specified in Refs. [28, 29]. An improved version suitable for very asymmetric disintegration processes was presented in Refs. [30, 31].

The two center potential in cylindrical coordinates is splitted [28, 30] into several parts which are treated separately

$$V(\rho,z,\varphi) = V_0(\rho,z) + V_{\rm as}(\rho,z) + V_{\rm n}(\rho,z) + V_{Ls}(\rho,z,\varphi) + V_{L^2}(\rho,z,\varphi) - V_c \tag{12.21}$$

where $V_0(\rho, z)$ represents the two center semi-symmetric harmonic potential whose eigenvectors can be analytically obtained by solving the Schrödinger equation. It is given by the relation

$$V_0(\rho,z) = \begin{cases} \frac{1}{2}m_0\omega_{z1}^2(z+z_1)^2 + \frac{1}{2}m_0\omega_\rho^2\rho^2, & z<0 \\ \frac{1}{2}m_0\omega_{z2}^2(z-z_2)^2 + \frac{1}{2}m_0\omega_\rho^2\rho^2, & z\geq 0 \end{cases} \tag{12.22}$$

where m_0 is the nucleon mass, z_1 and z_2 (positive) represent the distance between the centers of the spheroids and their intersection plane, ω_i are stiffnesses along the coordinate i. Other terms in the relation (12.21) $V_{\rm as}(\rho,z)$, $V_{\rm n}(\rho,z)$, $V_{Ls}(\rho,z,\varphi)$,

$V_{L^2}(\rho, z, \varphi)$ and V_c are the mass asymmetry, the necking, the spin-orbit coupling, the L^2 correction and the depth of the potential, respectively. All these terms are treated as corrections and will be diagonalized in the eigenfunction basis of the potential (12.22).

The orthogonal system of eigenfunctions used to diagonalize the total Hamiltonian is obtained by solving the Schrödinger equation in cylindrical coordinates:

$$\left[-\frac{\hbar^2}{2m_0}\Delta + V_0(\rho, z)\right]\Psi = E\Psi \qquad (12.23)$$

The ansatz $\Psi = Z(z)R(\rho)\Phi(\varphi)$ leads to the next solutions for the eigenvectors:

$$\begin{aligned}
\Phi(\varphi) &= \frac{1}{\sqrt{2\pi}}\exp(im\varphi) \\
R_{nm}(\rho) &= \sqrt{\frac{2n!}{(n+m)!}}\alpha_\rho \exp\left(-\frac{\alpha_\rho^2\rho^2}{2}\right)(\alpha_\rho\rho)^m L_n^m(\alpha_\rho^2\rho^2) \\
Z_\nu &= \begin{cases} C_{\nu_1}\exp\left(-\frac{\alpha_{z1}^2(z+z_1)^2}{2}\right)\mathbf{H}_{\nu_1}[-\alpha_{z1}(z+z_1)], & z<0, \\ C_{\nu_2}\exp\left(-\frac{\alpha_{z2}^2(z-z_2)^2}{2}\right)\mathbf{H}_{\nu_2}[-\alpha_{z2}(z-z_2)], & z\geq 0, \end{cases}
\end{aligned} \qquad (12.24)$$

where L_n^m is the Laguerre polynomial, $\mathbf{H}_\nu$ is the Hermite function, $\alpha_i = (m_0\omega_i/\hbar)^{1/2}$ are length parameters, and C_{ν_i} denote the normalization constants. The quantum numbers n and m are integers while the quantum number ν along the z-axis is real and has different values in the two intervals $(-\infty, 0]$ and $[0, \infty)$. The solutions (12.24) represent an orthogonal system of eigenvectors for the semi-symmetric two center harmonic potential that give the eigenvalues

$$\begin{aligned}
E_{\nu,n,m} &= \hbar\omega_{z1}\left(\nu_1 + \frac{1}{2}\right) + \hbar\omega_\rho(2n+m+1) \\
&= \hbar\omega_{z2}\left(\nu_2 + \frac{1}{2}\right) + \hbar\omega_\rho(2n+m+1).
\end{aligned} \qquad (12.25)$$

As an example, the TCSM can be used to obtain energy level diagrams starting from an initial state of a parent nucleus up to the asymptotic configuration of two fragments separated at infinity. Such a level diagram is displayed in Fig. 12.1 for the ^{14}C emission from ^{222}Ra. In this example, the nuclear shape parametrization is given by the intersection of two spheres of different radii R_1 and R_2. Here, R_2 is equal to the radius of the spherical ^{14}C emitted fragment and is kept constant. R_1 is obtained from the volume conservation condition.

The single particle levels are labeled by their spectroscopic notations on both sides of the pictures. For small values of R, the levels exhibit a behavior similar to that obtained in the frame of the Nilsson model. After the split of the system, $R > R_1 + R_2$, a superposition of the single particle states of the daughter ^{208}Pb and the emitted ^{14}C can be recognized.

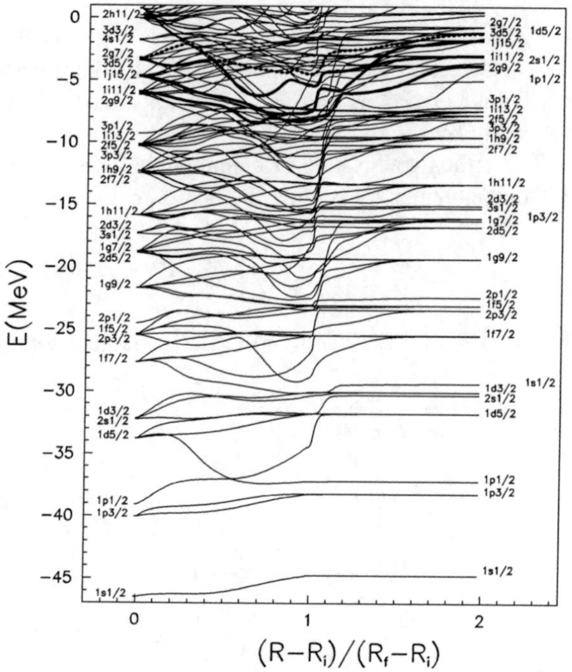

Fig. 12.1 Level scheme for the ^{14}C spontaneous emission from ^{222}Ra versus the distance between the centers of fragments. On the *left* and *right* sides on the figure, the spherical orbitals are labeled with spectroscopic notations. The notations on the left side belong to ^{222}Ra, while on the right side to ^{14}C

In Fig. 12.2, it is shown the deformation energy of the nuclear system, i.e. the sum between the liquid drop energy and the shell effects, including pairing corrections. The macroscopic energy is obtained in the framework of the Yukawa-plus-exponential model extended for binary systems with different

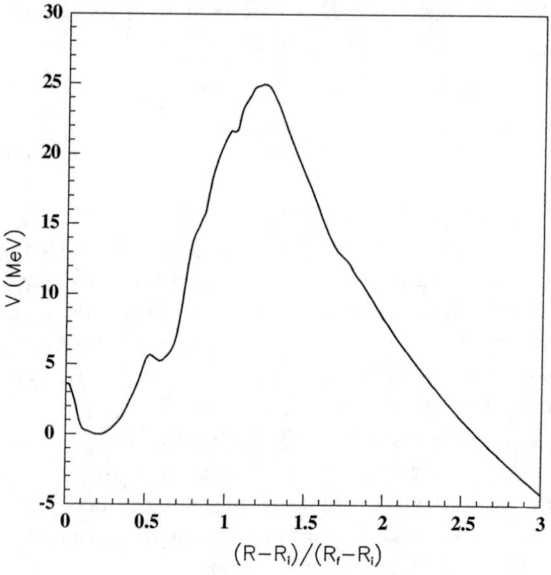

Fig. 12.2 The potential energy for the ^{208}Pb + ^{14}C system versus the distance between centers of emitted fragments. The energy scale is normalized to the *Q*-value

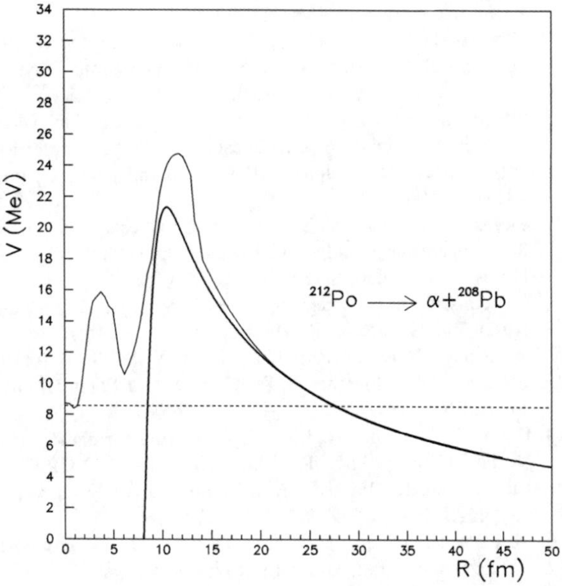

Fig. 12.3 The potential energy for the $^{208}Pb + \alpha$ system versus the distance between centers of emitted fragments (*thin line*). By a *thick line* we give the corresponding double folding interaction. The energy scale is normalized to zero at infinity. By a *dotted line* we plotted the Q-value

charge densities [32]. The Strutinsky prescriptions [33] were computed on the basis of the Superasymmetric TCSM [16].

It is interesting to investigate a similar potential for the α-decay. In Fig. 12.3 we plotted the energy of the $^{208}Pb + \alpha$ system by a thin line. One can see the occurrence of the pocket-like potential in the surface region due to shell effects. As a comparison, we give on the same figure the double folding interaction by a thick line. The barrier predicted by TCSM is higher than its double folding counterpart. This means that the half life becomes longer in comparison to the value predicted by the double folding procedure. Anyway, the precision in estimating the interaction within macroscopic–microscopic method is around 2 MeV, i.e. the half live error for α-decay is about two orders of magnitude. This is the reason why this approach was mainly applied for description of fission processes or heavy cluster decays, where the energy error is relative small in comparison with the Q-value.

References

1. Wildermuth, K., Kanellopulos, Th.: The "Cluster Model" of atomic nuclei. Nucl. Phys. **7**, 150–162 (1958)
2. Wildermuth, K., Kanellopulos, T.: Nucl. Phys. **8**, 449 (1958)
3. Wildermuth, K., Tang, Y.: A Unified Theory of the Nucleus, Braunschweig, Vieweg Verlag (1977)
4. Lovas, R.G., Liotta, R.J., Insolia, A., Varga, K., Delion, D.S.: Microscopic theory of cluster radioactivity. Phys. Rep. **294**, 265–362 (1998)
5. Iriondo, M., Jerrestam, D., Liotta, R.J.: Cluster radioactivity and clustering formation in nuclei. Nucl. Phys. A **454**, 252–266 (1986)

6. Blendowske, R., Fliessbach, T., Walliser, H.: Microscopic calculation of the ^{14}C decay of Ra nuclei. Nucl. Phys. A **464**, 75–89 (1987)
7. Blendowske, R., Fliessbach, T., Walliser, H.: Systematics of cluster-radioactivity-decay constants as suggested by microscopic calculations. Phys. Rev. Lett. **61**, 1930–1933 (1988)
8. Shi, Y.-j., Swiatecki, W.J.: Estimates of the influence of nuclear deformations and shell effects on the lifetimes of exotic radioactivities. Nucl. Phys. A **464**, 205–222 (1987)
9. Florescu, A., Holan, S., Săndulescu, A.: On the preformation probability for emission of clusters heavier than alpha-particle in the independent particle model. Rev. Roum. Phys. **33**, 131–143 (1988)
10. Florescu, A., Holan, S., Săndulescu, A.: Overlap integrals for ^{12}C, ^{14}C and ^{16}O emission in the independent particle model. Rev. Roum. Phys. **33**, 243–295 (1988)
11. Florescu, A., Holan, S., Săndulescu, A.: R matrix microscopic calculation for ^{12}C, ^{14}C and ^{16}O emission in the independent particle model. Rev. Roum. Phys. **34**, 595–619 (1989)
12. Ivaşcu, M., Silişteanu, I.: The microscopic approach to the rates of radioactive decays by emission of heavy clusters. Nucl. Phys. A **485**, 93–110 (1988)
13. Poenaru, D.N., Gherghescu, R., Greiner, W.: Potential energy surfaces for cluster emitting nuclei. Phys. Rev. C **73**, 014608/1–9 (2006)
14. Delion, D.S., Insolia, A., Liotta, R.J.: Pairing correlations and quadrupole deformation effects on the ^{14}C decay. Phys. Rev. Lett. **78**, 4549–4552 (1997)
15. Blendowske, R., Fliessbach, T., Walliser, H.: From α-decay to exotic decays—a unified model. Z. Phys. A **339**, 121–128 (1991)
16. Mirea, M.: Landau-Zenner effect in cluster decay. Phys. Rev. C **57**, 2484–2492 (1998)
17. Dumitrescu, O.: Fine structure of cluster decays. Phys. Rev. C **49**, 1466–1481 (1994)
18. Holzer, P., Mosel, U., Greiner, W.: Double-centre oscillator and its application to fission. Nucl. Phys. A **138**, 241–252 (1969)
19. Scharnweber, D., Greiner, W., Mosel, U.: The two-center shell model. Nucl. Phys. A **164**, 257–278 (1971)
20. Badralexe, E., Rizea, M., Săndulescu, A.: Symmetric two-centre model wave function. Rev. Roum. Phys. **19**, 63–80 (1973)
21. Mirea, M., Tassan-Got, L., Stephan, C., Bacri, C.O.: Dissipation effects in low energy fission. Nucl. Phys. A **735**, 21–39 (2004)
22. Mirea, M., Tassan-Got, L., Stephan, C., Bacri, C.O., Bobulescu, R.C.: Landau-Zenner effect in fission. Phys. Rev. C **76**, 064608/1–11 (2007)
23. Mirea, M.: Fine structure of α decay in odd nuclei. Phys. Rev. C **63**, 034603/1–5 (2001)
24. Gherghescu, R.A., Poenaru, D.N., Greiner, W.: Fission channels of $^{304}120$. J. Phys. G **24**, 1149–1163 (1998)
25. Gherghescu, R.A., Poenaru, D.N., Greiner, W.: Synthesis of $^{286}114$ and $^{290}114$ using low-energy fusion channels. J. Phys. G **32**, L73–L84 (2006)
26. Park, J.Y., Greiner, W., Scheid, W.: Signatures of molecular single-particle states by level crossing in heavy ions collisions. Phys. Rev. C **21**, 958–962 (1980)
27. Thiel, A.: The Landau-Zener effect in nuclear molecules. J. Phys. G **16**, 867–910 (1990)
28. Maruhn, J.A., Greiner, W.: The asymmetric two-center shell model. Z. Phys. **251**, 431–457 (1972)
29. Maruhn, J.A., Greiner, W.: Theory of fission-mass distribution demonstrated for ^{226}Ra, ^{236}U, ^{258}Fm. Phys. Rev. Lett. **32**, 548–551 (1974)
30. Mirea, M.: Superasymmetric two-center shell model for spontaneous heavy-ion emission. Phys. Rev. C **54**, 302–314 (1996)
31. Mirea, M.: Realistic orbital momentum operators for the superasymmetric two-center shell model. Nucl. Phys. A **780**, 13–33 (2006)
32. Poenaru, D.N., Ivascu, M., Mazilu, D.: Folded Yukawa-Plus-Exponential Model PES for nuclei with different charge densities. Comput. Phys. Commun. **19**, 205–214 (1980)
33. Brack, M., Damgaard, J., Jensen, A., Pauli, H., Strutinsky, V., Wong, W.: Funny hills: the shell correction approach to nuclear shell effects and its application to the fission process. Rev. Mod. Phys. **44**, 320–405 (1972)

Chapter 13
Conclusions

Particle and cluster emission from deformed nuclei is a field of research that only recently has become important. This late interest is partly due to the relatively rare occurrence of various forms of radioactivity as proton and two proton emission or heavy cluster decays, which hindered their study previously. But the main reason of the present interest is the possibility that emission processes offers the almost unique tool to study exotic nuclear systems close to the proton/neutron drip lines, or superheavy nuclei. Another reason consists in the increased precision of experimental tools, able to detect fine structure in proton or α decays and in cold fission.

Together with the recent experimental efforts concentrated in various types of radioactivity, an important theoretical activity has followed. Many methods were already developed long time ago, but other approaches have been introduced in order to study the decay processes and the related structures of the nuclei involved in the decay processes. We mention here ternary decay processes like two proton emission or ternary cold fission.

In this book, we have tried to describe all these methods to analyze emission processes, as detailed as possible, as they have been given in the literature. The notation and the terminology were unified in the presentation, explaining clearly the concepts as well as the theories themselves. We have thus explained the meaning of "wide" in contrast with "narrow" resonances in emission processes in various approaches like scattering state description, R-matrix method, coupled channels technique, reaction theory. We have also tried to show the extend of validity of these theories, designed to be applied in the complex energy plane in the framework of the Gamow resonances.

A large part of presentation was devoted to the description of the fine structure in various emission processes, like proton or α-decay and cold fission, within the coupled channels approach. This is one of the most promising directions to investigate the structure of very exotic and unstable nuclear systems produced by modern radioactive beams.

D. S. Delion, *Theory of Particle and Cluster Emission*, Lecture Notes in Physics, 819,
DOI: 10.1007/978-3-642-14406-6_13,

The conclusion of this presentation is that indeed emission processes offer an unique tools to investigate the structure of rare nuclei, which due to their short lifetime are otherwise very difficult to explore.

Chapter 14
Appendices

List of Used Symbols

$a_{jm}^{\dagger}$	Single particle creation operator
A	Mass number
$\mathscr{A}$	Antisymmetrization operator
$\mathscr{A}_{\lambda\mu}^{\dagger}(\tau_1 j_1 \tau_2 j_2)$	Creation quasiparticle pair operator
$\mathscr{A}_{cc'}$	Coupling matrix
$\alpha_{jm}^{\dagger}$	Quasi-particle creation operator
$\alpha_{2\mu}$	Surface quadrupole coordinate
$b_{2\mu}^{\dagger}$	Quadrupole boson creation operator
B	Binding energy
$B_{\tau}(n_1 l_1 j_1 n_2 l_2 j_2; L)$	B-coefficient of the α-particle preformation amplitude
β	Harmonic oscillator constant
β_{λ}	Multipole deformation parameter
β_c	Channel logarithmic derivative
c	Channel index
c_n	Single particle expansion coefficients
C_l	Centrifugal barrier factor
d_n	Single particle deformation expansion coefficients
D	Daughter nucleus
$D_{MK}^{J}(\omega)$	Wigner function
$\mathscr{D}_{MK}^{J}(\omega) = \sqrt{\frac{2J+1}{8\pi^2}} D_{MK}^{J}(\omega)$	Normalized Wigner function
δ	Derivative of the reduced width
δ_c	Channel phase shift
E	Emission energy (Q-value)
ε	Single particle energy
f	Focal distance of spheroidal coordinates

D. S. Delion, *Theory of Particle and Cluster Emission*, Lecture Notes in Physics, 819, DOI: 10.1007/978-3-642-14406-6_14, © Springer-Verlag Berlin Heidelberg 2010

$f_c(r)$	Radial wave function in the laboratory system of coordinates
$F_l(\chi, \rho)$	Regular Coulomb function
$\mathscr{F}(\mathbf{r})$	Preformation amplitude
$\phi^{(\beta)}_{nlm}(\mathbf{r}) = \mathscr{R}^{(\beta)}_{nl}(r)Y_{lm}(\hat{r})$	Radial harmonic oscillator wave function
$\varphi^{(\beta)}_{nljm}(\mathbf{x}) = [\phi^{(\beta)}_{nl}(\mathbf{r}) \otimes \chi_{\frac{1}{2}}(\mathbf{s})]_{jm}$	Spin-radial harmonic oscillator wave function
$\Phi(t, \mathbf{r})$	Time dependent wave function
$\Psi(\mathbf{r})$	Coordinate wave function
$\psi_{\tau\epsilon ljm}(\mathbf{x}) = u_{\tau\epsilon}(r)\mathscr{Y}^{(l)}_{jm}(\hat{r}, \mathbf{s})$	Single particle wave function
$g_c(r)$	Radial wave function in the intrinsic system of coordinates
$G_l(\chi, \rho)$	Irregular Coulomb function
$G_\tau(NL)$	G-coefficient of the α-particle preformation amplitude
γ	Reduced width
Γ	Decay width
$\Gamma(x)$	Euler Gamma function
$\Gamma_{\lambda\mu}(n)$	Phonon creation operator
$\mathbf{H}$	Hamiltonian
$H^{\pm}_l(\chi, \rho)$	Coulomb–Hankel function
$\mathscr{I}^{(\beta)}_{nn'}$	Overlap integral between radial ho wave functions
j	Single particle spin
J	Spin
$\mathscr{J}(\hat{r})$	Probability flux
κ	Momentum
χ	Coulomb parameter
$\chi_{\frac{1}{2}}(\mathbf{s})$	Spin wave function
$l,\ L$	Angular momentum
$L^{\alpha}_{\nu}(x)$	Laguerre polynomial
λ	Decay constant
λ	Angular momentum
λ_0	Preformation factor
m	Single particle spin projection
M	Spin projection
μ	Reduced mass
μ	Angular momentum projection
n	Assault frequency
N	Neutron number
N_c	Channel scattering amplitude
P	Parent nucleus
P_c	Channel penetrability
$\hat{P}$	Permutation operator

$\mathscr{P}(\mathbf{r})$	Internal probability
$Q_{\lambda\mu}$	Multipole operator
$\mathbf{r} = (r, \hat{r})$	Daughter-cluster relative radius
$\mathbf{R} = (R, \hat{R})$	Relative radius between comparable fragments
R	Matching radius
$R_{cc}{}'$	R-matrix
$\mathscr{R}_{nl}^{(\beta)}(r)$	Radial harmonic oscillator wave function
$\rho = \kappa r$	Reduced radius
ρ_k	Nuclear density
$S_{cc'}$	S-matrix
$T_{1/2}$	Half life
τ	Isospin coordinate
$u_c(r) = f_c(r)/r$	Radial wave function
$U_{\lambda\mu}^{(n)}(ljl'j')$	Deformed spin-orbit multipoles
v_c	Asymptotic channel velocity
$v(\mathbf{r}_{12})$	Effective nucleon–nucleon interaction
$V_0(r)$	Spherical potential
$V_d(\mathbf{r})$	Deformed potential
$V_\lambda(r)$	Multipole component of the potential
$W(N, L)$	W-coefficient of the α-particle preformation amplitude
$\mathbf{x}$	Internal coordinate
ξ	radial spheroidal coordinate
η	Angular spheroidal coordinate
$\mathscr{X}_\lambda^{(n)}(\tau_1 j_1 \tau_2 j_2)$	Forward QRPA amplitude
$\mathscr{Y}_\lambda^{(n)}(\tau_1 j_1 \tau_2 j_2)$	Backward QRPA amplitude
$Y_{lm}(\hat{r})$	Angular harmonics
$\mathscr{Y}_{JM}^{(c)}(\mathbf{x}_1, \mathbf{x}_2, \hat{r})$	Channel core-angular harmonics
$\mathscr{Y}_{jm}^{(l)}(\hat{r}, \mathbf{s}) = [Y_l(\hat{r}) \otimes \chi_{\frac{1}{2}}(\mathbf{s})]_{jm}$	Spin-orbit wave function
ω	Euler angle
Ω	Euler angle for comparable fragments
Z	Charge number

14.1 Single Particle Mean Field

The single particle deformed potential in the intrinsic system, defined by the major axes of the nuclear ellipsoid, has the following general form

$$V(\mathbf{r}, \mathbf{s}) = V_0(\mathbf{r}) + V_{so}(\mathbf{r}, \mathbf{s}), \tag{14.1}$$

where **r** denotes intrinsic coordinates and V_0 is the central part of the interaction, containing the nuclear and proton Coulomb interaction

$$V_0(\mathbf{r}) = V_N(\mathbf{r}) + V_C(\mathbf{r}). \tag{14.2}$$

14.1.1 Nuclear Potential

The nuclear part of the interaction is given by

$$V_N(\mathbf{r}) = -V_N^{(0)} f(\mathbf{r}, r_{0c}, a_c), \tag{14.3}$$

with a Woods–Saxon shape defined as the Fermi distribution

$$f(\mathbf{r}, r_0, a) = \frac{1}{1 + exp\left[\frac{r - R(\hat{r})}{a}\right]}, \tag{14.4}$$

where $\hat{r} \equiv (\theta, \phi)$ and the radius $R(\hat{r})$ of the nuclear surface is given by

$$R(\hat{r}) = cR_0\left[1 + \sum_{\lambda \geq 2\mu} \beta_{\lambda\mu} Y_{\lambda\mu}(\hat{r})\right], \quad R_0 = r_0 A_d^{1/3}. \tag{14.5}$$

The constant c is determined by the volume conservation condition

$$c^{-3} = \int \left[1 + \sum_{\lambda\mu} \beta_{\lambda\mu} Y_{\lambda\mu}(\hat{r})\right]^3 \frac{d\hat{r}}{4\pi}. \tag{14.6}$$

The most important parameters are quadrupole deformation

$$\beta_{20} = \beta_2 \cos\gamma, \quad \beta_{2\pm2} = \beta_2 \frac{\sin\gamma}{\sqrt{2}}, \tag{14.7}$$

and hexadecapole deformation

$$\beta_{40} = \frac{1}{6}\beta_4(5\cos^2\gamma + 1), \quad \beta_{4\pm2} = \frac{\sqrt{30}}{12}\beta_4 \sin 2\gamma, \quad \beta_{4\pm4} = \frac{\sqrt{70}}{12}\beta_4 \sin^2\gamma. \tag{14.8}$$

The expansion of the nuclear potential

$$V_N(\mathbf{r}) = V_N(r) + \sum_{\lambda > 0, \mu} V_{N,\lambda\mu}(r) Y_{\lambda\mu}(\hat{r}), \tag{14.9}$$

where

$$V_{N,\lambda\mu}(r) \equiv -V_N^{(0)} f_{\lambda\mu}(r), \tag{14.10}$$

contains the multipoles of the Fermi distribution defined as follows

$$f(r, \hat{r}, r_0, a) = \sum_{\lambda\mu} f_{\lambda\mu}(r, r_0, a) Y_{\lambda\mu}(\hat{r}). \tag{14.11}$$

The coefficients are given by the inverse transformation

$$f_{\lambda\mu}(r, r_0, a) = \int Y^*_{\lambda\mu}(\hat{r}) f(\mathbf{r}, r_0, a) d\hat{r}. \tag{14.12}$$

The axially spherical symmetry corresponds to vanishing angles $\gamma = 0$ and therefore one has $\mu = 0$. All above relations become simpler by doing the replacement

$$d\hat{r} = 2\pi dt, \quad t \equiv \cos\theta. \tag{14.13}$$

14.1.2 Coulomb Potential

The Coulomb potential between a distributed charge $Z_D e$, inside the deformed surface (14.5), and a proton charge e is given by

$$V_C(\mathbf{r}) = \int \frac{\rho(\mathbf{r}')}{|\mathbf{r} - \mathbf{r}'|} d\mathbf{r}', \tag{14.14}$$

where the charge density is

$$\rho(\mathbf{r}') = Z_D e^2 N f(\mathbf{r}', r_{0c}, a_c). \tag{14.15}$$

The inverse of the normalization constant

$$N^{-1} = \int f(\mathbf{r}', r_{0c}, a_c) d\mathbf{r}' = \sqrt{4\pi} \int_0^\infty f_{00}(r') r'^2 dr', \tag{14.16}$$

is rather close to the geometric volume $4\pi R_{0c}^3/3$. By using an expansion of the density in multipoles

$$\rho(\mathbf{r}') = Z_D e^2 \sum_{\lambda\mu} \rho_{\lambda\mu}(r') Y_{\lambda\mu}(\hat{r}'), \tag{14.17}$$

and the well-known relation

$$\frac{1}{|\mathbf{r} - \mathbf{r}'|} = \sum_{\lambda\mu} \frac{4\pi}{2l+1} \frac{r_<^l}{r_>^{l+1}} Y_{\lambda\mu}(\hat{r}) Y^*_{\lambda\mu}(\hat{r}'), \tag{14.18}$$

where $r_< = \min(r, r'), r_> = \max(r, r')$, the multipole expansion of the Coulomb potential becomes

$$V_C(\mathbf{r}) = \sum_{\lambda\mu}\left[Z_D e^2 \frac{4\pi}{2\lambda+1}\int_0^\infty \rho_{\lambda\mu}(r')\frac{r_<^\lambda}{r_>^{\lambda+1}} r'^2 dr'\right] Y_{\lambda\mu}(\hat{r}) \equiv \sum_{\lambda\mu} V_{C\lambda\mu}(r) Y_{\lambda\mu}(\hat{r}). \tag{14.19}$$

At large distances $r >> R(\hat{r}')$, where the density has vanishing values, one obtains

$$V_C(\mathbf{r}) \approx e \sum_{\lambda\mu}\frac{1}{r^{\lambda+1}}\left[Z_D e \frac{4\pi}{2\lambda+1}\int_0^\infty \rho_{\lambda\mu}(r') r'^{\lambda+2} dr'\right] Y_{\lambda\mu}(\hat{r}) \equiv e \sum_{\lambda\mu}\frac{Q_{\lambda\mu}}{r^{\lambda+1}} Y_{\lambda\mu}(\hat{r}), \tag{14.20}$$

where $Q_{\lambda\mu}$ denotes the intrinsic multipole moment.

Notice that the role of the diffusivity a_c in the density distribution is rather week. The replacement of the Fermi distribution by a sharp one

$$\rho(\mathbf{r}, r_{0c}, 0) = \frac{3Z_D e^2}{4\pi R_{oc}^3}\Theta[R(\hat{r}) - r], \tag{14.21}$$

gives an error less that 5% in the internal region and practically leads to the same results for large distances. In particular the Coulomb potential of an uniformly charged sphere is given by

$$V_C(r) = \begin{array}{ll} \frac{Z_D e^2}{2R_{0c}}\left(3 - \frac{r^2}{R_{0c}^2}\right), & r \leq R_{0c} \\ \frac{Z_D e^2}{r}, & r > R_{0c}. \end{array} \tag{14.22}$$

14.1.3 Spin-Orbit Potential

Spherical symmetric interaction is obtained when $\beta_2 = \beta_4 = 0$ in (14.7). In this case the spin-orbit potential is given by the ansatz

$$V_{so}(r, \mathbf{s}) = -V_{so}^{(0)} \frac{1}{r}\frac{df(r, r_{0so}, a_{so})}{dr}\mathbf{l}\cdot\boldsymbol{\sigma} \equiv V_{so}(r)\mathbf{l}\cdot\boldsymbol{\sigma}, \quad \sigma = 2s, \tag{14.23}$$

where

$$V_{so}^{(0)} \equiv -\Lambda\left(\frac{\hbar c}{2Mc^2}\right)^2 \bar{V}_{so}^{(0)}. \tag{14.24}$$

The spherical spin-orbit interaction can be written in an invariant form by using the momentum operator $\hat{\mathbf{k}} = -i\nabla$

$$\frac{1}{r}\frac{df(r)}{dr}\mathbf{l}\cdot\boldsymbol{\sigma} = \frac{1}{r}\frac{df(r)}{dr}[\mathbf{r}\times\hat{\mathbf{k}}]\cdot\boldsymbol{\sigma} = \frac{df(r)}{dr}\frac{\mathbf{r}}{r}\cdot[\hat{\mathbf{k}}\times\boldsymbol{\sigma}], \tag{14.25}$$

or, by using cyclic permutation

$$\frac{1}{r}\frac{df(r)}{dr}\mathbf{l}\cdot\boldsymbol{\sigma} = \nabla f(\mathbf{r})\cdot[-\mathbf{i}\nabla\times\boldsymbol{\sigma}]. \tag{14.26}$$

Therefore the deformed spin-orbit interaction and its multipole expansion is given by

$$\begin{aligned} V_{so}(\mathbf{r},\mathbf{s}) &= -V_{so}^{(0)}\nabla f(\mathbf{r}, r_{0so}, a_{so})\cdot[-\mathbf{i}\nabla\times\boldsymbol{\sigma}], \\ &= -V_{so}^{(0)}\sum_{\lambda\mu}\nabla[f_{\lambda\mu}(r, r_{0so}, a_{so})Y_{\lambda\mu}]\cdot[-i\nabla\times\sigma]. \end{aligned} \tag{14.27}$$

The gradient theorem allows one to write

$$\nabla\left[f(r)Y_{\lambda\mu}(\hat{r})\right] = \eth_{\lambda-1}f(r)\mathbf{Y}_{\lambda\mu}^{(\lambda-1)}(\hat{r}) + \eth_{\lambda+1}f(r)\mathbf{Y}_{\lambda\mu}^{(\lambda+1)}(\hat{r}), \tag{14.28}$$

where there are introduced the vectorial harmonics

$$\mathbf{Y}_{\lambda\mu}^{(\lambda\pm1)}(\hat{r}) = [Y_{\lambda\pm1}(\hat{r})\mathbf{e}]_{\lambda\mu}, \tag{14.29}$$

With the differential operators written as

$$\eth_{\lambda-1} \equiv \sqrt{\frac{\lambda}{2\lambda+1}}\left(\frac{d}{dr}+\frac{\lambda+1}{r}\right), \quad \eth_{\lambda+1} \equiv -\sqrt{\frac{\lambda+1}{2\lambda+1}}\left(\frac{d}{dr}-\frac{\lambda}{r}\right). \tag{14.30}$$

and with

$$\mathbf{Y}_{\lambda\mu}^{(\lambda\pm1)}(\hat{r})\cdot\mathbf{A} = [Y_{\lambda\pm1}(\hat{r})\mathbf{A}]_{\lambda\mu}, \tag{14.31}$$

where $\mathbf{A}$ is a vector, one obtains the multipole component of the spin-orbit interaction

$$\begin{aligned} &\nabla[f_{\lambda\mu}(r)Y_{\lambda\mu}(\hat{r})]\cdot[-i\nabla\times\sigma] \\ &= \left[\eth_{\lambda-1}f_{\lambda\mu}(r)T_{\lambda\mu}^{(\lambda-1)}(\hat{r},\boldsymbol{\sigma}) + \eth_{\lambda+1}f_{\lambda\mu}(r)T_{\lambda\mu}^{(\lambda+1)}(\hat{r},\boldsymbol{\sigma})\right], \end{aligned} \tag{14.32}$$

in terms of the following differential spin-orbit operators

$$T_{\lambda\mu}^{(\lambda\pm1)}(\hat{r},\boldsymbol{\sigma}) \equiv [Y_{\lambda\pm1}(\hat{r})(-i\nabla\times\boldsymbol{\sigma})]_{\lambda\mu}. \tag{14.33}$$

Thus, the expansion in multipoles of the spin-orbit potential is given by

$$V_{so}(\mathbf{r},\mathbf{s}) = V_{so}(r)\mathbf{l}\cdot\sigma + \sum_{\lambda>0,\mu}\left[V_{so,\lambda\mu}^{(\lambda+1)}(r)T_{\lambda\mu}^{(\lambda+1)}(\hat{r},\boldsymbol{\sigma}) + V_{so,\lambda\mu}^{(\lambda-1)}(r)T_{\lambda\mu}^{(\lambda-1)}(\hat{r},\boldsymbol{\sigma})\right], \tag{14.34}$$

where

$$V_{so,\lambda\mu}^{(\lambda\pm 1)}(r) \equiv -V_{so}^{(0)}\partial_{\lambda\pm 1}f_{\lambda\mu}(r, r_{0so}, a_{so}). \tag{14.35}$$

The so-called universal parameters [1, 2] for both proton and neutron mean fields are given respectively by

$$V_N^{(0)} = \bar{V}_{so}^{(0)} = 49.6\left(1 + \chi_\tau \frac{N-Z}{N+Z}\right)(\text{MeV}), \quad \chi_{p,n} = \pm 0.86, \tag{14.36}$$

where

$$\begin{aligned}
r_{0c}(p) &= 1.275\,(\text{fm}), \quad r_{0c}(n) = 1.347\,(\text{fm}),\\
r_{0so}(p) &= 1.320\,(\text{fm}), \quad r_{0so}(n) = 1.310\,(\text{fm}),\\
a_c(p) &= 0.70\,(\text{fm}), \quad a_{so}(n) = 0.70\,(\text{fm}),\\
\Lambda_p &= 36.0\,(\text{MeV}), \quad \Lambda_n = 35.0\,(\text{MeV}).
\end{aligned}$$

14.2 WKB for Coulomb Functions

The WKB solution for the radial equation

$$\begin{aligned}
&\left[-\frac{d^2}{d\rho^2} + \frac{V_l(\rho)}{E} - 1\right] f_l(r) = 0\\
&\frac{V_l(\rho)}{E} = \frac{l(l+1)}{\rho^2} + \frac{\rho}{\chi},
\end{aligned} \tag{14.37}$$

where the Coulomb parameter χ is defined by Eq. 2.67, has the following form

$$\begin{aligned}
H_l^{(\pm)}(\rho) &= \left(\frac{V_l}{E} - 1\right)^{-1/4} exp\left[\pm\int_\rho^{\rho_3}\left(\frac{V_l}{E} - 1\right)^{1/2} d\rho\right], \quad \frac{V_l}{E} > 1\\
&= \left(1 - \frac{V_l}{E}\right)^{-1/4} exp\left[\pm\int_\rho^{\rho_3}\left(1 - \frac{V_l}{E}\right)^{1/2} d\rho\right], \quad \frac{V_l}{E} < 1,
\end{aligned} \tag{14.38}$$

where the argument χ is dropped. The turning points are defined as the solutions $\rho_1 < \rho_2 < \rho_3$ of the equation

$$\frac{V_l(\rho)}{E} - 1 = 0. \tag{14.39}$$

The WKB solution approximation is valid if the condition

$$\left|\frac{d}{d\rho}\left|\frac{V_l}{E} - 1\right|^{-1/2}\right| << 1, \tag{14.40}$$

is fulfilled. A particular case is the Coulomb plus centrifugal term V_l^C. Apart from small regions around these points, one obtains good analytical WKB estimates in the case of large Coulomb barriers.

Let us concentrate upon the decay (outgoing) solution $H^{(+)}(\rho)$.

(A) For the *internal region*

$$\cos^2\alpha \equiv \frac{\rho}{\chi} = \frac{E}{V_0} < 1, \tag{14.41}$$

one has the following solution

$$\begin{aligned} H_l^{(+)}(\rho) &\approx (ctg\alpha)^{1/2} exp[\chi(\alpha - \sin\alpha\cos\alpha)] C_l \\ C_l &\equiv exp\left[\frac{l(l+1)}{\chi}\sqrt{\frac{\chi}{\rho} - 1}\right], \end{aligned} \tag{14.42}$$

under the condition

$$2\chi\sin^3\alpha\cos\alpha \gg 1. \tag{14.43}$$

Inside a large Coulomb barrier the outgoing Coulomb wave practically coincides with the irregular function

$$H_l^{(+)}(\rho) \approx G_l(\rho), \quad \frac{E}{V_0(\rho)} << 1. \tag{14.44}$$

(B) For the *external region*

$$ch^2 a \equiv \frac{\rho}{\chi} = \frac{E}{V_0} > 1, \tag{14.45}$$

one obtains the following oscillating solution

$$H_l^{(+)}(\rho) \to_{\rho\to\infty} exp\left[i\chi(sh\ a\ ch\ a - a) + i\frac{\pi}{4}\right] exp\left[-i\frac{l(l+1)}{\chi}\right], \tag{14.46}$$

which can be written as follows

$$H_l^{(+)}(\rho) = \cos(\phi_0 - \phi_l) + i\sin(\phi_0 - \phi_l), \tag{14.47}$$

in terms of the phases, defined respectively by

$$\phi_0 = \chi(sh\ a\ ch\ a - a) + \frac{\pi}{4}, \quad \phi_l = \frac{l(l+1)}{\chi}. \tag{14.48}$$

The WKB condition is similar, i.e.

$$2\chi sh^3 a\ ch\ a \gg 1. \tag{14.49}$$

The WKB approximation for most decay processes gives very close values with respect to the exact Coulomb function.

14.3 Rotations

In order to avoid any confusion connected with conventions we will give here the Wigner rotation functions used in this book. For this, let us consider the wave function $|JM\rangle$ with angular momentum J and projection M in the laboratory system of coordinates. One can "rotate" this wave function in the intrinsic system of coordinates according to

$$\hat{R}(\omega)|JM\rangle = \sum_{M'} |JM'\rangle\langle JM'|\hat{R}(\omega)|JM\rangle. \tag{14.50}$$

By introducing the coordinate representation of the wave function

$$\psi_{JM}(\mathbf{r}) = \langle \mathbf{r}|JM\rangle, \tag{14.51}$$

and by denoting the rotation (Wigner) matrix in the Rose convention

$$D^J_{M'M}(\omega) \equiv \langle JM'|\hat{R}(\omega)|JM\rangle, \tag{14.52}$$

one obtains that

$$\hat{R}(\omega)\psi_{JM}(\mathbf{r}) = \sum_{M'} D^J_{M'M}(\omega)\psi_{JM'}(\mathbf{r}) = \psi_{JM}(\mathbf{r}'). \tag{14.53}$$

The most important properties of the rotation matrix are

$$\sum_M D^J_{MM_1}(\omega)D^{J*}_{MM_2}(\omega) = \sum_M D^J_{M_1M}(\omega)D^{J*}_{M_2M}(\omega) = \delta_{M_1M_2}, \tag{14.54}$$

$$\sum_K D^J_{MK}(\omega)D^J_{KM'}(\omega') = D^J_{MM'}(\omega\omega'). \tag{14.55}$$

$$D^{J*}_{MM'}(\omega) = (-)^{M-M'} D^J_{-M-M'}(\omega), \tag{14.56}$$

$$D^J_{MM'}(\omega^{-1}) = D^{J*}_{M'M}(\omega). \tag{14.57}$$

By using the first relation one can invert the rotation (14.53)

$$\hat{R}(\omega^{-1})\psi_{JM}(\mathbf{r}') = \sum_{M'} D^{J*}_{MM'}(\omega)\psi_{JM'}(\mathbf{r}') = \psi_{JM}(\mathbf{r}). \tag{14.58}$$

One gets the well known factorization in terms of Euler angles $\omega = (\phi, \theta, \psi)$

$$D^J_{MK}(\phi, \theta, \psi) = e^{-iM\phi} d^J_{MK}(\theta) e^{-iK\psi}. \tag{14.59}$$

The rotation matrix

$$\mathscr{D}^J_{MK}(\omega) = \sqrt{\frac{2J+1}{8\pi^2}} D^J_{MK}(\omega). \tag{14.60}$$

is normalized to unity

$$\int \mathscr{D}^{J*}_{MK}(\omega)\mathscr{D}^{J'}_{M'K'}(\omega)d\omega = \delta_{JJ'}\delta_{MM'}\delta_{KK'}. \tag{14.61}$$

In the Rose representation the eigenfunction of the angular momentum squared is given by $\mathscr{D}^{J*}_{MK}(\omega)$. Denoting the laboratory third axis z and the intrinsic similar axis ζ one has

$$\mathbf{J}^2\mathscr{D}^{J*}_{MK}(\omega) = J(J+1)\mathscr{D}^{J*}_{MK}(\omega), \tag{14.62}$$

$$\mathbf{J}_z\mathscr{D}^{J*}_{MK}(\omega) = M\mathscr{D}^{J*}_{MK}(\omega), \tag{14.63}$$

$$\mathbf{J}_\zeta\mathscr{D}^{J*}_{MK}(\omega) = K\mathscr{D}^{J*}_{MK}(\omega). \tag{14.64}$$

Some particular values of projections are important

$$D^l_{m0}(\phi,\theta,\psi) = \sqrt{\frac{4\pi}{2l+1}} Y^*_{lm}(\theta,\phi), \tag{14.65}$$

$$D^l_{0m}(\phi,\theta,\psi) = (-)^m \sqrt{\frac{4\pi}{2l+1}} Y^*_{lm}(\theta,\psi), \tag{14.66}$$

$$D^l_{00}(0,\theta,0) = P_l(\cos\theta). \tag{14.67}$$

The product of two rotation D-functions can be expressed in terms of the sum over single D-functions as

$$D^{J_1}_{M_1K_1}(\omega)D^{J_2}_{M_2K_2}(\omega) = \sum_{J=|J_1-J_2|}^{J_1+J_2} \langle J_1M_1;J_2M_2|JM\rangle\langle J_1K_1;J_2K_2|JK\rangle D^J_{MK}(\omega), \tag{14.68}$$

For the orbital harmonics Y_{lm} this relation gives

$$Y^*_{lm}(\hat{r})Y_{l'm'}(\hat{r}) = (-)^m \sum_{L=|l-l'|}^{l+l'} \frac{\hat{l}\hat{l'}}{\sqrt{4\pi}\hat{L}} \langle l,0;l'0|L,0\rangle\langle l,-m;l'm'|LM\rangle Y_{LM}(\hat{r}). \tag{14.69}$$

The spin-orbital harmonics

$$\mathscr{Y}^{(ls)}_{jm}(\hat{r},\mathbf{s}) \equiv \langle\hat{r},\mathbf{s}|ljm\rangle, \tag{14.70}$$

defined as

$$\mathscr{Y}^{(ls)}_{jm}(\hat{r},\mathbf{s}) = \left[i^l Y_l(\hat{r}) \otimes \chi_s(\mathbf{s})\right]_{jm} = \sum_{m_1+m_2=m} \langle lm_1; sm_2|jm\rangle i^l Y_{lm_1}(\hat{r})\chi_{sm_2}(\mathbf{s}). \quad (14.71)$$

describe the angular behaviour of a fermion in a static nuclear field. By considering the orthonormality of spin functions

$$\chi^{\dagger}_{s,m}\chi_{s,m'} = \delta_{mm'}, \quad (14.72)$$

one obtains a relation similar to (14.69)

$$\begin{aligned}\mathscr{Y}^{(ls)\dagger}_{jm}(\hat{r},\mathbf{s})\mathscr{Y}^{(l's)}_{j'm'}(\hat{r},\mathbf{s}) = (-)^{m+\frac{1+l'-l}{2}} \sum_{L=|l-l'|}^{l+l'} \frac{\hat{j}\hat{j}'}{\sqrt{4\pi}\hat{L}} \\ \times \langle j,s;j',-s|L,0\rangle\langle l,-m;l'm'|LM\rangle Y_{LM}(\hat{r}).\end{aligned} \quad (14.73)$$

14.4 Reduced Matrix Elements

The Wigner Eckart theorem defines the reduced matrix element of spherical operators $Q_{\lambda\mu}$, as follows

$$\langle JM|Q_{\lambda\mu}|J'M'\rangle = \langle J||Q_\lambda||J'\rangle \frac{\langle J',M';\lambda,\mu|J,M\rangle}{\hat{J}}. \quad (14.74)$$

The reduced matrix element of spherical harmonics is given by

$$\langle Y_l||Y_\lambda||Y_{l'}\rangle = \frac{\hat{\lambda}\hat{l}}{\sqrt{4\pi}}\langle l,0;\lambda,0|l',0\rangle. \quad (14.75)$$

The following reduced matrix element is important for coupled channels calculations from axially symmetric emitters

$$\left\langle \mathscr{Y}^{(J_flj)}_{J_iM_i}|Q_\lambda \cdot T_\lambda|\mathscr{Y}^{(J'_fl'j')}_{J_iM_i}\right\rangle = (-)^{\lambda+J_f-J'_f}\left\{\begin{matrix} J_f & J'_f & \lambda \\ j' & j & J_i\end{matrix}\right\}\langle J_f||Q_\lambda||J'_f\rangle\langle lj||T_\lambda||l'j'\rangle, \quad (14.76)$$

Here, Q_λ acts on the collective wave function $|J_fM_f\rangle = |\Phi_{J_fM_f}\rangle$. For core excitations described by a rotational ground state band, one obtains the reduced matrix element (14.75), i.e

$$\langle J_f||D^\lambda_0||J'_f\rangle = \frac{\sqrt{4\pi}}{\hat{\lambda}}\langle Y_{J_f}||Y_\lambda||Y_{J'_f}\rangle = \hat{l}\langle l,0;\lambda,0|l',0\rangle. \quad (14.77)$$

The operator T_λ acts on proton spin-orbital wave function $|ljm\rangle \equiv |\mathscr{Y}^{(l\frac{1}{2})}_{jm}\rangle$. For a central potential the expansion is performed in terms of angular harmonics $T_{\lambda\mu} = Y_{\lambda\mu}$, which have the following reduced matrix element

$$\langle lj||Y_\lambda||l'j'\rangle = i^{l'-l}\frac{1+(-)^{l+l'+\lambda}}{2}\frac{\hat{\lambda}\hat{j}}{\sqrt{4\pi}}\left\langle j,\frac{1}{2};\lambda,0|j',\frac{1}{2}\right\rangle. \tag{14.78}$$

For the spin-orbit interaction there are two operators $T^{(\lambda\,\pm\,1)}_{\lambda\mu}$, defined by (14.33). A direct calculation shows that their reduced matrix elements are given by

$$\langle lj||T^{(\lambda\pm1)}_\lambda||l'j'\rangle\frac{f_{l'j'}(r)}{r} = \sqrt{36}(-)^{\lambda+j+j'+1}\hat{\lambda}\hat{j}'\sum_{l_aj_a}\hat{j}_a\begin{Bmatrix}\lambda\pm1 & 1 & \lambda\\ j' & j & j_a\end{Bmatrix}$$
$$\times\begin{Bmatrix} l_a & l' & 1\\ \frac{1}{2} & \frac{1}{2} & 1\\ j_a & j' & 1\end{Bmatrix}\langle lj||Y_{\lambda\pm1}||l_aj_a\rangle\langle l_a||\nabla||l'\rangle\frac{f_{l'j'}(r)}{r}, \tag{14.79}$$

where the first reduced matrix element can be evaluated by using (14.78) and

$$\langle l_a||\nabla||l'\rangle\frac{f_{l'j'}(r)}{r} = i^{l'-l_a}\hat{l}'\langle l,0;1,0|l_a,0\rangle$$
$$\times\left\{[l'(l'+1)-l_a(l_a+1)]\frac{f_{l'j'}(r)}{2r^2}+\frac{1}{r}\frac{df_{l'j'}(r)}{dr}\right\}. \tag{14.80}$$

One finally obtains the expression

$$\langle lj||T^{(\lambda\pm1)}_\lambda||l'j'\rangle\frac{f_{l'j'}(r)}{r} = T^{(0)}_{\lambda\pm1}(l,j,l',j')\frac{f_{l'j'}(r)}{r}+T^{(1)}_{\lambda\pm1}(l,j,l',j')\frac{1}{r}\frac{df_{l'j'}(r)}{dr}, \tag{14.81}$$

where the following short-hand notations are introduced

$$T^{(0)}_{\lambda\pm1}(l,j,l',j') = \frac{1}{2r}\sum_{l_aj_a}A_{\lambda\pm1}(l_a,j_a)[l'(l'+1)-l_a(l_a+1)], \tag{14.82}$$

$$T^{(1)}_{\lambda\pm1}(l,j,l',j') = \sum_{l_aj_a}A_{\lambda\pm1}(l_a,j_a), \tag{14.83}$$

$$A_L(l_a,j_a) \equiv \sqrt{\frac{9}{\pi}}i^{l'-l}(-)^{\lambda+j+j'+1}\hat{L}\hat{\lambda}\hat{j}\hat{j}'\hat{j}_a\begin{Bmatrix}L & 1 & \lambda\\ j' & j & j_a\end{Bmatrix}\begin{Bmatrix} l_a & l' & 1\\ \frac{1}{2} & \frac{1}{2} & 1\\ j_a & j' & 1\end{Bmatrix}$$
$$\times\frac{1+(-)^{l+l_a+L}}{2}\left\langle j,\frac{1}{2};L,0|j_a,\frac{1}{2}\right\rangle\hat{l}',0;1,0|l_a,0\rangle. \tag{14.84}$$

In our derivations we often have multipole operators acting on the radial coordinates which, by using the equation above, acquire the form

$$U^{(k)}_{\lambda\mu}(l,j,l',j') \equiv V^{(\lambda-1)}_{so,\lambda\mu}(r)T^{(k)}_{\lambda-1}(l,j,l',j')+V^{(\lambda+1)}_{so,\lambda\mu}(r)T^{(k)}_{\lambda+1}(l,j,l',j'), \tag{14.85}$$

where $k=0$ indicates just the radial function, while $k=1$ its derivative. The multipole formfactors $V^{(\lambda\pm1)}_{so,\lambda\mu}(r)$ are defined by (14.34). Thus, the matrix element of the spin-orbit interaction valid for the general triaxial case can be expressed in terms of a multipole expansion as follows

$$\langle ljm|V_{so}(\mathbf{r},\mathbf{s})|l'j'm'\rangle\frac{f_{l'j'}(r)}{r} = \delta_{ll'}\delta_{jj'}V_{so}(r)\langle \mathbf{l}\cdot\sigma\rangle + \sum_{\lambda>0,\mu}\frac{\langle j',m';\lambda\mu|j,m\rangle}{\hat{j}}$$
$$\times\left[U^{(0)}_{\lambda\mu}(l,j,l',j')\frac{f_{l'}(r)}{r} + U^{(1)}_{\lambda\mu}(l,j,l',j')\frac{1}{r}\frac{df_{l'}(r)}{dr}\right]. \tag{14.86}$$

14.5 Numerov Integration Method

Here, we present the Numerov scheme to integrate a second order linear system of differential equations

$$\frac{d^2Y}{dr^2} = \mathscr{A}Y, \tag{14.87}$$

where Y is a column vector and $\mathscr{A}$ is a square matrix.

The integration region is divided into equal intervals

$$r, r+h, r+2h, \ldots, r+nh, r+(n+1)h. \tag{14.88}$$

The values of the vector y at initial coordinates r, $r+h$ are known. The values at final coordinates are connected with the initial values by a "propagator" matrix, i.e.

$$\begin{bmatrix} Y(r+(n+1)h) \\ Y(r+nh) \end{bmatrix} = \begin{bmatrix} \mathscr{Z}_{n+1,1} & \mathscr{Z}_{n+1,0} \\ \mathscr{Z}_{n,1} & \mathscr{Z}_{n,0} \end{bmatrix}\begin{bmatrix} Y(r+h) \\ Y(r) \end{bmatrix}. \tag{14.89}$$

The "propagator" matrix is given by a recursive relation

$$\begin{bmatrix} \mathscr{Z}_{n+1,1} & \mathscr{Z}_{n+1,0} \\ \mathscr{Z}_{n,1} & \mathscr{Z}_{n,0} \end{bmatrix} = \begin{bmatrix} Z_1(r+nh) & Z_0(r+nh) \\ 1 & 0 \end{bmatrix}\begin{bmatrix} \mathscr{Z}_{n,1} & \mathscr{Z}_{n,0} \\ \mathscr{Z}_{n-1,1} & \mathscr{Z}_{n-1,0} \end{bmatrix}. \tag{14.90}$$

The initial values of the matrices Z_k are given by the so-called Numerov integration scheme

$$\begin{aligned} Z_1(r+h) &= 2 + h^2[\alpha_1\mathscr{A}(r+h) + 2\alpha_2\mathscr{A}(r+2h)] + h^4\alpha_2\alpha_1\mathscr{A}(r+2h)\mathscr{A}(r+h), \\ Z_0(r+h) &= -1 + h^2[\alpha_0\mathscr{A}(r) - \alpha_2\mathscr{A}(r+2h)] + h^4\alpha_2\alpha_0\mathscr{A}(r+2h)\mathscr{A}(r), \end{aligned} \tag{14.91}$$

with

$$\alpha_0 = \alpha_2 = \frac{1}{12}, \quad \alpha_1 = \frac{5}{6}. \tag{14.92}$$

In this way by changing the initial conditions $y(r)$, $y(r + h)$ one obtains the final values by a matrix multiplication (14.89). The "propagator" matrix (14.90) is given in terms of the system matrix $\mathcal{A}$.

14.6 Runge–Kutta Integration Method

Let us solve the linear system of equations

$$\frac{dY}{dr} = \mathcal{A}Y. \tag{14.93}$$

First the integration region is divided into N equal intervals

$$r, r+h, r+2h, \ldots, r+Nh. \tag{14.94}$$

The vector Y_{n+1} can be expressed in terms of the vector Y_n as follows

$$Y_{n+1} = \mathcal{P}_n Y_n, \tag{14.95}$$

where

$$\mathcal{P}_n = I + \frac{1}{6}K_n^{(1)} + \frac{1}{3}K_n^{(2)} + \frac{1}{3}K_n^{(3)} + \frac{1}{6}K_n^{(4)}, \tag{14.96}$$

with

$$\begin{aligned}
K_n^{(1)} &= h\mathcal{A}(r_n), \\
K_n^{(2)} &= h\mathcal{A}\left(r_n + \frac{1}{2}h\right)\left(I + \frac{1}{2}K_n^{(1)}\right), \\
K_n^{(3)} &= h\mathcal{A}\left(r_n + \frac{1}{2}h\right)\left(I + \frac{1}{2}K_n^{(2)}\right), \\
K_n^{(4)} &= h\mathcal{A}(r_n + h)\left(I + K_n^{(3)}\right),
\end{aligned} \tag{14.97}$$

with

$$r_n = r + (n-1)h. \tag{14.98}$$

One can obtain directly the final vector acting on the initial value

$$Y_N = \mathcal{P}(N,1)Y_1, \tag{14.99}$$

where

$$\mathcal{P}(N,1) = \prod_{n=1}^{N-1} \mathcal{P}_n. \tag{14.100}$$

14.7 Spheroidal System of Coordinates

The ternary emission process can be described as the binary emission of the light cluster from a hyperdeformed configuration of the heavy fragments. An efficient way to describe the fissioning shape of heavy fragments is the use of the so-called spheroidal coordinates. The prolate spheroidal coordinates is defined by using the sum and difference of distances to two foci r_1, r_2, i.e.

$$\begin{aligned} \xi &= \frac{r_1 + r_2}{d}, \quad 1 \leq \xi \leq \infty, \\ \eta &= \frac{r_1 - r_2}{d}, \quad -1 \leq \eta \leq 1, \end{aligned} \tag{14.101}$$

where $d = 2f$ is the interfocal distance. It will be shown below that the first coordinate ξ is the analog of the usual radial coordinate r and the second one η of the angular coordinate $\cos\theta$ in the spherical system.

The major and minor axes are defined respectively as

$$\rho = f\xi, \quad r = f\sqrt{\xi^2 - 1}. \tag{14.102}$$

For the hyperdeformed configuration of heavy fragments in Fig. 7.2 in Chap. 7, the best choice of the focal parameter f is defined by the following system of equations for the major and minor axes

$$\begin{aligned} f\xi_m &= \frac{R}{2} + R_m, \\ f\sqrt{\xi_m^2 - 1} &= R_m, \end{aligned} \tag{14.103}$$

where R is the distance between heavy fragments and $R_m = 1.2(A_1^{1/3} + A_2^{1/3})/2$ their mean radius. One gets the following relation

$$f = R_m\sqrt{\left(\frac{R}{2R_m} + 1\right)^2 - 1}. \tag{14.104}$$

The Schrödinger equation for the light fragment in spheroidal coordinates was derived in Ref. [3] for the angular momentum projection $K = 0$. According to Ref. [4] it is given the system of coupled equations in the basis of spheroidal harmonics $Y_{lK}(\eta, \phi)$ for any value of K. Let us expand in terms of spheroidal harmonics the axially symmetric potential

$$v(\xi, \eta) \equiv \frac{2\mu_3}{\hbar^2} V(\xi, \eta) = v_0(\xi) + \sum_{L \geq 1} v_L(\xi) Y_{L0}(\eta), \tag{14.105}$$

and the wave function

$$\psi_K(\xi,\eta) = \sum_l \frac{g_l^{(3)}(\xi)}{r} Y_{lK}(\eta), \tag{14.106}$$

where r is given by (14.102). By performing standard manipulations one obtains the following system, replacing the first line in Eq. 7.21

$$\begin{aligned}\frac{d^2 g_l^{(3)}}{d(f\xi)^2} &= \left[(v_0-\kappa^2)\left(1+\frac{2}{3}\frac{1}{\xi^2-1}\right)+\frac{1}{f^2(\xi^2-1)}\left(l(l+1)+\frac{K^2-1}{\xi^2-1}\right)\right]g_l^{(3)} \\ &+\sum_{l'}\frac{2}{3}\sqrt{\frac{2l'+1}{2l+1}}\frac{\kappa^2-v_0}{\xi^2-1}\langle l'0;20|l0\rangle\langle l'K;20|lK\rangle g_{l'}^{(3)} \\ &+\sum_{l'}\sum_{L\geq 1} v_L\sqrt{\frac{2l'+1}{2l+1}}\left[\left(1+\frac{2}{3}\frac{1}{\xi^2-1}\right)\sqrt{\frac{2L+1}{4\pi}}\langle l'0;L0|l0\rangle\langle l'K;L0|lK\rangle\right. \\ &\left.-\frac{2}{3}\frac{1}{\xi^2-1}\sum_{L'}\sqrt{\frac{2L'+1}{5}}\langle L0;20|L'0\rangle^2\langle l'0;L'0|l0\rangle\langle l'K;L'0|lK\rangle\right]g_{l'}^{(3)},\end{aligned} \tag{14.107}$$

where brackets stands for the Clebsch–Gordan coefficient. For a pure spheroidal interaction $v_0(\xi)$, one obtains a tridiagonal system of equations, instead of a diagonal one. This quadrupole coupling makes the main difference between spherical and spheroidal coordinates, but it is the price payed for the drastic decreasing of the dimension in describing the dynamics.

Let us stress on the fact that this is a deformed system of coordinates describing a deformed shape and therefore one has an explicit dependence on K. Except for a small region in the case $K = 0$ the term proportional to $K^2 - 1$ represents a strong repulsive barrier.

The system of differential equations for the radial components (14.107) in terms of the major radius $\rho = f\xi$

$$\frac{d^2 g_l}{d\rho^2} = \sum_{l'=1}^{N} \mathscr{A}_{l;l'} g_{l'}, \quad l = 1,\ldots,N, \tag{14.108}$$

can be solved in a standard way, as described in Sect. 4.1 in Chap. 4. In particular for small distances the components of the wave function have the following asymptotics

$$g_l \xrightarrow{\xi\to 1} = \varepsilon_l \equiv \begin{bmatrix} (\xi^2-1)^{(K+1)/2}, & K \neq 1 \\ (\xi-1)^{l+1}, & K = 1 \end{bmatrix}. \tag{14.109}$$

For large distances the spheroidal coordinates approach the spherical ones, i.e.

$$\rho \xrightarrow{\xi\to\infty}, \quad \eta \xrightarrow{\xi\to\infty} \cos\theta, \tag{14.110}$$

and the system of equations becomes decoupled.

It is also important to give the expression of the scalar product in spheroidal coordinates, necessary to estimate the norm of the wave function

$$\langle\psi^{(1)}\psi^{(2)}\rangle \equiv 2\pi f^3 \int_1^{\infty} d\xi \int_{-1}^{1} d\eta(\xi^2-\eta^2)\psi^{(1)}(\xi,\eta)\psi^{(2)}(\xi,\eta)$$
$$= \sum_l \int_1^{\infty} fd\xi g_l^{(1)} g_l^{(2)} + \frac{2}{3}\sum_{ll'} \int_1^{\infty} fd\xi \frac{g_l^{(1)} g_{l'}^{(2)}}{\xi^2-1}\left[\delta_{ll'} - \sqrt{\frac{2l+1}{2l'+1}}\langle l0;20|l'0\rangle^2\right]. \tag{14.111}$$

To find resonant states of the light fragment in the highly deformed potential of the heavy binary system, is a very difficult task. Let us mention that in order to reproduce this potential, it is necessary to use a very large spherical basis for the coupled channel procedure, i.e. more than 20 spherical partial waves.

It turns out that it is enough to consider up to 10 partial spheroidal waves to achieve less than 0.1% precision. To this purpose, for each inter-fragment distance $R = R_1 + R_2$ the focal distance can be adjusted in order to obtain the best description of the hyperdeformed binary shape.

14.8 Spherical Harmonic Oscillator

The Schrödinger equation for the ho potential

$$\left[-\frac{\hbar^2}{2\mu}\Delta + \frac{1}{2}\mu\omega^2 r^2\right]\phi(\mathbf{r}) = \hbar\omega\left[-\frac{\Delta^2}{2\beta} + \frac{\beta r^2}{2}\right]\phi(\mathbf{r}) = E_\phi(\mathbf{r}), \tag{14.112}$$

where the ho parameter is introduced

$$\beta = \frac{\mu\omega}{\hbar}, \tag{14.113}$$

has the eigenstates given by

$$\phi_{nlm}^{(\beta)}(\mathbf{r}) = \mathscr{R}_{nl}^{(\beta)}(r)Y_{lm}(\hat{r}),$$
$$\mathscr{R}_{nl}^{(\beta)}(r) = (-)^n\left[\frac{2\beta^{l+3/2}n!}{\Gamma(n+l+3/2)}\right]^{1/2} r^l e^{-\frac{\beta r^2}{2}} L_n^{l+1/2}(\beta r^2), \tag{14.114}$$

where L_n^α denotes the Laguerre polynomial.The radial ho wave function satisfies the following equation

$$\hbar\omega\left[-\frac{1}{2\beta}\left(\frac{1}{r}\frac{d^2}{dr^2}r - \frac{l(l+1)}{r^2}\right) + \frac{\beta r^2}{2}\right]\mathscr{R}_{nl}^{(\beta)}(r) = E_N \mathscr{R}_{nl}^{(\beta)}(r). \tag{14.115}$$

The eigenvalues are

$$E_N = \hbar\omega\left(N + \frac{3}{2}\right), \qquad N = 2n + l. \tag{14.116}$$

The sum of two ho potentials can be recoupled into the relative and cm terms as follows

$$\frac{1}{2}m\omega^2 r_1^2 + \frac{1}{2}m\omega^2 r_2^2 = \frac{1}{2}\hbar\omega\beta\left(r^2 + R^2\right), \tag{14.117}$$

where

$$\mathbf{r} = \frac{\mathbf{r}_1 - \mathbf{r}_2}{\sqrt{2}}, \quad \mathbf{R} = \frac{\mathbf{r}_1 + \mathbf{r}_2}{\sqrt{2}}. \tag{14.118}$$

The Talmi-Moshinsky transformation transforms the product of the ho wave function into the product between the relative and cm functions

$$\left[\phi^{(\beta_1)}_{n_1 l_1}(\mathbf{r}_1)\phi^{(\beta)}_{n_2 l_2}(\mathbf{r}_2)\right]_{\lambda\mu} = \sum_{nlNL}\langle nlNL;\lambda|n_1 l_1 n_2 l_2;\lambda\rangle\left[\phi^{(\beta)}_{nl}(\mathbf{r})\phi^{(\beta)}_{NL}(\mathbf{R})\right]_{\lambda\mu}. \tag{14.119}$$

The spherical ho wave functions can be defined by using the bracket notations as follows

$$\begin{aligned}\varphi^{(\beta)}_{nljm}(\mathbf{x}) &\equiv \langle\mathbf{x}|\beta nljm\rangle = [\phi^{(\beta)}_{nl}(\mathbf{r})i^l\chi_{\frac{1}{2}}(s)]_{jm},\\ \phi^{(\beta)}_{nlm}(\mathbf{r}) &\equiv \langle\mathbf{r}|\beta nlm\rangle = \mathscr{R}^{(\beta)}_{nl}(r)Y_{lm}(\hat{r})\ ,\end{aligned} \tag{14.120}$$

where

$$\mathscr{R}^{(\beta)}_{nl}(r) \equiv \langle r|\beta nl\rangle \tag{14.121}$$

are the radial spherical ho functions.

14.8.1 Spherical Shifted Harmonic Oscillator

Let us consider the radial equation for the shifted ho

$$\left\{-\frac{\hbar^2}{2\mu}\left[\frac{d^2}{dr^2} - \frac{l(l+1)}{r^2}\right] + \frac{\mu\omega^2}{2}(r - r_0)^2\right\}\psi(r) = E\psi(r), \tag{14.122}$$

where $\psi(r) = r\phi(r)$ and μ is the reduced mass. For large values of the equilibrium radius r_0, the solution is concentrated around r_0, so that $|r - r_0| << r_0$ and the equation acquires the following approximate form

$$\left[-\frac{\hbar^2}{2\mu}\frac{d^2}{dr^2}+\frac{\mu\omega^2}{2}(r-r_0)^2\right]\psi(r)=\left[E-\frac{\hbar^2 l(l+1)}{2\mu r_0^2}\right]\psi(r). \tag{14.123}$$

It can be written in terms of the new coordinate $\bar{r}=r-r_0$ as an ordinary one-dimensional ho equation

$$\left[-\frac{\hbar^2}{2\mu}\frac{d^2}{d\bar{r}^2}+\frac{\mu\omega^2}{2}\bar{r}^2\right]\psi(\bar{r})=\left[E-\frac{\hbar^2 l(l+1)}{2\mu r_0^2}\right]\psi(\bar{r}). \tag{14.124}$$

The eigenvalues are "rotational bands", built on top of ho eigenvalues

$$\begin{aligned} E_{nl} &= \hbar\omega\left(n+\frac{1}{2}\right)+\frac{\hbar^2 l(l+1)}{2_0^2} \\ &= \hbar\omega\left[n+\frac{1}{2}+\frac{l(l+1)}{2_0^2}\right], \end{aligned} \tag{14.125}$$

in terms of the ho parameter (14.113).

14.9 Diagonalization Procedure in a Non-Orthogonal Basis

In this Appendix we will remind the method to solve the eigenvalue problem for an hermitean matrix in a non-orthogonal basis. Let us consider the general eigenvalue problem

$$\mathbf{H}\Psi_k = E_k\Psi_k, \tag{14.126}$$

where the eigenfunction Ψ_k, corresponding to the eigenvalue E_k, is expanded in a non-orthogonal basis ϕ_n

$$\Psi_k=\sum_n C_{kn}^T\phi_n=\sum_n C_{nk}\phi_n, \tag{14.127}$$

T denoting the matrix transposition. Inserting (14.126) into (14.127), multiplying to left by ϕ_m and integrating one obtains

$$\sum_n\langle\phi_m|\mathbf{H}|\phi_n\rangle C_{kn}^T=E_k\sum_n C_{kn}^T\langle\phi_m|\phi_n\rangle. \tag{14.128}$$

This is just the general eigenvalue problem to be solved given by Eq. 9.21 in Chap. 9, with the metric matrix

$$I_{mn}\equiv\langle\phi_m|\phi_n\rangle. \tag{14.129}$$

Let us first find the eigenvalues F_j and eigenvectors Y_{lj} of the symmetric metric matrix

$$\sum_l I_{ml} Y_{lj} = F_j Y_{mj}, \tag{14.130}$$

where the orthonormality conditions for the eigenvectors holds

$$\sum_i Y_{mi} Y^T_{in} = \delta_{mn}. \tag{14.131}$$

The system of functions

$$\psi_i = \sum_k \frac{Y^T_{ik}}{\sqrt{F_i}} \phi_k, \tag{14.132}$$

is orthonormal, because from (14.130) and (14.131) one obtains

$$\langle \psi_i | \psi_j \rangle = \sum_k \frac{Y^T_{jk}}{\sqrt{F_i}} \sum_l I_{kl} \frac{Y_{lj}}{\sqrt{F_j}} = \sum_k \frac{Y^T_{ik}}{\sqrt{F_i}} \sqrt{F_j} \cdot Y_{kj} = \delta_{ij}. \tag{14.133}$$

Expanding the initial eigenfunctions Ψ_k (14.127) in terms of this basis

$$\Psi_k = \sum_i X^T_{ki} \psi_i, \tag{14.134}$$

one obtains an eigenvalue problem

$$\sum_i \langle \psi_l | H | \psi_i \rangle X^T_{ki} = E_k X^T_{ki}, \tag{14.135}$$

for the symmetric matrix

$$\langle \psi_l | \mathbf{H} | \psi_i \rangle = \sum_{mn} \frac{Y^T_{lm}}{\sqrt{F_l}} \langle \phi_m | \mathbf{H} | \phi_n \rangle \frac{Y_{ni}}{\sqrt{F_i}}. \tag{14.136}$$

Using (14.132) and (14.134) one finally obtains the following expression for the expansion coefficients in (14.127)

$$C^T_{kn} = \sum_i X^T_{ki} \frac{1}{\sqrt{F_i}} Y^T_{in}. \tag{14.137}$$

Here the eigenvalues F_i and the eigenvectors Y^T_{in} are solutions of the system (14.5), (14.6) and the eigenvectors X^T_{ki} the solutions of (14.135) and (14.136).

14.10 Four-Particle Metric Matrix

The metric matrix (9.22 in Chap. 9) is given by the following relations

$$\langle 0|\left[P_{\nu\nu\beta_2 b_2}P_{\pi\pi\alpha_2 a_2}\right]_{\alpha_4}\left[P^{\dagger}_{\pi\pi\alpha'_2 a'_2}P^{\dagger}_{\nu\nu\beta'_2 b'_2}\right]_{\alpha_4}|0\rangle = \delta_{\alpha_2\alpha'_2}\delta_{\beta_2\beta'_2}\delta_{a_2 a'_2}\delta_{b_2 b'_2}, \tag{14.138}$$

$$\begin{aligned}&\langle 0|\left[P_{\nu\pi\beta_2 b_2}P_{\nu\pi\alpha_2 a_2}\right]_{\alpha_4}\left[P^{\dagger}_{\pi\nu\alpha'_2 a'_2}P^{\dagger}_{\pi\nu\beta'_2 b'_2}\right]_{\alpha_4}|0\rangle\\ &= \delta_{\alpha_2\alpha'_2}\delta_{\beta_2\beta'_2}\delta_{a_2 a'_2}\delta_{b_2 b'_2} + \delta_{\alpha_2\beta'_2}\delta_{\beta_2\alpha'_2}\delta_{a_2 b'_2}\delta_{b_2 a'_2}(-)^{\alpha_2+\beta_2-\alpha_4}\\ &\quad - \sum_{ijkl}[A(ijkl;\alpha_2 a_2\beta_2 b_2\alpha'_2 a'_2\beta'_2 b'_2) + A(ijkl;\alpha_2 a_2\beta_2 b_2\beta'_2 b'_2\alpha'_2 a'_2)(-)^{\alpha'_2+\beta'_2-\alpha_4}],\end{aligned} \tag{14.139}$$

$$\langle 0|\left[P_{\nu\nu\beta_2 b_2}P_{\pi\pi\alpha_2 a_2}\right]_{\alpha_4}\left[P^{\dagger}_{\pi\nu\alpha'_2 a'_2}P^{\dagger}_{\pi\nu\beta'_2 b'_2}\right]_{\alpha_4}|0\rangle = -\sum_{ijkl}B(ijkl;\alpha_2 a_2\beta_2 b_2\alpha'_2 a'_2\beta'_2 b'_2) \tag{14.140}$$

where

$$\begin{aligned}A(ijkl;\alpha_2 a_2\beta_2 b_2\alpha'_2 a'_2\beta'_2 b'_2) \equiv& \hat{\alpha}_2\hat{\beta}_2\hat{\alpha}'_2\hat{\beta}'_2\begin{Bmatrix} i & j & \alpha_2\\ l & k & \beta_2\\ \alpha'_2 & \beta'_2 & \alpha_4\end{Bmatrix}\\ \times(-)^{k+l-\beta_2+k+j-\beta'_2}X^*_{\pi\nu}(ij;\alpha_2 a_2)&X^*_{\pi\nu}(kl;\beta_2 b_2)X_{\pi\nu}(il;\alpha'_2 a'_2)X_{\pi\nu}(kj;\beta'_2 b'_2),\end{aligned} \tag{14.141}$$

$$\begin{aligned}B(ijkl;\alpha_2 a_2\beta_2 b_2\alpha'_2 a'_2\beta'_2 b'_2) \equiv& \hat{\alpha}_2\hat{\beta}_2\hat{\alpha}'_2\hat{\beta}'_2\begin{Bmatrix} i & j & \alpha_2\\ k & l & \beta_2\\ \alpha'_2 & \beta'_2 & \alpha_4\end{Bmatrix}\\ \times\bar{X}^*_{\pi\pi}(ij;\alpha_2 a_2)\Delta_{ij}\bar{X}^*_{\nu\nu}(kl;\beta_2 b_2)&\Delta_{kl}X_{\pi\nu}(ik;\alpha'_2 a'_2)X_{\pi\nu}(jl;\beta'_2 b'_2)\end{aligned} \tag{14.142}$$

with the following notation

$$\bar{X}_{\tau\tau}(ij;\alpha_2 a_2) \equiv \begin{pmatrix} & X_{\tau\tau}(ij;\alpha_2 a_2), & i\le j\\ (-)^{i+j+1+\alpha_2} & X_{\tau\tau}(ji;\alpha_2 a_2), & i>j\end{pmatrix}. \tag{14.143}$$

14.11 Two Quasiparticle Preformation Amplitude

The expansion coefficient of the amplitude describing transitions between proton 2qp and ground states Eq. 9.77 in Chap. 9 is given by

$$W(N_\alpha L_\alpha K_\pi) \equiv 8 \sum_{N_\pi N_\nu} \sum_{L_\pi L_\nu} G_\pi(N_\pi L_\pi K_\pi) G_\nu(N_\nu L_\nu)_\pi, K_\pi; L_\nu, 0|L_\alpha, K_\pi\rangle \tag{14.144}$$

The definition of the new G-coefficient for protons is the following

$$G_\pi(N_\pi L_\pi K_\pi) = \sum_{12} B_\pi(n_1 l_1 j_1 n_2 l_2 j_2; L_\pi K_\pi) \langle (l_1 l_2) L_\pi (\frac{1}{2}\frac{1}{2})0; L_\pi | (l_1 \frac{1}{2}) j_1 (l_2 \frac{1}{2}) j_2; L_\pi \rangle$$
$$\times \sum_{n_\pi} \pi 0 N_\pi L_\pi; L_\pi | n_1 l_1 n_2 l_2; L_\pi \rangle \mathscr{I}^{(\beta\beta_\alpha)}_{n_\pi 0}. \tag{14.145}$$

The expression of the proton B-coefficient is

$$B_\pi(n_1 l_1 j_1 n_2 l_2 j_2; L_\pi K_\pi) = \sqrt{2} u_{\pi K_a} u_{\pi K_b} \langle j_1, K_a; j_2, K_b | L_\pi, K_\pi \rangle$$
$$\times c_{n_1}(\pi l_1 j_1 K_a) c_{n_2}(\pi l_2 j_2 K_b), \tag{14.146}$$

while for the neutron part is the same as Eq. 9.57, i.e.

$$B_\nu(n_3 l_3 j_3 n_4 l_4 j_4; L_\nu 0) = \sqrt{2} \sum_{K>0} X^\nu_K (-)^{j_4 - K} \langle j_3, K; j_4, -K | L_\nu, 0 \rangle$$
$$\times c_{n_3}(\nu l_3 j_3 K) c_{n_4}(\nu l_4 j_4 K). \tag{14.147}$$

References

1. Dudek, J., Szymanski, Z., Werner, T., Faessler, A., Lima, C.: Description of high spin states in ^{146}Gd using the optimized Woods-Saxon potential. Phys. Rev. C **26**, 1712–1718 (1982)
2. Cwiok, S., Dudek, J., Nazarewicz, W., Skalski, J., Werner, T.: Single-particle energies, wave functions, quadrupole moments and g-factors in an axially deformed Woods-Saxon potential with applications to the two-centre-type nuclear problem. Comput. Phys. Commun. **46**, 379–399 (1987)
3. Delion, D.S., Săndulescu, A., Mişicu, S., Cârstoiu, F., Greiner, W.: Quasimolecular resonances in the binary cold fission of ^{252}Cf. Phys. Rev. C **64**, 041303(R)/1–5 (2001)
4. Delion, D.S., Săndulescu, A., Greiner, W.: Self-consistent description of the ternary cold fission: tri-rotor mode. J. Phys. G **28**, 2921–2938 (2002)

Index